Topography and the Environment

We work with leading authors to develop the
strongest educational materials in geography,
bringing cutting-edge thinking and best
learning practice to a global market.

Under a range of well-known imprints, including
Prentice Hall, we craft high quality print and
electronic publications which help readers to understand
and apply their content, whether studying or at work.

To find out more about the complete range of our
publishing, please visit us on the World Wide Web at:
www.pearsoneduc.com

Topography and the Environment

Richard Huggett and Jo Cheesman

Dear Colen,
With love
Cheesman.

Prentice
Hall

An imprint of **Pearson Education**

Harlow, England · London · New York · Reading, Massachusetts · San Francisco
Toronto · Don Mills, Ontario · Sydney · Tokyo · Singapore · Hong Kong · Seoul
Taipei · Cape Town · Madrid · Mexico City · Amsterdam · Munich · Paris · Milan

Pearson Education Limited
Edinburgh Gate
Harlow
Essex CM20 2JE

and Associated Companies throughout the world

Visit us on the World Wide Web at:
www.pearsoneduc.com

First published 2002

ISBN 0-582-41857-7

British Library Cataloguing-in-Publication Data
A catalogue record for this book is available from the British Library

Library of Congress Cataloging-in-Publication Data
Huggett, Richard J.
    Topography and the environment / Richard Huggett and Jo Cheesman.
        p. cm.
    Includes bibliographical references (p.).
    ISBN 0–582–41857–7
        1. Geomorphology.   2. Environmental conditions.   I. Cheesman, Jo.
    II. Title.
    GB406 .H84 2002
    551.41—dc21   2001052367

10 9 8 7 6 5 4 3 2 1
05   04   03   02

Typeset in $9\frac{1}{2}$/11pt Times by 35
Printed in Malaysia, LSP

To our families

Contents

Preface

Topography is at once a fundamental and a subtle environmental factor. It wields large and small influences over all other factors of the ecosphere – climate, water storage and movement, soils, animals and plants, and human activities. Several conspicuous topographic influences are well known. Most readers will be aware that a hike up a mountain is the climatic equivalent of a trek from a lower to a higher latitude. They will also be familiar with the stark climatic and botanical contrasts between north-facing and south-facing slopes in the European Alps. Subtle influences are not so well known but are equally significant. In the French Alps, slope exposure affects alpine marmots' habitat preferences and growth rates. Problems of air pollution in Mexico City are associated with poor ventilation and local and regional wind systems that are largely under topographic control. Sea-floor topography may subtly alter marine processes. On 17 July 1998, a magnitude 7.1 earthquake spawned three tsunamis (tidal waves) whose size would normally be associated with a larger quake. The 15-m waves killed more than 2200 people living along Papua New Guinea's north coast. The waves may have grown abnormally large due to the three-dimensional topography of the nearby sea-floor. Equally, seismic vibrations may have triggered submarine landslides or gas explosions that could have given the waves a boost.

In the present book we investigate the basic and refined effects of topography upon other environmental factors – climates, surface water, soils and sediments, animals and plants, and humans. In doing so, we pinpoint four aspects of topography as an environmental factor. First, we acknowledge a corpus of work that limits topography to the physical ground surface. Workers in this field are largely physical geographers, and in particular geomorphologists who are principally interested in the characterisation and evolution of land form. Second, we recognise a renaissance of the venerable idea of topography as 'all the features of the Earth's surface', including human-made features. The language of the landscape ecologists now focuses around patch, corridor, and matrix. These are all crucial topographic elements that affect microclimates, the abundance and distribution of animals and plants, and conservation strategies. Third, we note a rapid emergence of topographically based modelling, developed in association with geographical information systems. With software packages and powerful computing facilities readily available, topographically based modelling is increasingly applied in a range of disciplines concerned with the Earth's surface environment. Fourth, we perceive a thriving interest in topographic location – place – as a human construct. Some researchers are interested not just in the character of a particular place (its location and structure), but also in how people perceive it and respond to it (its perceptual and cultural meaning). The book explores all four aspects of topography, with the first three taking the lion's share of the text. The idea of place as a human construct is an immense topic and is touched upon in relation to cultural, social, and economic influences upon settlement location and in relation to agroforestry and ecosystem management.

The book is divided into two parts. Part I introduces topography and Part II explores individual topographic influences on various environmental factors. Part I comprises two chapters. Chapter 1 discusses the nature of topography, the role of topography as an environmental factor, and the nature of the toposphere. Chapter 2 looks at the description of topography, discussing topographic elements (locational parameters, digital elevation data, and physical and ecological landscape units), topographic structures (networks, mosaics, and regions), and topographic connections (local and regional flows, global flows, and isolation in a connected world). Part II consists of five chapters. Chapter 3 examines topographic influences upon climates. It considers the modelling of climatic factors using topographic data; climatic variations with altitude; local climates (topoclimates), including urban heat-islands, small-scale topographic alteration of airflow, and local winds; synoptic-scale climates, including fall winds and compensating winds; and global topographic connections, including the effects of mountain chains and plateaux. Chapter 4 investigates topographic influences on water. It studies the hydrology of hillslopes, drainage basins, and channel networks at local, regional, and global scales, as well as the automated derivation of catchment characteristics. Chapter 5 probes topographic influences upon soils, examining altitude and soils (altitudinal climosequences, geomorphic zones on mountains), aspect and soils, soils and slopes, soil landscapes, and soil erosion. Chapter 6 considers topographic influences upon life. It looks at altitude and life, including life zones, tree-lines, and altitudinal species ranges; land form and life, including aspect, slope, microtopography, and surfaces; landscape patches (patch size, patch shape, and patch edges), corridors (roads and trails, powerlines, hedgerows, and streams and rivers), networks (tree networks, corridor and circuit networks), and mosaics (landscape structure and connectivity and landscape properties). Chapter 7 examines how topography exerts an influence on the location of settlements, routes, agriculture, environmental destruction and conservation, and human individuals and populations, which includes terrain and military campaigns.

Finally, we should like to thank several people without whose assistance this book might never have been written. Joint thanks go to Michael Bradford, who has always supported 'armchair research' projects, to Nick Scarle, who has worked wonders with the diagrams, to the colleagues who kindly let us use their photographs, and to Matthew Smith at Pearson Education for boldly going where no publisher has gone before. We would also record the following personal thanks:

Jo Cheesman: I would like to acknowledge my former colleagues at the School of Geography at Manchester University for the support that I received while still there and the continued support I received once I had left which enabled me to complete my sections of this book. Much of the research for my contributions to this book has come from my doctoral and postdoctoral research. During the years spanned by this research I have received unwavering support from my family, especially my parents, friends, and in particular, Steve. An enormous 'thank you' to all of you.

Richard Huggett: I thank Derek Davenport for continued beery discussions about all manner of things. I also thank Shelley, Zoë, and Ben who, despite their best endeavours, never manage to keep me off my PC for very long.

Richard Huggett
Jo Cheesman

Manchester
May 2001

Acknowledgements

We are grateful to the following for permission to reproduce copyright material:

Figure 2.6 after Figure 4.2 (p. 117) from *Land Mosaics: The Ecology of Landscapes and Regions*, Cambridge University Press, (Forman, R.T.T., 1995); Figure 3.12 after Figures 9 and 10 from 'Observations and Numerical Modelling of Mountain Waves over the Southern Alps of New Zealand' in *Quarterly Journal of the Royal Meteorological Society* (**vol. 126**, 2765–88), Royal Meteorological Society, (Lane, T.P., Reeder, M.J., Morton, B.R., and Clark, T.L., 2000); Figure 5.14 after Figure 1 from 'Hydric Conditions and Hydromorphic Properties within a Mollisol Catena in South-eastern Minnesota' in *Soil Science of America Journal* (**No. 62**, 1116–25), Soil Science of America, (Thompson, J.A. and Bell, J.C., 1998); Figure 5.16 after Figures 2 and 5 from *Slickspot Soil Genesis in the Carrizo Plain, California*' in *Soil Science of America Journal* (**No. 57**, 162–8), Soil Science of America, (Reid, D.A., Graham, R.C., Southard, R.J., and Amrheim, C., 1993); Figure 5.17 after Figure 2 from 'Pedogenesis of a Vernal pool Entisol-Alfisol-Vertisol Catena in Southern California' in *Social Science of America Journal* (**No. 60**, 316–23), Soil Science of America, (Weitkamp, W.A., Graham, R.C., Anderson, M.A., and Amrheim, C., 1996); Figure 6.20 after Figure 6 from 'A High-Andean Toposequence: The Geoecology of Caulescent Paramo Rosettes' in *Mountain Research and Development* (**No. 15**, 133–52), reproduced by permission of the publishers, International Mountain Society and United Nations University, (Pérez, F.L., 1995); Figure 7.1 after figure on p. 17 from *Rural Settlement*, Macmillan Education (Aspects of Geography series), (Roberts, B.K., 1987), reproduced by kind permission of Professor Brian K. Roberts; Figure 7.2 after Figure 2.6 (p. 31) from *Landscapes of Settlement, Prehistory to the Present*, Routledge, (Roberts, B.K., 1996).

Colour Plate 1 reproduced from the Land-Form PANORAMA Ordance Survey® map with the permission of Ordnance Survey on behalf of The Controller of Her Majesty's Stationery Office, © Crown Copyright 2001; Colour Plate 2 from the National Centre for Environmental Data and Surveillance, Bath, BA2 9ES; Colour Plates 3 and 4 © John Ogden; Colour Plate 5 © Joan Ehrenfeld; Colour Plates 6, 7 and 8 © Marcela Ferreyra.

Whilst every effort has been made to trace the owners of copyright material, in a few cases this has proved impossible and we take this opportunity to offer our apologies to any copyright holders whose rights we may have unwittingly infringed.

PART I   **Introducing topography**

CHAPTER 1

Topography, the toposphere, and the environment

What is topography?

Topography is the lie of the land, or the general configuration of the land surface, including its relief and the location of its features, natural and human-made. It is also the lie of the sea floor and may be used in describing submarine relief features. In truth, it may be applied to any feature that bears an uneven surface: the surface topography of metals, the topography of the brain, and corneal topography are examples. Several other terms convey a similar meaning to topography. Terrain is the natural physical features and configuration of a land area, and usually excludes artificial features. Physiography, at least according to those who subscribe to Thomas Henry Huxley's (1877) catholic definition that embraces all natural phenomena and their interdependencies, is very similar to topography but has an even wider compass. It is also used in a narrow sense to mean the physical ground surface.

Topography is a primary concern of several groups of researchers – archaeologists and historians, geomorphologists, and landscape ecologists. To the archaeologist and historian, topography includes the geography and natural resources of regions, the evolving structural forms of cities and the countryside, and functional areas within cities and the countryside, including churches, market places, workshops, and private houses. This sweeping definition of topography has tradition on its side. It was the subject of books by such old topographers as Gerald of Wales (1146–1243), who wrote a *History and Topography of Ireland* (1983 edition), and John Thomas Smith, who wrote *Ancient Topography of London* (1815). Modern 'topographers' still pen books on the history and topography of specific regions. Titles include *Studies in Ancient Greek Topography* (Pritchett 1969–1992) and *Discovering Kerry: Its History, Heritage and Topography* (Barrington 1976). Landscape ecologists, and latter-day European physiographers, follow tradition in taking topography to include all landscape features as well as the purely physical ground-surface features. Geomorphologists, and most twentieth-century American physiographers, are inclined to take topography as simply the ground surface (see Davis 1924; Fenneman 1939; Stoddart 1975). These different meanings of topography partly reflect the artificiality of disciplinary boundaries, but they are real differences and cannot be ignored. Confusion might arise from using topography in its traditional comprehensive sense to include all surface features, and some readers might prefer another term, such as landscape or terrain, to be used for this purpose. Nevertheless, topography is the term chosen in this book. One reason for this choice is that its comprehensive definition has an integrative role – it stresses the unity of landscape components (soils, slopes, rivers, lakes, animals, plants, and human-made features). A second reason is that, when broadly defined, topography embodies two potent geographical ideas that are applicable to all components of landscapes – place and space.

Place is a simple idea if it is used purely as a named locality – the Kalahari desert, the Scilly Isles, Manchester. But place has come to mean more than just a specific location (e.g. Relph 1996). It includes the character of that location (what is there) and how humans perceive it and respond to it – it has locational, structural, and perceptual and cultural aspects. Perceptual aspects are many and varied. They include the idea of 'scenery', or the overall features of a place considered from an aesthetic angle (e.g. Downs and Stea 1976; Jackle 1987). Perceptual aspects include the ideas of topophilia and topophobia, which try to capture the personal bonds, positive and negative, between people and place. Topophilia is the 'affect-ive bond between people and place or setting', the 'human love of place . . . diffuse as a concept, vivid and concrete as personal experience' (Tuan 1974, 92). Topophobia is its opposite, literally 'the fear of place'. These potent ideas are seen in 'place attachment', that is, the extent to which individuals value or iden-tify with particular environmental settings. Take the case of the people who use 'rail-trails', which are multi-use recreation trails established along disused railways. A study of such users suggested that place attachment has at least two dimensions: a place dependence, reflecting the importance of the place in facilitating a user's activity; and a more affective place identity, reflecting an individual's valuing of a set-ting for more symbolic or emotional reasons (Moore and Graefe 1994). Cultural aspects are nicely cap-tured in the following quotation: '. . . natural land-scapes are the result of biophysical processes that shape the land and create the unmistakable differ-ences between one place and another . . . similarly, human landscapes and settlements are the consequence of culture modifying and imposing its needs on nat-ural or wild places' (Hough 1990). In Australia, Aboriginal people have many sacred sites – rock clefts, valleys, water holes, rocks – that serve cultural and religious functions (Chatwin 1998). Uluru (Ayer's Rock) is perhaps the most famous example, but, to the Aboriginals, the Australian landscape is laden with a cultural significance that is impalpable to people from other parts of the world. Many cultures name places after topographic features. In England, such place-name endings as -ford, -field, and -ley betray a topographic origin. Bradford was the site of a broad ford. Family names may become associated with the topographically derived names. Macclesfield was founded by Anglo-Saxon settlers, whose head-man may have been called Macca, which was a Saxon personal name, the place becoming known as 'Macca's field'. Cuffley is probably derived from a personal name, Cuffa, and leah, which is a clearing in a wood.

Space is an inherently structural idea, focusing on relative location. It examines how flows, slopes, surfaces, networks, migration routes, trails, lines of communication, and so on connect places. Some geographers have identified space as the core of their discipline. One geographer was bold enough to articulate a general spatial-systems theory (Coffey 1981). His idea was that a few concepts, rooted in spatial structure and spatial dynamics, underpin the explanations of a wide range of phenomena in human and physical geography: spatial structure, comprising geometry, topology, and dimensionality; and move-

ments in space. Spatial structure and movements in space are themselves expression of spatial process, which involves growth and organisation as key ele-ments. However, such schemes, stimulating though they be, overlook the cultural and social spaces so treasured by human geographers. Cultural and social spaces are part of inhabited space and do not lend themselves to a geometrical analysis (e.g. Hubert 1998, 1999).

What is the role of topography as an environmental factor?

Several locational and structural topographic factors affect the ecosphere. The influences of latitude, lon-gitude, altitude, and terrain structure are particularly potent and merit some preliminary discussion.

Locational factors

Latitude, longitude, and altitude are locational topo-graphic factors that have basic, if indirect, ecospheric influences. A place's location indirectly determines, or at least constrains, the climate that it will experience and, to some degree, what other environmental con-ditions will be like. To appreciate the all-important yet too often taken-for-granted influences of locational factors upon the ecosphere, it may be helpful to start by considering a simplified planet.

Imagine the Earth were all land of uniform height with no oceans. On this monotonous Earth, the only topographic influence exerted on the ecosphere would be the indirect effect of location. The tropics would be hot and the poles cold, so latitude, acting through climate, would be an influential environmental fac-tor. There would be no climatic differences within latitudinal zones, and climates would not vary with longitude.

Longitudinal climatic differences would arise if the planetary surface were divided into oceans and land areas. The nature of the longitudinal variations would depend on the precise arrangement of land and sea. Whatever the arrangement, land and sea have differ-ent thermal properties that would lead to a distortion of latitudinal zones – longitudinal climatic gradients would appear. Even greater distortion would be pos-sible if the planetary surface had varied relief, and in particular if there were mighty mountain ranges that ran from north to south. The circulation of warm and cold ocean waters would almost certainly lead to fur-ther distortions.

Geographical gradients

Latitudinal, longitudinal, and altitudinal climatic gradients are all present on the real planet Earth. Altitudinal gradients and their effects on the ecosphere will be discussed in later chapters. The effects of latitude and longitude will not be explored fully in the present book. However, because latitudinal and longitudinal climatic patterns provide a kind of back-cloth against which to view topographic influences, it seems sensible to scrutinise them in a little depth at this juncture.

Latitude wields an enormous indirect influence upon the ecosphere. Many driving climatic variables follow latitudinal gradients. Solar radiation, net radiation, and temperature in particular all tend to decrease from 'highs' in the tropics to 'lows' at the poles. In addition, seasonal and daily climatic rhythms vary with latitude owing to the seasonal trend in the Sun's daily path, which itself varies with latitude. Significantly, seasonal changes of solar radiation, day length, and temperature are small in low latitudes but large in high latitudes. The result is that seasonal effects are more pronounced than diurnal effects in middle and high latitudes, but not in tropical latitudes. The tropics are therefore said to experience diurnal climates whereas middle and high latitudes experience seasonal climates. The latitudinal differences in temperature regime affect precipitation characteristics. Precipitation systems in low latitudes are primarily convective and small-scale, whereas those in middle and high latitudes are usually cyclonic and large-scale. Snow is confined to middle and high latitudes (and tropical mountains). The latitude of a place also determines where it lies within the global wind belts. An important effect of this is that the seaboards of continents, especially those with high mountain ranges, that lie in the path of either the trade winds or the westerly winds receive prodigious amounts of precipitation. Such places include the east coast of Central America and the north-east coast of South America, which lie in the path of trade winds, and the Pacific north-west coast of North America, which lies in the path of westerly winds. All these regions receive up to 5000 mm of precipitation per year.

Longitudinal influences are essentially distortions of latitudinal (and altitudinal) patterns owing to such structural topographic factors as the arrangement of land and sea, the presence of warm and cold ocean currents, and the presence of mountains and plateaux. For example, on a latitudinal basis, mean annual precipitation is highest in a band around the equator and in the bands in which cyclonic storms are common (40–50° N and S). However, parts of equatorial East Africa receive less than 200 mm of rain per year, while 12 000 mm fall on the slopes of Mount Waialeale, Hawaii, which lies in the subtropics (22°5′N). A major distortion of latitudinal gradients is seen in the annual range of mean monthly temperatures, which is a measure of continentality (p. 8). This distortion is caused by the different thermal properties of land and sea, and by the ameliorating influence of warm ocean currents. In brief, the deep interiors of continents, far removed from relatively warm oceans, tend to experience much colder winters and hotter summers than continental margins at the same latitude. The western shores of Europe are bathed in winter by the warm North Atlantic Drift. This ocean current reduces the likelihood of cold temperatures in south-west Britain and allows frost-sensitive plants and animals to survive. Several plants and animals that have their main range in south-west Europe manage to live in south-west Britain – the strawberry tree (*Arbutus unedo*), the Cornish heath (*Erica vagans*), and the pale butterwort (*Pinguicula lusitanica*) are botanical examples; Barrett's marbled coronet moth (*Hadena luteago barrettii*), the hoary footman moth (*Eilema caniola*), and the Kerry spotted slug (*Geomalacus maculosus*) are zoological examples. Cold ocean currents also help to produce longitudinal climatic gradients. The shores of southern South America and southern Africa are cooled by air passing over the cold Humboldt Current and Benguela Current, respectively, which helps to sustain the Atacama and Kalahari deserts. Longitudinal climatic differences between the western and eastern seaboards of the South American subtropics are great. Under the Tropic of Capricorn, west-coast annual rainfall in the Atacama Desert is less than 100 mm, while in east-coast Brazil it is more than 1000 mm and supports tropical forest.

Geographical patterns

Latitude, longitude, and altitude combine to produce four fundamental, climatically based, global patterns that are crucial to understanding the structure and function of the ecosphere – a thermal pattern, a moisture pattern, a throughput pattern, and an accumulation pattern (Figure 1.1). These patterns are largely dictated by zonal and regional climatic gradients (Box and Meentemeyer 1991).

1. *Thermal pattern*   The thermal pattern is related to the amount of incoming energy. It mirrors energy availability and is largely determined by the Earth's surface radiation or energy balance. Insolation,

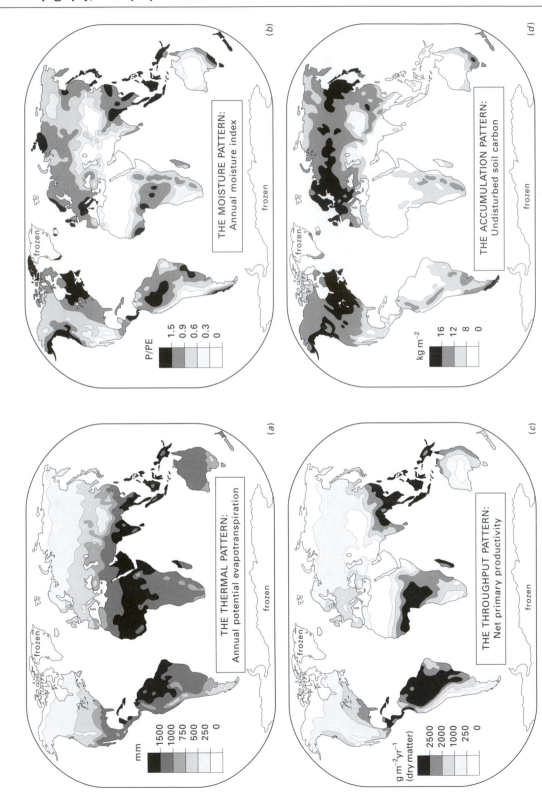

Figure 1.1 Four basic global patterns in climate-related physical and ecological phenomena. (a) Thermal pattern, as represented by annual potential evapotranspiration. (b) Moisture pattern, as represented by the ratio of annual rainfall to potential evapotranspiration (P/PE), or moisture index. (c) Throughput pattern, as represented by net primary productivity. (d) Accumulation pattern, as represented by carbon stored in undisturbed soils. (Adapted from Box and Meentemeyer, 1991.)

temperature levels, annual potential evapotranspiration, and community respiration rate follow the thermal pattern. All these variables are driven by solar energy and are not generally limited by moisture availability. Consequently, they show their highest levels at the equator and lowest levels at the poles.

2. *Moisture pattern*   The moisture pattern is related to the precipitation amounts. It mirrors water availability. Available moisture is far more important in influencing ecological processes than precipitation totals alone, since it is a measure of the water that passes through the landscape (roughly the precipitation less the evapotranspiration). This point is readily understood with an example: a mean annual rainfall of 400 mm might support a forest in Canada, but a dry savannah in Tanzania. Precipitation, the ratio of precipitation to potential evaporation, vegetative cover, and solar energy efficiency of net primary production adhere to the moisture pattern. They tend to have high or medium values in the tropics and temperate belt and low values in the arid subtropics and continental temperate belts.

3. *Throughput pattern*   The throughput pattern mirrors the simultaneous availability of heat and moisture. Actual evapotranspiration, soil texture, the field capacity of soils, net primary production, gross primary production, litter production, the decomposition rates of wood and litter, soil acidity, the base saturation of soils, and possibly plant and animal species richness conform with this pattern. They tend to have high or medium values in the tropics, warm temperate zone, and typical temperate zone, and low values in the arid subtropics and continental temperature regions.

4. *Accumulation pattern*   The accumulation pattern mirrors gains of material over long time spans. Standing biomass, litter accumulation, and soil carbon content comply with this pattern. For instance, soil carbon tends to increase to a steady-state value when gains from leaf-fall balance losses from decomposition. Standing phytomass tends to accumulate so long as annual net primary production exceeds annual litterfall.

Structural factors

Topography can act directly upon the ecosphere, affecting climate, water flow and storage, soils and sediments, and living things. Several structural topographic attributes influence meteorological elements and climate. Essentially, topography has a three-dimensional character that alters, first, airflow patterns, so modifying precipitation, cloud, and so on; and second, the surface radiation balance, so modifying the thermal characteristic of the near-surface air and soil. On a large scale, the size and orientation of such features as mountains and plateaux are powerful factors, as is the disposition of land and sea. On a regional scale, relative relief and terrain shape tend to be particularly influential properties. Surprisingly, perhaps, vegetation appears to wield an influence over some regional climates. A simulation using global climate models showed that boreal forests keep winter and summer air temperatures warmer than would be the case if the forest were replaced by bare ground or tundra vegetation, largely by reducing the albedo (reflectivity) of the surface (Bonan *et al.* 1992). Another simulation suggested that during the mid-Holocene epoch, when boreal forest extended into the present tundra zone, temperatures would have been 2°C higher than now due to variations in the Earth's orbit, and an additional 4°C higher in spring and 1°C higher in other seasons due to the reduction in albedo (Foley *et al.* 1994). And high-latitude forest in the Late Cretaceous period may help to explain the global warmth prevailing at the time (Otto-Bliesner and Upchurch 1997). On a local scale, slope angle, slope aspect, and albedo may produce remarkable variations of climate near the ground.

Land and sea

The disposition of continents and oceans has signal importance for the ecosphere. The distribution of land and sea affects climate, ocean currents, the degree of continentality, and the access between different oceans and different continents. Changing sea level and the rearrangement of continents alters the number of islands and island-continents, and consequently the insularity of terrestrial habitats. This, in turn, influences climate and species diversity.

The alignment of continents has a fundamental yet subtle influence on human culture (Barrow 1995, 122). The early spread of human influence after the evolution of agriculture was more readily achieved in Eurasia, which straddles lines of constant climate over vast distances in an east–west direction, than in the Americas, which run across the gamut of climatic zones. It was, and still is, more difficult for animals and crops to spread in the New World than in the Old World because adaptation to different climates was required. Temperate climates span Eurasia from

Ireland to China, and domesticated animals and cereals are fairly universal across the continent. In the New World, the tropical region separating North and South America severely hinders the passage of animals and crops between the two continents. Had the Americas been turned through 90° and aligned east–west, then the rise of agriculture in the New World might have been faster, and its civilisations might have evolved and spread more rapidly, than those in the Old World.

Continentality (or its inverse, oceanicity) is a very important factor. It arises from differences in the thermal properties of land and water. Briefly, land gains and loses heat more quickly than does water. Moreover, the oceans (and other water bodies) are far more reluctant to give up their stored heat than is the land, so the oceans tend to stay relatively warm during cold seasons. In consequence, the annual and diurnal ranges of surface and air temperatures are much larger over continents than over oceans. The ranges are greatest deep in continental interiors, where the ameliorating effects of oceans are far removed. The effects of continentality are more marked in the Northern Hemisphere, owing to the preponderance of land there. Continentality affects cloud and precipitation regimes – the interiors of the northern continents have winters dominated by cold anticyclones, generally with little cloud, and summers dominated by heat lows and convective storms and cloud regimes. Even at a regional scale the effects of continentality are detectable. The north-west seaboard of the British Isles has a strongly oceanic temperature regime, whereas the south-east has a more continental regime. These differences are reflected in Conrad's (1946) continentality index, k:

$$k = \left(\frac{1.7A}{\sin(\phi + 10)} \right) - 14 \qquad (1.1)$$

where A is the annual range of mean temperature (°C) and ϕ is the latitude angle in degrees. Values of k in the British Isles range from 1.3 at Cape Wrath, which lies on the very exposed north-western tip of the Scottish mainland, to 12.5 at Heathrow, near London (Tout 1976). To put these figures in a continental perspective, the value of k at Thorshavn, in the Faroes, is around zero and for the extreme continental climate of Verkhoyansk, Siberia, it is about 100.

Monsoons, which are large-scale reversals of wind regimes, are partly created by the differential heating of land and sea. The Asiatic monsoon, which affects a sizeable portion of Asia, is perhaps the best known. The classic explanation of monsoons had summer winds, in the manner of a gigantic sea breeze, being 'pulled' into a low pressure cell sitting over a continent and formed by intense heating. This explanation, though attractively simple, has been superseded by more complicated explanations that recognise the involvement of interacting global and regional factors acting through the full depth of the troposphere in the formation of most monsoon systems.

Land and sea breezes and onshore and offshore lake breezes are caused by the differential heating of land and sea.

Relief, ruggedness, and aspect

The three-dimensional form of the land surface on a regional scale affects climates and life. Some mountains develop their own weather systems and characteristic climates. All topographic features disrupt airflow patterns and may generate distinctive atmospheric circulation patterns. Some effects are global, as when north–south trending mountain ranges interfere with planetary waves and instigate the genesis of cyclones. Regional and local effects arise under clear skies and calm conditions, when valleys tend to develop distinctive circulations of air.

The regional aspect (orientation) of large topographic features determines exposure to prevailing winds. Leeward slopes, especially on large hills and mountains, normally lie within a rain shadow. Rain shadow effects on vegetation are pronounced in the Basin and Range Province of the United States: the climates of the Great Basin and mountains are influenced by the Sierra Nevada, and the climates of the prairies and plains are semi-arid owing to the presence of the Rocky Mountains. In the Cascades, the eastern, leeward slopes are drier than the western, windward slopes. Consequently, the vegetation changes from western and mountain hemlocks (*Tsuga heterophylla* and *T. mertensiana*) and lovely and alpine firs (*Abies amabilis* and *A. lasiocarpa*) to western larch (*Larix occidentalis*) and ponderosa pine (*Pinus ponderosa*), and finally to sagebrush (*Artemisia tridentata*) desert (Billings 1990).

Animals also appear to respond to some regional topographic properties. In western North America, mountain goats (*Oreamnos americanus*) and peregrine falcons (*Falco peregrinus*) prefer to live in steep, rugged terrain, whereas pronghorn antelopes (*Antilocapra americana*) and greater prairie-chickens (*Tympanuchus cupido*) are confined to flat and gently rolling terrain (Beasom *et al.* 1983).

Local topography

On a local scale, topographic factors that influence climate include the aspect and slope of the ground surface, the orientation of barriers (such as hedge-rows, walls, and buildings), the vertical structure of vegetation and human-made features (the top of which defines the uppermost surface of the landscape and may be quite distinct from the ground surface), the presence of rivers and lakes, and the distribution of vegetation and land use types. The microclimates and local climates resulting from such topographic influences in turn affect local water balances, soil evolution, geomorphic processes, and the distribution of animals and plants.

In areas where forest lies next to grassland, it is not uncommon for differential heating during the day to produce a flow of air from the relatively cool forest to the relatively warm grassland – a sea breeze without a sea, as it has been called. Similarly, urban areas often have climates that differ from surrounding rural areas, chiefly owing to thermal effects.

Slope aspect produces microclimatic differences large enough to have a considerable impact on ecosystems. In the Tanana watershed region of Alaska, south-facing slopes are mantled by a subarctic brown-forest soil supporting mature white spruce (*Picea glauca*); north-facing slopes are covered by a half-bog soil in which grows non-marketable black spruce (*P. mariana*) (Krause *et al.* 1959). On the south-facing slopes of the Kullaberg Peninsula, south-west Sweden, several arthropod species are found far to the north of their main range. The species include the silky wave moth (*Idaea dilutaris*), the beetle *Danacea pallipes*, and the spider *Theridion conigerum* (Ryrholm 1988).

Slope gradient exerts a considerable influence on ecosystems. In the Derbyshire Dales, England, a sequence of soils and vegetation is strung along a typical catena (Balme 1953). Incipient podzols form on the gentle summit slopes and support heather (*Calluna vulgaris*), bilberry (*Vaccinium myrtillus*), wavy hair-grass (*Deschampsia flexuosa*), and, in poorly drained areas, mat grass (*Nardus stricta*). Brown earths form on the less steep valley-side slopes and support a calcifuge grassland community dominated by common bent (*Agrostis tenuis*), sheep's fescue (*Festuca ovina*), and red fescue (*F. rubra*), with calcicolous species restricted to limestone outcrops. The steeper valley-side slopes are characterised by rendzinas with incipient leaching that support a close sward of sheep's fescue and red fescue, though in areas where the sward is densest, such calcifuges as mountain pansy (*Viola lutea*) are found. Calcareous rendzinas with dark alkaline humus occur in the lower valley slopes and valley floor. Here, the vegetation is either calcareous grassland dominated by sheep's fescue growing in scattered tufts with such calcicolous herbs as common rock-rose (*Helianthemum chamaecistus*) and rough hawkbit (*Leontodon hispidus*), or dense thickets dominated by ash (*Fraxinus excelsior*) and field maple (*Acer campestre*).

Slope curvature, as well as slope gradient, affects ecosystems. In the central Appalachian Mountains, pine forest tends to grow on convex noseslopes, oak forest on sideslopes, and northern hardwood species (beech, maple, and birch) in concave hollows (Hack and Goodlett 1960).

What is the toposphere?

The toposphere sits at the interfaces of the pedosphere, atmosphere, and hydrosphere. The totality of the Earth's physical land surface was called the 'relief sphere' by the German geomorphologist Julius Büdel (1982). A more euphonious alternative is the 'toposphere' (Huggett 1995, 7). As topography is more than just the ground surface, so the toposphere embodies more than just the planet's physical relief features. It includes all the surface features of the globe – ground surface, vegetation, and human-made objects. Even animals might be regarded as moving parts within the toposphere. The toposphere is not a two-dimensional surface; in many places it has a vertical dimension. Its lower boundary is the ground surface and its upper boundary is defined by the height of vegetation (the vegetation canopy) or the tops of buildings (the urban canopy). Caverns, grottoes, and tunnels (made by burrowing animals and humans) extend the lower boundary of the toposphere into the ground, often through a network of complex passages. Under special circumstances animals may affect the toposphere – huge flocks of birds or large insect swarms may temporarily raise the top of the toposphere above the vegetation canopy and buildings. Where vegetation is absent, as in sand seas or the deep ocean floor, the toposphere is restricted to the ground surface. But in many environments, the toposphere is a three-dimensional body that nurtures its own microclimate within it.

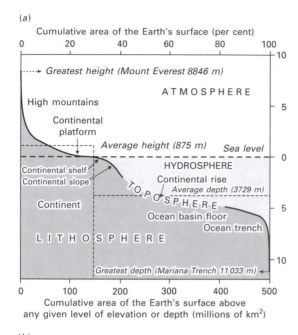

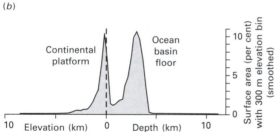

Figure 1.2 Hypsometric curves of the toposphere. (a) Cumulative hypsometric curve with relief features indicated. (b) Differential hypsometric curve showing the area of the toposphere between successive levels using a 300-m elevation bin. (The hypsometric curves are adapted from Rosenblatt *et al.*, 1994; otherwise adapted from Huggett, 1997, 223.)

The physical toposphere

The continents, extended to the edge of the presently submerged continental shelves, and the oceans are the major topographical features of the Earth (Figure 1.2). It is convenient to discuss the continental toposphere and the oceanic toposphere separately. The two are linked but form a major topographic division at the Earth's surface that has an enormous influence on the ecosphere. A big difference is obviously that the continental toposphere sits at the bottom of the atmosphere, whereas the oceanic toposphere sits at the bottom of the oceans. The average height of the land

surface is 875 m, whereas the average depth of the oceans is −3729 m. Interestingly, though, the highest point on land, about 9 km, is comparable to the deepest point in the oceans, about 11 km. But the differences are much more profound than such simple contrasts suggest. Continents and oceans have very different origins. No ocean floor appears to be older than about 240 million years, and the ocean floor is created at mid-ocean ridges and destroyed at subduction sites. Continents, if not everlasting or static, have been afloat on the underlying asthenosphere since they were first formed some 2.5 billion years ago.

Continental toposphere

Large-scale topographic features of continents are ancient crystalline shields, old and young mountains, and upland and lowland plains. Upland plains may form plateaux (tablelands). The dominant feature of the continental toposphere is the continental platform (Figure 1.2).

The distribution of large-scale continental topographic forms mirrors the distribution of the crustal types. Continental shields and platforms tend to have low to moderate elevation. The sediments on the platforms either are horizontal or undulate a little. The mountain belts produce distinct topographic features. Palaeozoic mountain belts are often expressed as curved belts of eroded upland. Mesozoic and Cenozoic mountain belts are seen as eroded uplands, or as eroded upland areas separated by basins, as in the Basin and Range Province in the Great Basin, western United States. Young mountain belts, or active orogens, are commonly expressed as ribbons of high elevation (over 3 km) with signs of intense folding and thrusting. In places, they contain high plateaux, which are relatively undeformed and uplifted regions. An example is the Colorado Plateau.

The highest mountain is Mount Everest, in Nepal and China, standing at 8846 m; the second highest is K2 (Mount Godwin Austen), in Pakistan and China, at 8611 m. These two mountains are in the Himalayas, a mountain range some 2400 km long and 200–400 km wide. The Himalayas include 96 peaks higher than 7315 m (24 000 feet). Other lofty mountains are Aconcagua, Argentina (6959 m), Mount McKinley, Alaska (6194 m), and Kilimanjaro, Tanzania (5895 m). Smaller-scale structural topographic features of continents are associated with folds, fault-blocks, rift systems, and other lineaments (see Summerfield 1991). Other features are associated more directly with erosional and depositional geomorphic processes –

features such as hillslopes, sink-holes, caves, river valleys, deltas, and desert pavements.

Rivers and lakes lie upon the ground surface and form patterns that commonly follow local geological structures. Rivers range in length from a few metres to thousands of kilometres. The Nile extends for 6670 km, the Amazon for 6437 km, and the Yangtze (Chang Jiang) for 6300 km. In places, rivers plunge over sheer slopes to form waterfalls. These range from minor falls in rivers to such impressive drops as the 979 m Angel Falls, Venezuela; the 604 m Giessbach Falls, Switzerland; the 580 m Sutherland Falls, New Zealand; the 491 m Ribbon Falls, United States; and the 422 m Gavarnie Falls, France. Lower but broad drops, as in the Niagara and Victoria Falls, are equally impressive. The Boyoma Falls of the Democratic Republic of the Congo (formerly Zaïre) boast the highest annual water flow.

Freshwater lakes occupy about 825 000 km^2, or 0.55 per cent of the land surface. They range from small seasonal ponds to such giants as Lake Superior, North America (83 270 km^2), and Lake Victoria, Africa (69 400 km^2). Forty large lakes hold about 80 per cent of the world's lake-water (Nace 1969). The bulkiest and deepest lake is Lake Baikal in central Asia. This lake has an area of almost 22 000 km^2, but owing to its great depth, contains nearly as much water as all the North American Great Lakes.

Oceanic toposphere

For a long time, the topography of the ocean floor was largely a mystery. Even now, the deep ocean bed is not well explored, but the main topographic features have been mapped. Continental shelves run from the low water mark on beaches to a depth of about 200 m, and are part of continental land masses. In places, continental shelves are crossed by ravines that carry sediment to the open ocean. At the edge of the continental shelf, the ocean floor pitches at an angle of 3–6° along the continental slope. This ends in the less steep continental rise, which falls at a rate of 1-in-100 to 1-in-700 and attains an average depth of about 4 km. The continental rise gives way to the ocean basin (abyssal plain). The extensive plains of the deep ocean floor are interrupted by seamounts. Seamounts are volcanoes, some of which break the ocean surface to form islands. Hawaii is the exposed top of an underwater mountain that at 9150 m is higher than Mount Everest. There are also flat-topped volcanoes, known as guyots. The other prominent feature of the deep ocean is mid-ocean ridges. These

are rugged underwater mountain chains that may be 1000 km across and may have peaks rising several kilometres above the ocean floor. In places, including Iceland and Tristan da Cunha, the ridges break the surface to form islands. The ridges are often traversed by transform faults. Very long and deep ocean trenches border some oceanic regions; they may have volcanoes on their flanks. The deepest point is the Mariana trench (−11 033 m).

Other major features of the oceanic toposphere are island arcs, marginal-sea basins, and inland-sea basins. Island arcs lie above subduction zones in the lithosphere. There are two kinds: continental-margin arcs, which include Japan and New Zealand; and island-arcs, which are collections of mainly andesitic volcanic peaks. Marginal-sea basins are portions of ocean crust, such as the Philippine Sea, that sit between island arcs. Inland-sea basins are partially or totally surrounded by tectonically stable continental crust. The salty Caspian Sea of Asia is the largest at 373 000 km^2. Its surface lies 28 m below sea level and its maximum depth is 980 m.

The biological toposphere

Of all living things, plants contribute most to the structure of the toposphere. Vegetation gives the natural toposphere its vertical dimension. Vegetation is tallest in woods and forests and shortest in deserts. Humans have transformed much of the land-surface vegetation cover for agricultural and silvicultural purposes, generally converting woodland and forest to monocultural fields of grass crops (wheat, maize, rice). The fields are in places separated by hedgerows.

Humans have not only remade the features of the biological toposphere, they have also created new features for their own ends – buildings, bridges, roads, railways, canals, paths, fences, and so on. The tallest structures are the CN Tower, Toronto, Canada (553 m), and the Petronas Tower, Kuala Lumpur, Malaysia (450 m). The Sears Tower, Chicago, which stands at 442 m not including the spires, is the tallest office block in the world.

References

Balme, O. E. (1953) Edaphic and vegetational zoning on the Carboniferous limestone of the Derbyshire Dales. *Journal of Ecology* **41**, 331–44.

Barrington, T. J. (1976) *Discovering Kerry: Its History, Heritage and Topography*. Dublin: Blackwater Press.

Barrow, J. D. (1995) *The Artful Universe: The Cosmic Source of Human Creativity*. Harmondsworth: Penguin.

Beasom, S. L., Wiggers, E. P., and Giardino, J. R. (1983) A technique for assessing land surface ruggedness. *Journal of Wildlife Management* **47**, 1163–6.

Billings, W. D. (1990) The mountain forests of North America and their environments. In C. B. Osmond, L. F. Pitelka, and G. M. Hidy (eds) *Plant Biology of the Basic and Range* (Ecological Studies, Vol. 80), pp. 47–86. Berlin: Springer.

Bonan, G. B., Pollard, D., and Thompson, S. L. (1992) Effects of boreal forest vegetation on global climate. *Nature* **359**, 716–18.

Box, E. O. and Meentemeyer, V. (1991) Geographic modeling and modern ecology. In G. Esser and D. Overdieck (eds) *Modern Ecology: Basic and Applied Aspects*, pp. 773–804. Amsterdam: Elsevier.

Büdel, J. (1982) *Climatic Geomorphology*. Translated by Lenore Fischer and Detlef Busche. Princeton, NJ: Princeton University Press.

Chatwin, B. (1998) *The Songlines*. London: Vintage.

Coffey, W. J. (1981) *Geography: Towards a General Spatial Systems Approach*. London and New York: Methuen.

Conrad, V. (1946) Usual formulas of continentality and their limits of validity. *Transactions of the American Geophysical Union* **27**, 663–4.

Davis, W. M. (1924) The progress of geography in the United States. *Annals of the Association of American Geographers* **14**, 159–215.

Downs, R. M. and Stea, D. (1976) *Image and Environment: Cognitive Mapping and Spatial Behavior*. Chicago, IL: Aldine Publishing.

Fenneman, N. M. (1939) The rise of physiography. *Bulletin of the Geological Society of America* **50**, 349–60.

Foley, J. A., Kutzbach, J. E., Coe, M. T., and Levis, S. (1994) Feedbacks between climate and boreal forests during the Holocene epoch. *Nature* **371**, 52–4.

Gerald of Wales (Cambrensis Giraldus) (1983 edn) *History and Topography of Ireland*. Translated by John J. O'Meara. Harmondsworth: Penguin.

Hack, J. T. and Goodlett, J. C. (1960) *Geomorphology and Forest Ecology of a Mountain Region in the Central Appalachians* (US Geological Survey Professional Paper 347). Washington, DC: US Government Printing Office.

Hough, M. (1990) *Out of Place: Restoring Identity to the Regional Landscape*. New Haven, CT, and London: Yale University Press.

Hubert, J.-P. (1998) À la recherche d'une géométrie de l'espace habité chez Camille Vallaux, Jean Gottmann et Gilles Ritchot. *L'Espace Géographique* 1998 (3), 217–27.

Hubert, J.-P. (1999) L'aménagement et le concept de structure. In A. Fischer and J. Malezieux (eds) *Industrie et Aménagement*, pp. 129–44. Paris: Éditions L'Harmattan.

Huggett, R. J. (1995) *Geoecology: An Evolutionary Approach*. London and New York: Routledge.

Huggett, R. J. (1997) *Environmental Change: The Evolving Ecosphere*. London: Routledge.

Huxley, T. H. (1877) *Physiography: An Introduction to the Study of Nature*, 1st edn. London: Macmillan.

Jackle, J. A. (1987) *The Visual Elements of Landscape*. Amherst, MA: University of Massachusetts Press.

Krause, H. H., Rieger, S., and Wilde, S. A. (1959) Soils and forest growth in the Tanana Watershed of interior Alaska. *Ecology* **40**, 492–5.

Moore, R. L. and Graefe, A. R. (1994) Attachments to recreation settings: the case of rail-trail users. *Leisure Sciences* **16**, 17–31.

Nace, R. L. (1969) World water inventory and control. In R. J. Chorley (ed.) *Water, Earth, and Man: A Synthesis of Hydrology, Geomorphology, and Socio-economic Geography*, pp. 31–42. London: Methuen.

Otto-Bliesner, B. L. and Upchurch, G. R. (1997) Vegetation-induced warming of high-latitude regions during the Late Cretaceous period. *Nature* **385**, 804–7.

Pritchett, W. K. (1969–1992) *Studies in Ancient Greek Topography, Parts 1–8*. University of California Publications in Classical Studies. Berkeley, CA, and London: University of California Press.

Relph, E. (1996) Place. In I. Douglas, R. J. Huggett, and M. E. Robinson (eds) *Companion Encyclopedia of Geography: The Environment and Humankind*, pp. 906–22. London and New York: Routledge.

Rosenblatt, P., Pinet, P. C., and Thouvenot, E. (1994) Comparative hypsometric analysis of Earth and Venus. *Geophysical Research Letters* **21**, 465–8.

Ryrholm, N. (1988) An extralimital population in a warm climatic outpost: the case of the moth *Idaea dilutaria* in Scandinavia. *International Journal of Biometeorology* **32**, 205–16.

Smith, J. T. (1815) *Ancient Topography of London: Containing Not Only Views of Buildings . . . but Some Account of Places and Customs either Unknown, or Overlooked by the London Historians*. London: published and sold by the Proprietor, John Thomas Smith.

Stoddart, D. M. (1975) 'That Victorian science': Huxley's physiography and its impact on geography. *Transactions of the Institute of British Geographers* **66**, 17–40.

Summerfield, M. A. (1991) *Global Geomorphology*. Harlow: Longman.

Tout, D. G. (1976) Temperature. In T. J. Chandler and S. Gregory (eds) *The Climate of the British Isles*, pp. 96–128. London and New York: Longman.

Tuan, Y.-F. (1974) *Topophilia: A Study of Environmental Perception, Attitudes, and Values*. Englewood Cliffs, NJ: Prentice Hall.

CHAPTER 2

Characterising topography

This chapter deals with the definition and description of topographic elements, including locational factors (latitude, longitude, and altitude) and compositional factors (landform and landscape units). It looks at the topographic structures that are constructed from these elements – networks, mosaics, and regions. It also considers the varied flows of materials and energy that link landscapes at local, regional, and global scales.

Topographic elements

Locating places

Longitude and latitude

Latitude and longitude define a geographical grid for locating points on the Earth's surface. Parallels of latitude, sometimes simply called parallels, run east–west. Meridians are north–south lines and define longitude.

Latitude is prescribed by the angular distance of a point from the equator – it indicates how far a point lies north or south of the equator. It is measured in degrees of arc from the equator (0°) towards either pole (90°). Manchester Airport, England, lies at latitude 53°21′N; Hobart International Airport, Australia, lies at latitude 42°84′S.

Longitude measures the position of a point eastward or westward from the reference prime meridian (the line joining the north pole and south pole that passes through the old location of the Greenwich Royal Observatory, near London). It is expressed as the angle of arc between the meridian of the point and the reference meridian. Manchester Airport lies at longitude 2°17′W; Hobart International Airport lies at longitude 147°50′E.

Altitude

Altitude, also called elevation, is the height of a place above (or below) a reference level, such as mean sea level. Manchester Airport has an altitude of 69 m above mean sea level; Hobart International Airport has an altitude of 4 m. Death Valley, in south-eastern California and south-western Nevada, has an altitude of –86 m. In seas and oceans, negative altitudes are usually referred to as depth. The Puerto Rico trench has a depth of 9200 m. Differences in altitude are used to define many other topographic attributes, such as slope gradient and slope curvature (p. 20).

Global Positioning System

In 1945, Arthur C. Clarke, the science-fiction writer, predicted the invention of geopositional satellites. The forerunner of the modern Global Positioning System (GPS) was originally developed in 1964 by the US Navy for tracking US submarines using a single satellite. The United States GPS, which is maintained by the Department of Defense, now uses a constellation of 24 satellites and associated control systems that allow a suitable radio receiver to find its location anywhere on Earth, by day or night. The Russian Global Navigation Satellite System (GLONASS) positioning system, set up in 1982, theoretically does the same thing, but it has experienced operational problems.

Originally, the US GPS was implemented for military purposes, but an increasing number of non-military applications have appeared. For instance, the GPS is proving to be a valuable tool in geomorphology, especially when it is linked to geographical information systems (e.g. Cornelius et al. 1994; Twigg 1998). It enables latitude, longitude, and altitude to be determined, although the accuracy in doing so depends on the method used to obtain a 'fix' (absolute, differential, or static).

Characterising places

The most common method of characterising topography in the past used contours and spot heights on paper maps. The information source for these data was either field-based surveying or extraction from overlapping stereoscopic aerial photographs. However, in the last 25 years there has been increasing development and use of Geographical Information Systems (GIS) that allow input, storage, and manipulation of digital data representing spatial and aspatial features of the Earth's surface. The digital representation of topography has probably attracted greater attention than that of any other feature of the Earth's surface. There are numerous ways of modelling the spatial form of surface topography. Digital representations are referred to as either Digital Elevation Models (DEMs) or Digital Terrain Models (DTMs). A DEM is 'an ordered array of numbers that represent the spatial distribution of elevations above some arbitrary datum in a landscape' (Moore *et al.* 1993, 8). DTMs are 'ordered arrays of numbers that represent the spatial distribution of terrain attributes' (Moore *et al.* 1993, 8). DEMs are, therefore, a subset of DTMs.

The structure or organisation of DEM data can be subdivided into (1) contours, (2) triangular irregular networks (TINs), and (3) regular grids (Figure 2.1). The structure used depends upon the initial raw data source and the proposed application. Contour-based methods use contour lines stored as digital line graphs (DLGs) made up of *x* and *y* co-ordinate pairs of specific elevation. TINs sample surface-specific points, such as ridges, peaks, and breaks in slope, the regular network of points being stored as sets of *x*, *y*, and *z* co-ordinates each with pointers to their neighbours in the net (Peuker *et al.* 1978; Mark 1975). TINs

are made up of facets, which are triangular planes created by joining three adjacent points. The triangles create a continuous, connected surface based upon Delauney triangulation (Weibel and Heller 1991). Regular grid methods (sometimes referred to as gridded DEMs or altitude matrices) may use a regularly spaced square, or a triangular or rectangular grid. The choice of cell size or data resolution depends upon the application and the study area being examined. Colour Plate 1 shows a DEM.

A GIS will store all three DEM formats and allows inter-transformation between them. Deciding which format to use depends upon the application and the type of data analysis required. Gridded DEMs are the most commonly used, since they allow additional topographic parameters to be calculated and have a relatively simple structure that is easy to integrate with other data when carrying out process modelling. With TINs, the irregularity of the structure makes computation of topographic attributes more complex. For instance, it is difficult to determine the upslope connection of a facet (Moore *et al.* 1993). Calculation of topographic parameters with DLGs is very difficult and they are, therefore, less popular. All the same, DLGs are often created as an initial step in creating the two more popular models.

Digital elevation data sources

Raw elevation data can be derived from photogrammetric methods, including stereo aerial photographs, satellite imagery, and airborne laser interferometry, or from field surveys using GPS or total stations. If stereo aerial photographs and satellite images are the sources for elevation data, there will be a complete coverage of the landscape at the resolution of the image or photographs. The pre-processing of aerial photographs or satellite imagery is essential. It includes geometric correction and removal of distortion

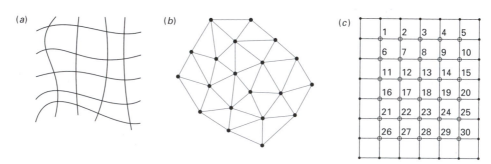

Figure 2.1 DEM data can be subdivided into (a) contours, (b) triangular irregular networks (TINs), and (c) regular grids.

with respect to the chosen spheroid, projection, and orientation. Other distortions, such as the tilt and wobble of the viewing platform (i.e. aeroplane or satellite) and atmospheric effects, also have to be removed to produce a quality product. When using aerial photographs, co-ordinate points are measured and their locations and elevations determined, a least-squares solution then being employed to produce a stereo model that can be processed to generate a DEM (Dixon *et al.* 1998; Gilvear *et al.* 1998). The DEM is created by extracting data to create a proper hypsometric surface (measurement of elevation at the Earth's surface with respect to sea level) at a resolution appropriate for the study application. Stereo images from SPOT, the French Earth observation satellite (Le Système pour l'Observation de la Terre), can be used to produce orthophotos and DEMs in a similar manner to the aerial photographs. An advantage of using satellite images is that they are already in digital format.

Airborne laser interferometry involves using scanners to provide high resolution surface measurements. An example is Light Detection and Ranging (LIDAR). Although LIDAR is a relatively young and complex technology, it provides a technique that is accurate, suitable for areas of rugged and difficult terrain, and increasingly affordable (Flood and Gutelius 1997). LIDAR works by measuring the laser-pulse travel time from a transmitter to a target and back to the receiver. The laser pulse travels at the speed of light (3×10^8 m/s), so very accurate timing is required to obtain fine vertical resolutions. As the aircraft flies over an area, a scanning mirror directs the laser pulses back and forth across-track. The collected data is a set of points arranged across the flight-line. The combination of multiple flight-line data provides coverage for an area. LIDAR data do not require orthorectification because every LIDAR measurement is georeferenced, which is achieved by collecting several measurements, including location of the aircraft, using GPS (Jensen 2000). Several studies have been carried out to determine the accuracy of LIDAR measurements (Vaughn *et al.* 1996; Ridgway *et al.* 1997; Krabill *et al.* 1995). These report that LIDAR provides measurements with vertical accuracies of around ±10 cm, although this can be improved to ±5 cm if care is taken in the various calibrations and corrections.

An extremely useful characteristic of LIDAR is its ability to penetrate the vegetation canopy and map the ground beneath. Some of the laser pulse is actually backscattered by the vegetation, but only a portion of the laser needs to reach the ground to pro-vide a ground measurement. LIDAR will record the returns both from the vegetation canopy and from the ground, so that the elevation of the vegetation canopy and of the ground can be measured. This characteristic has been exploited in a number of studies for estimating timber volume and forest biomass (Nilsson 1996; Nelson *et al.* 1988) and vegetation height (Weltz *et al.* 1994). Nonetheless, if the vegetation canopy is very dense, so that only a few pulses reach the ground, the data cannot be used for creating an accurate DEM, which is disadvantageous should DEM production be the objective. In addition, buildings will completely intercept the laser pulses. Work has been carried out to separate the ground measurements from measurements of buildings, vegetation, and other structures. In the United Kingdom, the Environment Agency is currently carrying out a programme of capturing LIDAR data in regions susceptible to coastal erosion and river flooding (Colour Plate 2).

Many users of DEMs may not have the necessary sophisticated equipment or expertise to capture raw elevation data using the above photogrammetric methods. A common alternative is to extract elevation information from paper topographic maps using digitising techniques. Digitising can be carried out manually using a digitising tablet, but this is a monotonous and laborious job, the results of which are subject to many human errors. Alternatively, a topographic map may be scanned and automatic tracing software used to capture the contour information. The resulting contours can be retained as contour DLGs or interpolated onto a TIN or regular grid (Ceruti 1980; Oswald and Raetzsch 1984). Spot heights can also be digitised and incorporated into a TIN or interpolated into a regular-gridded DEM. Most government mapping agencies, such as the United Kingdom's Ordnance Survey (OS) or the United States Geological Survey (USGS), have produced DEMs by digitising topographic maps or using aerial photographs, and the resulting DEMs are available as a commercial product. Britain, USA, and Australia are all completely covered by coarse DEMs at a 1:250 000 scale. Higher resolution DEMs, derived from 1:50 000 or 1:25 000 maps, are also available in many countries (e.g. the UK's Ordnance Survey provide Land-form PANORAMA 1:50 000 and Land-form PROFILE 1:10 000 DEMs, both as gridded DEMs and as digital contours). The USGS (United States Geological Survey) have produced DEMs from a number of different methods, including from contour maps, digitised elevations and photogrammetric stereomodels based on aerial photographs, and satellite remote sensing images. They produce a range of

Table 2.1   Example applications that have utilised digital elevation data

Application	Type of DEM	Reference
Landscape representation and hydrological simulations	Gridded	Zhang and Montgomery (1994)
Hydrological and ecological applications	Contours	Moore *et al.* (1988)
Forest gully erosion	Gridded	DeRose *et al.* (1998)
Predicting erosion hazard areas	Gridded	Vertessy *et al.* (1990)
Prediction of concentrated flow erosion in cultivated catchments	Gridded	Ludwig *et al.* (1996)
Soil erosion modellings using 'ANSWERS'	Gridded	De Roo *et al.* (1989)
Predicting topographic limits to a gully network	Gridded	Prosser and Abernethy (1996)
Analysis of erosion thresholds, channel networks and landscape morphology	Gridded	Dietrich *et al.* (1993)
Spatially distributed solar radiation modelling	Gridded	McKenney *et al.* (1999)
Modelling the surface and bed topography of glaciers	TIN	Theakstone and Jacobsen (1997)
Distributed snowmelt simulations of an alpine catchment	Gridded	Blöschl *et al.* (1991)
Snow accumulation and distribution in an alpine catchment	Gridded	Elder *et al.* (1991)
Viewshed and least-cost path analysis	Gridded	Lee and Stucky (1998)
River channel adjustments	Gridded	Downs and Priestnall (1999)
Source areas, drainage density, and channel initiation	Gridded	Montgomery and Dietrich (1989)
Recognition of geological and geomorphic patterns	Gridded	Chorowicz *et al.* (1989)
Modelling river-channel topography	TIN and Gridded	Milne and Sear (1997)

DEMs with different grid spacings and coverages, from 7.5-minute to 30-minute resolutions (Garbrecht and Martz 2000).

Square-gridded DEMs have become the most common format of DEM in use, owing to their ease of computer representation and manipulation (Moore *et al.* 1991; Wise 1998; Wilson and Gallant 2000). Originally, they were derived from stereoscopic aerial photographs, the aim being to create an orthophoto map with the DEM as a by-product, although they can be derived using any of the previously described methods. Owing to the simplicity of gridded DEMs and the ease with which this form of data can be handled within a computer, in particular within GIS software packages implementing raster functionality, they are frequently used within spatial analysis and process modelling of physical phenomena (Table 2.1) and to create such additional surface variables as slope, aspect, solar irradiance, and visibility.

Gridded DEMs, however, have a number of disadvantages, including (as reported by Moore *et al.* 1993; Burrough and McDonnell 1998; Wilson and Gallant 2000): (1) data redundancy in areas of uniform terrain; (2) they cannot easily handle abrupt changes or differing complexity in elevation due to resolution, so that important details of land surface in complex regions are often missed; (3) the resolution of the grid influences storage requirements, computational efficiency, and results obtained; and (4) the computed upslope flow-paths used in hydrological analysis tend to zigzag unrealistically, which hinders the accurate determination of such primary attributes as catchment area (Zevenbergen and Thorne 1987; Moore *et al.* 1991) (see p. 110). Recent developments have overcome some of these drawbacks; for example, new compression techniques have reduced storage requirements and improved computer efficiency (Kidner and Smith 1992; Smith and Lewis 1994).

Using a TIN structure often offers a more flexible and efficient alternative to overcome many of the above problems, and in particular to avoid data redundancy. The advantage of a TIN is that it allows extra information to be stored for areas of complex terrain without having to store additional redundant information for flatter, more homogeneous regions. The data capture procedure for a TIN can specifically follow ridges, stream lines, and other important topological features (even human-made features) and these can be digitised at the necessary accuracy. This data capture can be manual or automated. TINs allow efficient, accurate data storage representing topography because the density of the triangles can be varied to match the roughness of the terrain (Moore *et al.* 1991). The TIN structure allows the generation of aspect, slope, contour maps, profiles, horizons, and line-of-sight maps, although the derived outputs frequently display triangular patterns. TINs have been used by Tajchman (1981) to compute topographic characteristics of a mountainous catchment and by Jones *et al.* (1990) for watershed delineation. However, sometimes the triangular discretisation can hinder certain spatial analysis due to the complex computations required, as in the derivation of surface geometry.

Errors in DEMs

Since DEMs are scaled models of topography they can never be truly accurate. Nonetheless, accuracy of DEMs varies dramatically as a result of differences in accuracy of source data and the interpolation method used to create the DEM. It is generally considered that DEM accuracy can never be greater than that of the original source of data. The optimal DEM quality depends upon the application for which the DEM is to be used, and an important problem is that DEMs created for one application are often subsequently for different purposes. Sometimes a quantitative measure of the accuracy of a DEM, created by organisations such as National Mapping Agencies, will be provided with the DEM product. However, such accuracy statements can be misleading, since (1) the method for obtaining such an error value is not usually detailed; (2) they do not test for the spatial distribution of the error; and (3) the methods used to produce the error value do not evaluate the DEM for the different types of error (random or systematic). Systematic errors are artefacts of the DEM production process and are considered to be the most significant errors in a spatial or statistical analysis because they are not easily detected yet can introduce significant bias (Brown and Bara 1994). It is important to understand the nature of error within a DEM, since this error will be propagated into any further data sets derived directly from processing the DEM or produced through process modelling involving the DEM. Over the last 15 years, considerable research has been directed to the causes, detection, representation, and correction of DEM errors.

Many studies have considered the cause of errors in DEMs created by different methods (e.g. Carter 1988; Weibel and Heller 1991; Wise 1998, 2000).

Contours digitised from topographic maps are commonly the source of DEM errors. In the first instance, there is the false assumption that smooth contours are a true representation of topography and that contour accuracy is dependent upon the quality and scale of the aerial photographs originally used to produce them and the characteristics of the photogrammetric device. In addition, paper maps may be damaged by stretching and folding, and if the contours are digitised manually, human errors are likely to be added. Once the contours have been digitised, they are often interpolated to create a gridded DEM and errors may be incorporated because of the geometry of the search algorithms (Burrough and McDonnell 1998). Gridded DEMs created in this manner frequently display a 'padi' (or rice terrace) effect in the areas bordering each contour. This is because a digitised contour line produces sets of data points with the same z value. During the interpolation process a certain number of data points (from the source data set) within a given search radius are found and a weighted average calculated. When all the points have the same z value, usually in low-relief areas, the padi effect is produced.

Burrough and McDonnell (1998) suggest an alternative method, designed specifically for digitised contours. It requires thinning digitised contour points to a minimum, then adding important topographic points such as peaks, ridges, valley bottoms, and so on, and then interpolating with a large search window that incorporates many data points. While this is computationally demanding, the results are considered superior to those of simple interpolation methods. Regardless, DEMs derived from topographic maps are considered to be of poorer quality than DEMs derived from direct photogrammetric measurements (Burrough and McDonnell 1998).

Carrara *et al.* (1997) provide a comparison of several common methods for producing DEMs from digitised contours. Two grid-based contour interpolators and two TIN generators were compared for three sample areas of complex morphology. (The DEM generator methods are all available in commercial GIS software packages.) They found that all the DEMs produced were affected by one or more types of error or pitfall, although one TIN generator and one grid interpolator proved to be capable of effectively producing terrain models that largely reflected the ground morphology as expressed by the source contour lines. Wise (1998, 2000) also found that there is a great deal of variation between DEMs derived from the same source data but using different interpolation techniques.

There is no unique criterion or single technique for measuring the quality of DEMs, and many methods have been developed to detect and estimate the magnitude and spatial distribution of the errors (e.g. Polidori *et al.* 1991; Brown and Bara 1994; Felicísimo 1994; Fryer *et al.* 1994; Li 1994; Garbrecht and Starks 1995; López 1997). One of the most common quantitative methods is to compare elevations in the DEM with 'true' independent elevation values on the terrain. The true values may be collected from mapped spot-heights (Wise 1998; Monckton 1994; Evans 1980; Skidmore 1989), manual interpolation of contours (Clarke *et al.* 1982), or independent field measurement (Bolstad and Stowe 1994; Giles and Franklin 1996) that may be collected using GPS (Adkins and Merry 1994). Other quantitative methods have used simulated DEMs (Chang and Tsai 1991; Carter 1992; Hodgson 1995).

One of the earliest statistical methods used to produce an expression of the error was the root-mean-square error (RMSE):

$$\text{RMSE} = \sqrt{\left(\sum_{i=1}^{n} d_i^2\right)\Big/n} \qquad (2.1)$$

where $d_i = Z_{est} - Z_{obs}$, where Z_{est} is the DEM value and Z_{obs} is the 'true' elevation value for the same location. A number of summary statistics can be calculated including standard deviation, maximum, minimum, and mean error. Mean error (ME) measures whether the DEM is systematically underestimating or overestimating elevation:

$$\text{ME} = \left(\sum_{i=1}^{n} d_i^2\right)\Big/n \qquad (2.2)$$

The USGS use RMSE to classify their DEMs into three levels of increasing quality. Level 1 is for data derived from scanning National High-Altitude Photography Program, National Aerial Photography Program, or equivalent photography. They state that the vertical RMSE of 7 m is the targeted accuracy, and a RMSE of 15 m is the maximum permitted. Level 2 is for elevation data sets smoothed or processed for consistency and edited to remove identifiable systematic errors. One-half of the original map contour interval is the maximum RMSE permitted, and no errors greater than one contour interval are permitted at all (Garbrecht and Martz 2000). In Level 3, DEMs are produced from digital line-graph data using selected elements from both hypsography (i.e. contours, spot heights) and hydrography (lakes, shorelines, drainage). Ridgelines and major transportation features are sometimes included. The maximum permitted error is a RMSE of one-third of the contour interval and

there are no errors greater than two-thirds of the contour interval in magnitude. The majority of data produced within the last decade fall within Level 2 (Garbrecht and Martz 2000).

Such methods have been criticised for failing to investigate the spatial structure of error in digital elevation data (Monckton 1994). Monckton attempted to overcome this by using Moran's I spatial autocorrelation to evaluate the spatial error of DEMs employing the elevational differences between the DEM and spot heights taken from maps of a larger scale than that used to derive the DEM. Graphical output was produced to indicate the spatial autocorrelation at lag distances of 250 m and above, values below this distance being unobtainable owing to the arrangement of the spot-height data. Monckton was able to use this method to ascertain that there was a clear random spatial arrangement of error at all lags, indicating that the error was probably truly randomly generated rather than an artefact of the production process and was independent of the nature of the terrain itself. The drawback to this method is that the use of spot heights may introduce some bias through their location and this may be detrimental to their statistical purity.

López (1997) developed another method which locates randomly distributed, weakly spatially correlated errors by applying a method based on principal components analysis (PCA). The process involved the decomposition of a DEM into strips and carrying out PCA on each one. The method aids the process of identifying unlikely values which can then be replaced using some suitable interpolation method.

Brown and Bara (1994) presented two techniques for revealing the presence and form of systematic error, both based on the assumption that short-range anisotropy in spatial data is indicative of error. Anisotropy occurs when the general pattern of variation in one direction (e.g. east to west) is different from the pattern of variation in another direction (e.g. north to south). The two methods involved the calculation of semivariograms and fractal dimensions. As well as successfully detecting the presence and structure of errors, the methods provided a quantitative basis for applying corrections to the surface.

Visualisation, as a method of evaluation, has been focused upon by other researchers (Kraus 1994; Hunter and Goodchild 1995; O'Callaghan and Mark 1984; Wood and Fisher 1993; Bolstad and Stowe 1994). Simple visualisation checks include reasonableness, conformance to general knowledge of terrain shape, and geomorphic consistency (e.g. connected

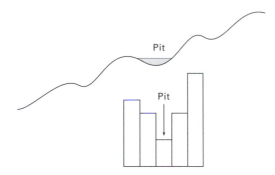

Figure 2.2 A pit in a DEM.

stream channels, ridges) (Bolstad and Stowe 1994). Other methods include shaded relief and isopleth maps that can be calculated from a DEM and allow a visual inspection for local anomalies that manifest themselves as bright or dark spots. These methods can be employed for random and systematic error location. Hunter and Goodchild (1995) found these methods, as well as probability mapping, useful for identifying edge-matching problems. Another graphical approach is to create contours from the DEM; this provides a sensitive assessment of terrain structure, since the position, aspect, and curvature of the contours depend directly on the elevation, aspect, and plan curvature, respectively, of the DEM (Wilson and Gallant 2000). Derived contours are very sensitive to elevation errors in the source data, particularly where small isolated contours exist on the source map.

It can be seen that there are numerous methods for evaluating the quality of a DEM. However, no individual method provides a comprehensive account of the DEM quality. The most prudent approach would be to adopt a range of methods, both graphical and numerical, which will determine the magnitude and spatial extent of both random and systematic errors.

Once the errors have been identified it may be necessary to correct them, and algorithms for the automation of this have been developed (e.g. Hannah 1981; O'Callaghan and Mark 1984; Jenson and Domingue 1988; Brown and Bara 1994). In particular there has been significant progress in the detection and correction of pits (sometimes known as sinks). Pits are cells which are surrounded by neighbouring cells that all have higher elevations (Figure 2.2). It is possible that some are actual closed depressions, that is natural features or excavations (Jenson and Domingue 1988); but it is possible that they are artefacts of the gridding process, that is systematic errors (O'Callaghan and

Mark 1984; Fairfield and Leymaine 1991). Artefact pits sometimes occur in narrow valleys where the width of the valley bottom is smaller than the cell size, and they occur at any resolution. They are also produced in areas of gentle topography caused by errors in interpolation, incorrect or insufficient data.

Artefact pits are particularly problematic in hydrological applications since they disrupt natural drainage topography such that flowlines become trapped in them and must be removed to allow construction of a continuous local drainage directions (ldd) net. This is known as a depressionless DEM and is necessary for the computation of further topographic attributes. The most common method of removal is to compare the elevation of the pit cell with those immediately around it. The pit cell is 'filled up' – its elevation increased – until it is equal to one or more of the neighbouring cells. Then that neighbouring cell must be examined to ensure that it drains downhill to another location. If this does not occur, the elevation is increased until a linkage is found. The removal of pits is therefore an automatic, iterative process. Another, less popular, method is smoothing the DEM data to remove the depressions (O'Callaghan and Mark 1984), but this has the drawback of adapting all elevations in the data and will not remove large depressions. Furthermore, smoothing simply masks the problems without actually improving the quality of the output. All these methods imply that all pits are the result of underestimation of elevation. However, some pits occur because of the obstruction of flowpaths by the overestimation of elevation (Garbrecht and Martz 2000). In such situations, Garbrecht and Martz suggest that breaching the obstruction is more effective than filling the sink created by the obstruction. In areas of low relief relative to the vertical resolution of the DEM, pits caused by obstruction of flow-paths frequently occur and breaching an obstruction is very effective.

Physical landscape units – landform elements

From the geomorphological viewpoint, the ground surface is composed of landform elements. Landform elements are recognised as simply-curved geometric surfaces lacking inflections (complicated kinks) and are considered in relation to upslope, downslope, and lateral elements. They go by a plethora of names – facets, sites, land elements, terrain components, and facies. The 'site' (Linton 1951) was an elaboration of the 'facet' (Wooldridge 1932), and involved altitude,

Table 2.2   Primary and secondary attributes that can be computed from DEMs and their applications

Attribute	Definition	Applications
Altitude	Height above mean sea level or local reference	Climate variables – pressure, temperature, vegetation and soil patterns, material volumes, cut and fill and visibility calculations, potential energy determination
Slope	Rate of change of elevation – gradient	Steepness of topography, overland and sub-surface flow, land capability/use classification, vegetation patterns, resistance to uphill transport, correction of satellite and aerial photographs, geomorphology, soil water content
Aspect	Compass direction of steepest downhill slope – azimuth of slope	Solar insolation/irradiance, evapotranspiration, flora and fauna attributes, distribution and abundance
Profile curvature	Rate of change of slope	Geomorphology – flow acceleration, erosion/deposition patterns and rate, soil and land evaluation indices, terrain unit classification
Plan curvature	Rate of change of aspect	Converging/diverging flow, soil water characteristics, terrain unit classification
Upslope slope	Mean slope of upslope area	Runoff velocity
Dispersal slope	Mean slope of dispersal area	Rate of soil drainage
Catchment slope	Average slope over the catchment	Time of concentration
Upslope area	Catchment area above a small length of contour	Runoff volume, steady-state runoff rate
Dispersal area	Area downslope from a small length of contour	Soil drainage rate
Catchment area	Area draining to catchment outlet	Runoff volume
Specific catchment area	Upslope area per unit width of contour	Runoff volume draining out of catchment, soil characteristics, soil-water content, geomorphology
Flow path length	Maximum distance of water flow to a point in the catchment	Erosion rates, sediment yield, time of concentration
Upslope length	Mean length of flow paths to a point in the catchment	Flow acceleration, erosion rates
Dispersal length	Distance from a point in the catchment to the outlet	Soil drainage impedance
Catchment length	Distance from highest point to outlet	Overland flow attenuation
Local drain direction (ldd)	Direction of steepest downhill flow	Calculation of catchment attributes as a function of stream topology. Computing lateral transport of materials over locally defined network
Stream length	Length of longest path along ldd upstream of a given cell	Flow acceleration, erosion rates, sediment yield
Stream channel	Cells with flowing water/cells with more than a given number of upstream elements	Location of flow, erosion/sedimentation, flow intensity
Tangential curvature	Plan curvature multiplied by slope	Alternative measure of local flow convergence and divergence
Elevation percentile	Proportion of cells in a user-defined circle lower than the centre cell	Relative landscape position, flora and fauna abundance and distribution
Ridge	Cells with no upstream contributing area	Watershed/drainage divides, soil erosion, connectivity
Wetness Index	$\ln\left(\dfrac{\text{specific catchment area}}{\tan(\text{slope})}\right)$	Index of moisture retention
Viewshed	Zones of intervisibility	Location of wind turbines, mobile phone aerials, hotels, military applications
Irradiance	Amount of solar energy received per unit area	Soil and vegetation studies, evapotranspiration

Source: Adapted from Moore *et al.* (1993), Burrough and McDonnell (1998), Wilson and Gallant (2000).

extent, slope, curvature, ruggedness, and relation to the water-table. The other terms were coined in the 1960s (see Speight 1974). Landform element seems a suitably neutral term and will be used here.

Landform elements are described by local geometry. Several parameters are derivatives of altitude – slope angle, slope profile curvature, and contour curvature. Further parameters go beyond local geometry, placing the element in a wider landscape setting – distance from the element to the crest, catchment area per unit of contour length, dispersal area (the land area down-slope from a short increment of contour). The classic work on landform elements and their descriptors has largely been superseded by digital elevation models.

Digitally derived landscape elements

Topographic elements of a landscape can be computed directly from a DEM and these are often classified into primary (or first-order) and secondary (or second-order) attributes (Moore *et al.* 1993; Evans 1980). Primary attributes are calculated directly from the digital elevation data; the most commonly derived include slope and aspect. Secondary attributes combine primary attributes and are 'indices that describe or characterise the spatial variability of specific processes occurring in the landscape' (Moore *et al.* 1993, 15); examples are soil erosion potential and a wetness index. Such methods allow the spatial variability of the processes to be represented whereas in the past it was only possible to model them as point processes.

Slope, aspect, plan curvature, and profile curvature are all examples of primary attributes (Table 2.2). These attributes are usually computed using directional derivatives of a topographic surface either by using second-order finite difference schemes or by fitting a bivariate interpolation function $z = f(x, y)$ to the DEM and then computing the derivatives of the function (Wilson and Gallant 2000; Moore *et al.* 1993). Secondary attributes are computed using two or more primary attributes and offer the ability to describe pattern as a function of process.

Most digital-terrain analysis methods are based on gridded DEM data structures since the methods are simpler and easier to implement, although methods have also been developed for TINs to calculate slope and aspect for each triangular facet (e.g. Tajchman 1981) and contour-based networks such as TAPES-C. TAPES-C (Topographical Analysis Program for the Environmental Sciences – Contours) was produced

by Moore *et al.* (1988), Moore (1988) and Moore and Grayson (1989, 1990). TAPES-C can calculate slope, aspect, and plan curvature for (1) irregularly shaped elements or patches that are bounded by equipotential lines on two sides (contour lines), and (2) streamlines that are orthogonal to the equipotential lines on the other two sides (no-flow boundaries).

There is an enormous amount of literature describing the use of DEMs, mainly of gridded DEM, to produce both primary and secondary attributes (e.g. Zevenbergen and Thorne 1987; Jenson and Domingue 1988), and then how such outputs can be incorporated into spatial models to simulate physical processes that are in some way influenced and controlled by the nature of topography (e.g. Band and Moore 1995; Skidmore 1990). Two excellent books which present many examples are those edited by Lane *et al.* (1998) and Wilson and Gallant (2000). There is also considerable research into the combination of terrain attributes to create some form of topographic classification of a region (e.g. Giles *et al.* 1994; Giles 1998; Giles and Franklin 1996; Chorowicz *et al.* 1989; Blaszczynski 1997; Falcidieno and Spagnuolo 1991; Graff and Usery 1993). Table 2.2 provides a list of the more commonly derived terrain attributes, and Table 2.1 lists a few examples of DEM applications; further details of selected applications are provided in subsequent chapters. A short review of the two most commonly used primary attributes, slope and aspect, is given here.

Slope and aspect

Slope and aspect are two of the most important topographic attributes. Slope is a plane tangent to the terrain surface represented by the DEM at any given point and is made up of two components: (1) gradient, the maximum rate of change of altitude; and (2) aspect, the compass direction of the maximum rate of change. Gradient is usually expressed in degrees or per cent, and aspect in degrees (and converted to a compass bearing). Since slope is the reason why gravity induces the flow of water and other materials, it is found in many environmental process models (Morgan 1986, 63–110) and in particular hydrological and geomorphological studies. For instance, slope and flowpath (i.e. slope steepness and length) are parameters in the dimensionless Universal Soil Loss Equation (USLE), which is designed to quantify sheet and rill erosion by water. Some researchers (e.g. Griffin *et al.* 1988; Mitasova *et al.* 1996, 1997; Desmet and Govers 1997) have attempted spatial application of

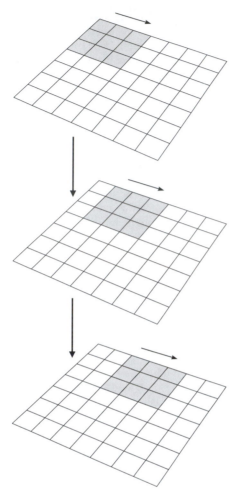

Figure 2.3 A moving 3 × 3 cell window or kernel used to compute derivatives of the hypsometric curve locally for every cell of a DEM grid.

this model using slope and slope length derived from DEMs.

Derivatives of the hypsometric curve can be computed locally for every cell of a DEM grid from a 3 × 3 cell window or kernel that is moved across the whole map (Figure 2.3). To calculate the gradient of each cell, the simplest finite difference in the x direction at a point i,j is the maximum downward gradient:

$$\left[\frac{\delta z}{\delta x}\right]_{i,j} = \max(z_{i+1,j} - z_{i-1,j}) \times 2\delta x \quad (2.3)$$

where δx is the distance between cell centres. For comparisons along diagonals, the $\sqrt{2}$ correction to δx is applied. The problem with this function is that

local errors in terrain elevation contribute significantly to errors in the slope estimates (Burrough and McDonnell 1998). Second-order finite-difference methods provide better results (Ritter 1987; Zevenbergen and Thorne 1987). A second-order finite-difference algorithm can be fitted to the four closest neighbours in the window, so slope is given by:

$$\tan S = \left[\left(\frac{\delta z}{\delta x}\right)^2 + \left(\frac{\delta z}{\delta y}\right)^2\right]^{0.5} \quad (2.4)$$

where z is altitude and x and y are the geographical co-ordinates of the cell.

Aspect is the orientation of the line of steepest descent. It is usually measured in degrees clockwise from north. Aspect can be important in some process models, for instance in the calculation of solar radiation. It may be calculated by:

$$\tan A = -\left(\frac{\delta z}{\delta y}\right)\bigg/\left(\frac{\delta z}{\delta x}\right) \quad (-p < A < p) \quad (2.5)$$

Zevenbergen and Thorne (1987) demonstrated the calculation of slope and aspect, as well as concavity and convexity, using a six-parameter quadratic equation fitted to a 3 × 3 window. Horn (1981) provided a third-order finite-difference estimate using all eight outer points of the window. Other methods fit multiple regression to the nine elevation points in the 3 × 3 window and derive slope and aspect from that.

Errors in terrain attributes

As with the DEM used to derive topographic attributes, these primary and secondary attributes are also prone to inaccuracies or differences in derived outputs. There are a number of sources of error in the estimation of surface derivatives: (1) errors in the original elevation measurements; (2) errors in the interpolation of elevation values on the DEM grid; (3) errors due to spatial sampling effects; and (4) inadequacies of the algorithms used to create the terrain attributes (Wise 1998; Srinivasan and Engel 1991). Systematic and random errors may affect the relationships between terrain variables and terrain-controlled conditions and processes. These errors increase in magnitude as first- and second-order derivatives are computed. The derivation of second-order attributes usually produces the most serious difficulties (Wilson and Gallant 2000). A number of studies have attempted to provide methods to detect and measure these errors and to assess the propagation of DEM errors in the derivatives.

Skidmore (1989) showed that different algorithms used to calculate gradient and aspect in different GIS packages produce different results from the same DEM and concluded that both second-order and third-order methods were superior to the above described in equation (2.3). Florinsky (1998) studied the accuracy of data on some local topographic attributes derived from a DEM. He produced formulae for calculating the root-mean-square errors of four primary topographic variables (gradient, aspect, and horizontal and vertical land-surface curvatures). Florinsky also demonstrated that mapping is the most convenient and pictorial implementation of the derived formulae and showed that high errors occurred particularly in flat areas. Wilson and Gallant (2000) suggested shaded relief as a graphical method for assessing the representation of slope and aspect in a DEM, while Bolstad and Stowe (1994) used field measurements.

In an interesting study with a slightly different slant to the above studies, Hodgson (1995) investigated what cell-size slope and aspect, derived from a DEM window, actually represented and how different was the actual surface angle at the central cell from the surface angle computed using a window of elevation values. What he found was that the slope/aspect angle derived from neighbouring elevation points best depicts the surface orientation for a larger cell – either 1.6 times or 2.0 times larger than the size of the central cell. He suggested that, for any study involving the combination of slope or aspect with other data sets, the elevation data used to derive the slope and aspect should be created at a resolution greater than that of the other data sets.

Some researchers have been investigating the use of terrain derivatives to test the accuracy of the DEMs used to produce them. Thus, Wise (1998) carried out accuracy assessments of slope, aspect, and curvature to produce a measure of DEM quality. In addition, graphical representation of other primary attributes, in particular profile curvature, can provide a check for the accuracy of the DEM in representing terrain shape. Bolstad and Stowe (1994) used slope and aspect to provide a comparison of two DEMs from different production methods. They used field measurements of slope and aspect, taken using a hand-held clinometer and field compass, to determine the accuracy produced using the third-order finite-difference method (Horn 1981). Standard descriptive statistics as well as paired *t*-tests and correlations were used to evaluate the terrain accuracies. They believed that this approach provided them with an adequate test for the DEMs.

Ecological landscape units – landscape elements

From an ecological viewpoint, the landscape may be regarded as a non-uniform (heterogeneous) and non-random mosaic consisting of three basic elements – patches, corridors, and background matrixes. These landscape elements are made of individual plants (trees, shrubs, herbs), small buildings, roads, small water bodies, and the like.

Patches

Patches are fairly uniform (homogeneous) areas that differ from their surroundings. Woods, fields, ponds, rock outcrops, and houses are all patches. Patches may be characterised by their size (large or small), number (few or many), boundary or edge attributes (width and 'wiggliness'), and shape (roundness or elongation):

1. *Patch size and number* These attributes are easy to measure. They are topographic attributes of great significance for ecology. A lively debate in conservation revolves around whether it is better to have a large patch or a small patch (LOS); or, alternatively, whether it is better to have a single large patch or several small patches (SLOSS). Patch size and number also affect biodiversity. They have been widely studied in investigations of species–area relationships. All these issues will be discussed in Chapters 6 and 7.

2. *Patch boundaries* All landscape elements have boundaries. A boundary consists of a border (a line separating the edges of neighbouring landscape elements) and edges (Figure 2.4). An edge is the outer part of a patch or other landscape element. In vegetation patches, it is a zone exhibiting the 'edge effect', which is the tendency of the outer part of a landscape element to support high population densities and high species diversity. The

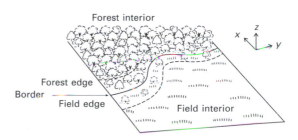

Figure 2.4 The structure of boundaries, showing borders and edges. (Adapted from Forman, 1995, 86.)

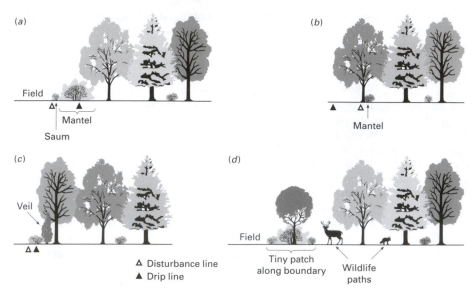

Figure 2.5 Forest edges: combinations of saum, mantel, and veil, (a) and (b) contrasting saum and mantel caused by the position of the drip line (the outermost extent of canopy-level branches) and disturbance line (the outermost extent of disturbance in the field from ploughing or grazing). (c) Veil growth after recent boundary formation (e.g. tree-felling). (d) A mosaic boundary. (Adapted from Forman, 1995, 94.)

interior of a patch, or core, contains species that live exclusively or predominantly away from the edge (interior species, as opposed to edge species). Edges and boundaries have a horizontal and vertical structure. A simple way of characterising the horizontal structure of boundaries is by their degree of sinuosity. Natural patches tend to have curvy boundaries with convexities (lobes) and concavities (coves). Patches created by humans often have straight boundaries that lack coves or lobes. Tiny-patch boundaries are common, in which a tiny parcel of one ecosystem type is surrounded by a second type on both sides, or on one side of the border. So, tiny patches of woodland might grow within grassland as diminutive outliers of an extensive woodland. Boundaries may also be designated hard or soft, to varying degrees. A hard boundary separates a field from a forest, whereas a soft boundary separating wet heath from dry heath is more gradual. Edges have a varied structure. Edges of vegetated patches and corridors normally possess any or all of three components – saum, mantel, and veil (Figure 2.5).

3. *Patch shape* This may be measured in many ways. Four types of measure exist: those based on length and width (long and short axes); those based on perimeter and area; those based on area alone;

and those based on area and length (Table 2.3). Natural patch shapes lack the straight edges of human-made patches. This is evident in Figure 2.6, which classifies some common patches according to the convolution of their edges (number of lobes) and their elongation (the ratio of length to width).

Corridors

Corridors are strips of land that differ from the land to either side. Examples are roads, hedgerows, and rivers. Corridors may be usefully grouped into three classes: trough corridors, wooded strip corridors, and stream and river corridors (Forman 1995, 158).

1. *Trough corridors* These are strips of vegetation lower in height than the neighbouring matrix. They include roads and roadsides, powerlines for electricity transmission, gas lines, oil pipelines, railways, dykes (e.g. Offa's Dyke), and trails. Trails are manifold – walking paths, bridle paths, rides, cross-country ski trails, bicycle trails, motorbike trails, snowmobile trails, drove roads (mainly for cattle but also for sheep, pigs, and geese), and animal trails. Streets in urban areas form canyons between rows of buildings and may be regarded as trough corridors, too.

Table 2.3 Measures of patch shape

Measure	Formula	Source
Form	$F_1 = a/b$	Davis (1986)
Form ratio	$F_2 = A/a^2$	Horton (1932)
Elongation	$E_1 = b/a$	Davis (1986)
Elongation ratio	$E_2 = \left\{ 2\sqrt{\dfrac{A}{p}} \right\} \Big/ a$	Schumm (1956)
Circularity	$C_1 = \sqrt{ab/a^2}$	Davis (1986)
Circularity	$C_2 = \dfrac{4A}{p^2}$	Griffith (1982), Davis (1986)
Circularity	$C_3 = \sqrt{\dfrac{A}{A_C}}$	Unwin (1981), Davis (1986)
Circularity ratio	$C_4 = \dfrac{A}{A_C}$	Stoddart (1965), Unwin (1981)
Circularity ratio	$C_5 = A \Big/ \left\{ \pi \left(\dfrac{p}{2\pi} \right)^2 \right\}$	Miller (1953)
Ellipticity ratio	$E_r = L \Big/ 2 \left\{ A \Big/ \left[\pi \left(\dfrac{L}{2} \right) \right] \right\}$	Stoddart (1965)
Ellipticity index	$E_i = \left[\dfrac{\pi a (0.5a)}{A} \right]$	Stoddart (1965), Davis (1986)
Compactness	$K_1 = \dfrac{2\sqrt{\pi A}}{p}$	Bosch (1978), Davis (1986)
Compactness index	$K_2 = A \Big/ \sqrt{2\pi \int_a R^2 \mathrm{d}x\,\mathrm{d}y}$	Blair and Biss (1967)
Grain shape index	$G_{SI} = \dfrac{p}{a}$	Davis (1986)
Mean radius	$\bar{R} = \sum R_j / n$	Boyce and Clark (1964) Stoddart (1965)
Shape factor	$S_F = p_C / p$	Bosch (1978), Davis (1986)

Notes: a = length of major axis; b = length of minor axis; A = area; p = perimeter; A_C = area of smallest circle enclosing the shape; n = number of sides, considered as a polygon; p_C = perimeter of circle with same area as the shape; R_j = jth radius of the shape from gravity centre to perimeter; R = radial axes from gravity centre or small area.

2. *Wooded strip corridors* These include hedgerows, fencerows, windbreaks, walls, and woodland corridors. They may be single-row, multiple-row, or wooded strips. Single-row strips include hedges, fences, walls, and some windbreaks. They may arise naturally, as when a hedgerow grows over an old ditch or fence, or may be planted. Multiple-row strips are always planted, usually as wind-breaks. Several combinations of trees and shrubs are possible, but strips with a separate row of trees and shrubs are commonly used as windbreaks for houses and farms. Wooded strips are remnants or regenerated woodland corridors.

3. *Stream and river corridors* These are also called riparian corridors and include streams, rivers, and canals.

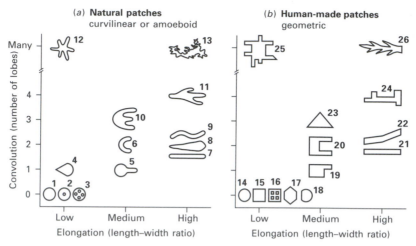

Figure 2.6 Patch shapes in landscapes. (a) Natural patches. **1** Bog, pingo, tarn, hummock in wetland. **2** Slope surrounding dry hilltop, wetland in karst terrain. **3** Gaps within a patch. **4** Delta, alluvial fan. **5** Landslide, avalanche, woods extending along an island in a river. **6** Oxbow lake, barchan. **7** Glacial wetland or lake. **8** Drumlin, dune, eyot. **9** Riparian woods. **10** Mountaintop with lava flows, headland around fjord. **11** Riparian woods along stream with tributaries. **12** Mountaintop vegetation extending along ridges, disturbance by mammal trampling around a waterhole. **13** Fire disturbance, pest outbreak disturbance. (b) Human-made patches. **14** Village around well or fort, central pivot irrigation. **15** City block, logged clear-cut in a chequerboard forest. **16** Woods with internal clear-cuts. **17** Geographical pattern of land use surrounding a central village. **18** Farm pond with a dam. **19** Woodlot in an agricultural area, city park. **20** Suburban woods with an encroaching housing estate. **21** Cultivated field. **22** Golf-course fairway, ski slope. **23** Field cut diagonally by a later road. **24** Woodlot lying within the intersection of several farms. **25** Town or city with development along roads and railways. **26** Dammed reservoir. (After Forman, 1995, 117.)

Greenways are hybrid corridors comprising parks, trails, waterways, scenic roads, and bike paths.

Matrixes

Matrixes are the background ecosystems or land-use types in which patches and corridors are set. It is often easy to identify the matrix in a landscape; it is simply the dominant ecosystem or land use – forest, grassland, heathland, arable, residential, greenhouses, or whatever. Identification is more problematic when two, three, or more ecosystems and land uses co-dominate in a landscape. In these situations, the matrix may be singled out by using area, connectivity, and 'control over dynamics' in sequence (Figure 2.7). Total area is the primary and easiest criterion to apply. Should one land-use type cover well over half the landscape, or is much more extensive than the second-largest land-use type, then it is the matrix. When the two most extensive land-use types cover the same area, then connectivity can be used to distinguish between them. The best connected land-use type is designated as the matrix. If, after considering area and connectivity, the matrix is still not determined, then a third characteristic – 'control over dynamics' – may be applied. What is being sought is the land-use element that has the greatest control over the landscape. 'Control over dynamics' is difficult to measure directly, and area and connectivity are really surrogate measures of it. The idea is to identify the land use that is, for example, a source of plant seeds, herds of herbivores, keystone predators, floodwaters, or humans (see Forman 1995, pp. 277–8).

In rare cases, the matrix is an uninterrupted area with other landscape types at its fringes. This happens in a remote woodland with settlements dotted around the edge, and in a low-lying wetland with dry slopes at its margins. Normally, the matrix is cut up by corridors and punctured by patches. Corridors produce a subdivided matrix, while patches lead to a

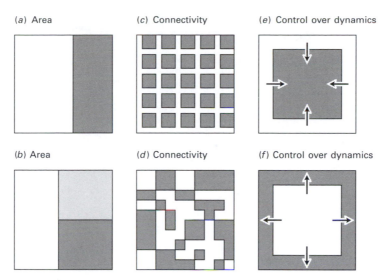

(a) Area   *(c)* Connectivity   *(e)* Control over dynamics

(b) Area   *(d)* Connectivity   *(f)* Control over dynamics

Figure 2.7 Characteristics used in identifying the matrix. White land-use types are the matrix. The matrix covers 60 per cent of (a), 45 per cent of (b), and 50 per cent of the others. Arrows show net direction of movement. The key factor identifying each matrix is indicated for each land-use pattern. (Adapted from Forman, 1995, 277.)

porous or perforated matrix (Forman 1995, 278). These differences are enormously important ecologically.

Topographic structures

Landform elements and landscape elements are the building blocks of landscapes and regions. In various combinations, they create spatial structures – networks, mosaics, and regions. These structures form a spatial hierarchy, ranging in size from a square metre and less to the entire planetary surface. It is standard practice (though with no agreed standard) to label the various levels in this hierarchy. Popular labels use the prefixes micro, meso, macro, and mega, producing the terms microscale, mesoscale, macroscale, and megascale. It is far simpler, but far less impressive, to use the terms small-scale, medium-scale, large-scale, and very large-scale. The boundaries between the categories are somewhat arbitrary. How small does large have to be before it becomes medium? Our current preferred scheme is shown in Figure 2.8. In this scheme, terrestrial space is sliced up like this: microscale is up to 1 km^2; mesoscale is 1–10 000 km^2; macroscale is 10 000–1 000 000 km^2; and megascale is from 1 000 000 km^2 to the entire surface area of the Earth (Huggett 1997).

Networks

Corridors may interconnect to form networks. One way of characterising networks is to reduce them to graphs, that is, arrays of points that are connected or not connected to one another by lines. The points are called vertices (also styled nodes, junctions, intersections, terminals, and zero-cells) and the lines are called edges (also styled links, sides, arcs, segments, branches, routes, and one-cells). Nodes represent the origin, intersection, or terminal of a corridor. Attached nodes are patches (e.g. woods and towns) that lie along a corridor. In transport systems, major nodes, commonly located at the fringes of a region, are 'gateways'; 'trunk lines' are the major routes connecting gateways; 'feeder lines' are smaller routes supplying gateways; and 'bridge lines' connect gateways in neighbouring regions.

Types of network

Classifying networks according to their topology produces two chief types: planar graphs and non-planar graphs (Haggett 1967a; Haggett and Chorley 1969). Non-planar graphs contain lines that cross in plan but do not actually intersect, as in air routes lying at differing heights. They are not of concern in this book, though they may be affected by topography. Air routes

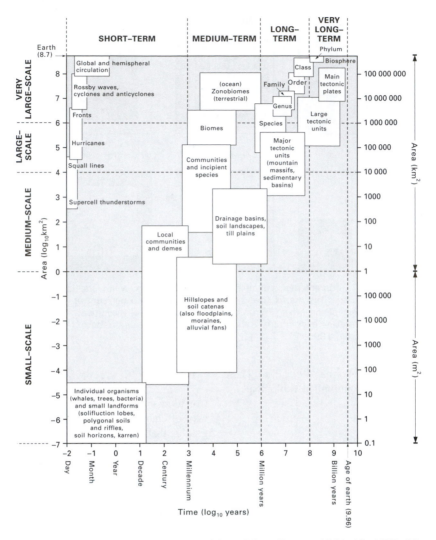

Figure 2.8 Scales of spatial structure. (Adapted from Huggett, 1991, 10; 1997, 5.)

tend to shun such large mountain ranges as the Asian Himalaya–Pamir–Nan Shan group owing to unpredictable weather and lack of emergency landing sites. Deserts also tend to be avoided because there are no places to land if difficulties should be encountered.

Planar graphs are real systems of lines on the Earth's surface – roads, rivers, and so forth. They comprise two groups: linear barriers and linear flow systems. Linear barriers form the edges of cells (patches or regions). Barrier networks may be isolated (as in the case of islands) or contiguous (as in the case of a set of administrative areas, such as the county boundary network in the United Kingdom).

Linear flow systems include corridors, trees, and circuits (Figure 2.9):

1. *Corridor networks*   In these, movement occurs along paths, such as individual roads and rivers. In some cases, paths are tightly aggregated networks. An example is a herd of mammals moving in a single, wide and fairly straight route. In other cases, the paths are loosely aggregated, as in a braided river or in a herd of mammals moving in a more dispersed fashion. Irregular networks are the loosest form and occur where there is no preferred direction of movement, as when an

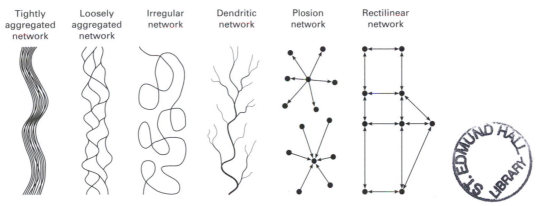

Figure 2.9 Linear systems of corridors and networks. (Adapted from Forman, 1995, 256.)

individual animal seeks food in uniform terrain. Loosely aggregated nets and irregular nets are sometimes called wavy nets (Forman 1995, 269).

2. *Tree (dendritic or branching) networks* These are exemplified by a river network. They contain no circuits and resemble a tree with a trunk dividing into branches, branches dividing into twigs, and so on. Flows in tree networks converge (as in the river case) but may also diverge (as in the case of the human body's arterial system).

3. *Circuit networks* These contain closed loops or circuits, as in railway networks and hedgerow networks. Two chief types are 'plosion' networks, where some nodes act as strong sources or sinks of movement, and rectilinear networks, where nodal sources and sinks are more or less balanced (Forman 1995, 256). In European countries, road networks are more wavy than rectilinear, largely because they are more 'organic' and less 'designed' than in old colonial territories.

Different types of network usually crisscross one another. However, because most corridors serve one purpose (walking, power transmission, motoring, or whatever), corridor networks tend to cover large areas. In the United States, for example, the powerline network covers about two million hectares, an area roughly the size of Israel (Dreyer and Niering 1986). Roadside verges also cover a large area. In the southwest agricultural region of Western Australia, a 60–200-m strip of natural vegetation is maintained along many of the 107 500 km of public roads (Hussey 1991). These natural roadside strips were created by State Government policy following expressions of

public concern for wildflower preservation in the late 1950s. The entire set of roadside strips forms a 'giant green network' (Forman 1995, 171). Similar green networks exist in South Africa (Dawson 1991), Great Britain (Way 1977), the United States (Voorhees and Cassell 1980), and elsewhere. In mainland Great Britain, there are about 231 000 ha of road reserves (the total strip of land reserved for transport purposes), covering about 0.9 per cent of the land area (Way 1977). In The United States the figure is 8 100 000 ha (Adams and Geis 1983; see also Bennett 1991).

Measuring networks

Several properties of networks can be measured quantitatively. Important properties include density and frequency, order, degree of integration, and accessibility.

Density is the length of links (roads, rivers, hedgerows, trails, and so forth) per unit area, while frequency is the number of links per unit area. For example, drainage density is the length of stream channels per unit area, and stream frequency is the number of streams per unit area (see p. 98). Drainage density varies from less than 5 km/km^2 on permeable sandstone to 500 km/km^2 or more on clay 'badlands'.

Order is a property of river networks. Each stream or link is assigned an order, starting with 1 for fingertip tributaries. Subsequent orders depend on the ordering system adopted. In the Strahler system, two first-order streams combine to form a second-order stream, but any first-order tributaries joining the second-order stream do not change its order. Only when two second-order streams join is a third-order stream created

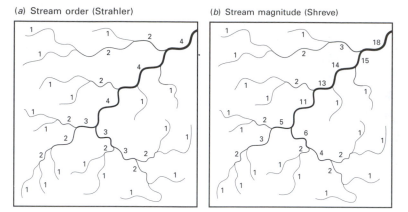

Figure 2.10   Stream ordering: (a) Strahler's system; (b) Shreve's system.

(Figure 2.10(a)). In Shreve's system, stream magnitude is the sum of all contributory streams. So, two first-order streams combine to form a second-order stream, and when this is joined by a first-order tributary, a third-order stream is created (Figure 2.10(b)).

The bifurcation ratio is a widely used topological property of river networks. It is defined as the ratio between the number of stream segments of a particular order and the number of stream segments of the next higher order. Within any drainage basin, the bifurcation ratio is constant. This fact gives rise to Horton's law of stream numbers, which states that the numbers of streams of different order in a given drainage basin tend closely to approximate an inverse geometric series in which the first term is unity and the ratio is the bifurcation ratio. Horton's law has subsequently been extended to nearly all natural branching systems, including trees and vascular systems, as the law of path numbers (Haggett 1967b). The bifurcation ratio must be at least 2, and in many natural river networks it varies between 3 and 5, depending on substrate, vegetation, and so on.

Network integration or network complexity depends on the pattern of links between nodes. The greater the integration, the more connected and more accessible (though not necessarily in terms of distance travelled) places are. A connectivity or adjacency matrix, **C**, is a starting point from which to explore network connectivity (Figure 2.11(c)). The nodes within a network are used as row and column headings. The elements of the matrix are then numbered as 0, if the nodes are not connected to adjacent nodes, or 1, if they are connected to adjacent nodes. Row totals then give the 'nodality' of each node. Several measures

can be extracted from a network graph or its connectivity matrix. Three numbers are basic to these calculations (Figure 2.11(b)):

e = the number of links (paths, edges) in the network;

v = the number of points (nodes, vertices) in the network; and

g = the number of separate, non-connecting subgraphs in the network.

Network complexity may be measured by the beta index and by the cyclomatic number. The cyclomatic number, μ, is the number of basic circuits in a network. It is derived from the number of edges, e, the number of nodes, v, and the number of separate (non-connecting) subgraphs, g:

$$\mu = e - v + g \qquad (2.6)$$

It ranges from zero where there are no subgraphs to 15 and above (the limit is set by the number of nodes) for richly interconnected networks with many subgraphs. To compare networks of differing size, the alpha or redundancy index, α, is helpful as it relates the cyclomatic number to the maximum number of circuits:

$$\alpha = \frac{\mu}{2v - 5} \qquad (2.7)$$

The alpha index ranges from 0 for non-connected paths to 1 for a fully connected network. It may also be expressed as

$$\alpha = \left(\frac{e - v + g}{2v - 5} \right) \times 100 \qquad (2.8)$$

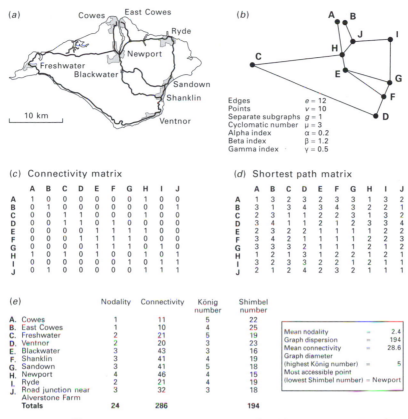

(a) Cowes, East Cowes, Ryde, Newport, Freshwater, Blackwater, Sandown, Shanklin, Ventnor — 10 km

(b)

Edges $e = 12$
Points $v = 10$
Separate subgraphs $g = 1$
Cyclomatic number $\mu = 3$
Alpha index $\alpha = 0.2$
Beta index $\beta = 1.2$
Gamma index $\gamma = 0.5$

(c) Connectivity matrix

	A	B	C	D	E	F	G	H	I	J
A	1	0	0	0	0	0	0	1	0	0
B	0	1	0	0	0	0	0	0	0	1
C	0	0	1	1	0	0	0	1	0	0
D	0	0	1	1	0	1	0	0	0	0
E	0	0	0	0	1	1	1	1	1	0
F	0	0	0	1	1	1	1	0	0	0
G	0	0	0	0	1	1	1	0	1	0
H	1	0	1	0	1	0	0	1	0	1
I	0	0	0	0	1	0	1	0	1	1
J	0	1	0	0	0	0	0	1	1	1

(d) Shortest path matrix

	A	B	C	D	E	F	G	H	I	J
A	1	3	2	3	2	3	3	1	3	2
B	3	1	3	4	3	4	3	2	2	1
C	2	3	1	1	2	2	3	1	3	2
D	3	4	1	1	2	1	2	3	3	4
E	2	3	2	2	1	1	1	1	2	2
F	3	4	2	1	1	1	1	2	2	3
G	3	3	3	2	1	1	1	2	1	2
H	1	2	1	3	1	2	2	1	2	1
I	3	2	3	3	2	2	1	2	1	1
J	2	1	2	4	2	3	2	1	1	1

(e)

	Nodality	Connectivity	König number	Shimbel number
A. Cowes	1	11	5	22
B. East Cowes	1	10	4	25
C. Freshwater	2	21	5	19
D. Ventnor	2	20	3	23
E. Blackwater	3	43	3	16
F. Shanklin	3	41	4	19
G. Sandown	3	41	5	18
H. Newport	4	46	4	15
I. Ryde	2	21	4	19
J. Road junction near Alverstone Farm	3	32	3	18
Totals	**24**	**286**		**194**

Mean nodality =	2.4
Graph dispersion =	194
Mean connectivity =	28.6
Graph diameter (highest König number) =	5
Most accessible point (lowest Shimbel number) =	Newport

Figure 2.11 The main-road network on the Isle of Wight, England, and various network measures associated with it.

The beta index, β, is the ratio of the number of edges or links, e, to the number of vertices or nodes, v:

$$\beta = \frac{e}{v} \qquad (2.9)$$

It differentiates between simple and complex topological network structures. Beta values below 1.0 indicate a partially connected network, as seen in branching stream networks and disconnected networks. Beta values of 1.0 indicate a network with just one possible circuit. Beta values above 1.0 indicate varying degrees of integration up to a maximum of $\beta = 3.0$ (for planar graphs). In the early 1960s, the beta index for the poorly developed Nigerian railway was 0.9, whereas for the more complex French rail network it was 1.4 (Kansky 1963). Like the cyclomatic number, the beta index varies according to the size of the network (the number of points it contains) and is not very useful for comparative purposes. A more helpful measure is the gamma index, γ, which is the ratio between the number of actual paths to the maximum number of possible paths in a network of a given number of points:

$$\gamma = \frac{e}{3(v - 2)} \qquad (2.10)$$

Network accessibility can be gauged in several ways from the connectivity matrix, **C**. A simple measure of the accessibility of each node is the nodality, which is the sum of the row elements (Figure 2.11 (e)). A more convoluted measure is computed by powering the connectivity matrix (multiplying it by itself). If 1s are used in the principal diagonal (to indicate that all places are connected to themselves), then at each step, n, of matrix powering, the matrix shows the number of paths of exactly n links between each pair of nodes. When all places are linked (no zero entries in the powered matrix), then n is the graph diameter.

A further measure that can be derived from the connectivity matrix, **C**, is the shortest path between pairs of points. The shortest-path matrix may be put together by inspecting the original graph (Figure 2.11(d)), or as a by-product of powering **C**, since each element in it is the power at which **C** becomes non-zero. Several indices may be obtained from the shortest-path matrix. The maximum number on each row gives the number of steps to reach the most distant point (in terms of topology, not absolute distance), or, to put it another way, the maximum number of paths in the shortest path from any one point to any other point in the network. These are called König numbers, and the largest of them specifies the graph diameter. König numbers indicate the centrality of points within the network: the larger the König number the less central the vertex. The row totals of the shortest-path matrix are Shimbel numbers. They are a guide to the accessibility of points within the network: the lower the Shimbel number the more accessible the point.

Mosaics and regions

Landform elements, as well as landscape elements, combine to form spatial mosaics. These mosaics are generally not random but have specific spatial configurations that are repeated in different places. Landform elements combine to form landform mosaics, within which two functional spatial units are ubiquitous – catenas and drainage basins. Landform regions also exist, although their identification is somewhat disputable. Landscape elements (patches, corridors, and matrixes) combine to form landscape mosaics, within which there is a range of topographic structures. These landscape structures are distinct spatial clusters of ecosystems or land uses or both. It might be thought that there would be considerable agreement between landform mosaics and landscape mosaics. To some extent there is, but there are also differences, largely because living things, and especially humans, do not have to follow fully the constraints imposed by the land surface. The common ground between the landform mosaic and the landscape mosaic needs researching. Obvious points of similarity lie in three spatial (and functional) structures – catenas, drainage basins, and networks. Connections between landform regions and landscape regions are, for the most part, far less clear.

Catenas and drainage basins

Catenas (toposequences) are repeated sequences of landscape units occurring along a hillslope, from summit to valley bottom (Milne 1935). Each catena comprises an eluvial, a colluvial, and an illuvial zone (Morison *et al.* 1948). The eluvial zone lies at the top of the catena and consists of a flat summit and shoul-

der; the colluvial zone lies in the middle of the catena and consists of a backslope and footslope; the illuvial zone lies at the bottom of the catena – it is the toeslope which, like the summit, is fairly flat. All landscapes may be viewed as collections of catenas (Scheidegger 1986). Vegetation communities may also display changes along catenas.

Catenas are part of a three-dimensional landscape, much of which is organised into erosional drainage basins. In moving down slopes, weathering products tend to move at right angles to land-surface contours. Flowlines of material converge and diverge according to the curvature of the land-surface contours. The pattern of vergency influences the amounts of water, solutes, colloids, and clastic sediments held in store at different landscape positions.

Landform regions

The mosaic of landform elements found in any terrain may have a discernible pattern. The search for landform patterns is the subject of terrain evaluation, the proponents of which believe that landform regions exist, each with a characteristic set of interrelated landform elements. These small-scale and medium-scale landscape regions are variously styled stows, recurrent landscape patterns, land systems, terrain patterns, landform systems, relief units, and landscapes. Most of these terms include soil and vegetation characteristics within their scope, but landform characteristics are essential to their definition. The 10-level 'terrain hierarchy' proposed by Mitchell (1991, 42–53) covers the gamut of terrain units, from the smallest to the largest (Table 2.4). Each of the smaller units is a 'genetically controlled assemblage of relief and drainage features', while the larger units depend on 'climatic, structural, and lithological criteria' (Mitchell 1991, 42).

Landscape regions

Patches, corridors, and matrixes combine in multifarious ways to create distinctive landscape mosaics. There appear to be six basic types of landscape (Figure 2.12): large-patch, small-patch, dendritic, rectilinear, chequerboard, and interdigitated landscapes (Forman 1990; 1995, 309). Large-patch landscapes have one or several large patches set in a matrix, as when large woods are surrounded by pasture. Small-patch landscapes have small patches set in a matrix, as when small woods are surrounded by pasture, or small bogs are surrounded by heathland. Dendritic landscapes are dominated by corridors in matrixes,

Table 2.4 A 10-level terrain hierarchy

Terrain unit	Definition	Examples within cool temperate oceanic land zone
Land zone	Major climatic region	Cool temperate oceanic
Land division	Major continent-sized structures	North-west American, north-east American, east Asian, and New Zealand
Land province	Second-order units bearing similar rock types and structures	Lowland Britain, highland Britain, lowland Ireland
Land region	Third-order units with similar surface forms, geological structures and rock types, and a comparable geomorphic evolution	Cumbrian lowlands, Lake District, central Pennines, south Pennines, Lancashire–Cheshire plain, Hampshire basin, the Weald
Land system (simple)	Recurrent assemblage of land facets that have a similar origin	North Downs, South Downs, Vale of Sussex, Greensand Ridge, Vale of Kent, Romney Marsh, High Weald
Land catena	Major repetitive component of a land system. A chain of geographically related land facets	Transect from hilltop to valley bottom in the High Weald
Land facet	Fairly homogeneous tract of land, differing from surrounding areas in terms of geology, water regime, and topography	Cliff, small river terrace, sand dune, alluvial fan, salt marsh
Land clump	Patterned repetition of two or more land elements too disparate to be a land facet	Badland, exposed bed of braided stream, area of hummock dunes
Land subfacet	Constituent part of a land facet caused by a differing intensity of the formative processes	Convex top and concave bottom sections of a hillslope too short to be deemed a catena
Land element	Smallest morphological landscape unit	Small rock outcrops, rills, swampy patches in fields

Source: Adapted from Mitchell (1991, 43).

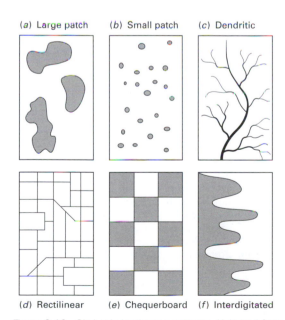

(a) Large patch   (b) Small patch   (c) Dendritic

(d) Rectilinear   (e) Chequerboard   (f) Interdigitated

Figure 2.12 Six basic landscape patterns. (Adapted from Forman, 1990; 1995, 309.)

such as wooded stream courses set in grassland. Rectilinear landscapes are dominated by a network of straight corridors, as in a hedgerow network. Chequerboard landscapes lack a dominant network and comprise two or three types of similar-sized patch that alternate across an area, such as fields of different crops. Interdigitated landscapes possess interpenetrating lobes of two different landscape types, for example uplands and lowlands, or built and agricultural areas in the urban margin.

Measuring landscape elements and mosaics

A large number of measures are available to measure landscape properties. In theory, these measures could be applied to both landform mosaics and landscape mosaics. In practice, they are largely applied to landscape mosaics as work on landform mosaics has stagnated over the last decade or so, or at least moved in other directions. Formulas are available for measuring mosaic diversity, edge properties, and patch properties (Table 2.5). Software packages are available for calculating landscape metrics. An example is FRAGSTATS (McGarigal and Marks 1995).

Table 2.5 Descriptive measures of landscape mosaics

Measure	Formula	Variables
Diversity measures		
Relative richness, R	$R = (m/m_{\max}) \times 100$	m = number of land-cover types
		$m_{\max}$ = maximum number of land-cover types
Relative evenness, E	$E = \left[\left(-\ln \sum\limits_{i=1}^{m} p_k^2\right)_j \middle/ \left(-\ln \sum\limits_{i=1}^{m} p_k^2\right)_{\max}\right] \times 100$	m = number of land-cover types
		$m_{\max}$ = maximum number of land-cover types
		p_k = fraction of area in land-cover type k
Diversity, H	$H = \sum\limits_{k=1}^{m} p_k \ln p_k$	m = number of land-cover types
		p_k = fraction of landscape in land-cover type k
Dominance, D_{o}	$D_0 = H_{\max} + \sum\limits_{k=1}^{m} p_k \ln p_k$	m = number of land-cover types
		p_k = fraction of area in land-cover type k
Boundary or edge measures		
Boundary length, B_{l}	$B_{\mathrm{l}} = L/A$	L = length of boundaries
		A = area
Boundary density, B_{d}	$B_{\mathrm{d}} = n_{\mathrm{b}}/A$	n_{b} = number of boundaries
		A = area
Edge number, $E_{i,j}$	$E_{i,j} = \Sigma e_{i,j} \times l$	$e_{i,j}$ = number of interfaces between grid cells of types i and j
		l = length of cell boundary
Fractal dimension, D	$D = \log P/\log A_{\mathrm{p}}$	P = perimeter of a patch at a specific length-scale
		A_{p} = area of a patch
Relative patchiness, P	$P = \left(\sum\limits_{i=1}^{N} D_i \middle/ n\right) \times 100$	n = number of boundaries between adjacent grid-cells
		D_i = dissimilarity value for the ith boundary between adjacent grid-cells
Patch-centred measures		
Patch isolation, r_i	$r_i = \dfrac{1}{n}\sum\limits_{j=1}^{j=n} d_{i,j}$	n = number of neighbouring patches considered
		$d_{i,j}$ = distance between patch i and any neighbouring patch j
Patch accessibility, a_i	$a_i = \sum\limits_{j=1}^{j=n} d_{i,j}$	n = number of neighbouring patches considered
		$d_{i,j}$ = distance along a linkage (e.g. a fencerow)
All-patch pattern measures		
Patch density, p_{d}	$p_{\mathrm{d}} = n/A_{\mathrm{l}}$	n = number of patches
		A_{l} = area of landscape
Patch dispersion, R_{c}	$R_{\mathrm{c}} = 2d_{\mathrm{c}}(\lambda/\pi)$	d_{c} = mean distance from a patch (or its centre or centroid) to its nearest neighbouring patch
		λ = mean density of patches
Nearest neighbour statistic, $q_{i,j}$	$q_{i,j} = n_{i,j}/n_i$	$n_{i,j}$ = number of grid-cells of type i adjacent to type j
		n_i = number of grid-cells of type i
Contagion, C	$C = 2m \log m + \sum\limits_{i=1}^{m}\sum\limits_{j=1}^{m} q_{i,j} \log q_{i,j}$	m = number of land-cover types
		$q_{i,j}$ = probability of land-cover type i being adjacent to land-cover type j

Source: Adapted from Forman (1995) and Turner (1989, 1990); see also Forman and Godron (1986).

Topographic connections

Landscape elements and regions (places) are linked by flows of energy and materials, seeds, spores, and individuals. These flows are in part regulated by the topographic attributes of adjacent elements and regions. In turn, the topographic attributes may be altered by the flows. Interconnections occur at all levels. This idea was encapsulated in the expression 'everything is connected to everything else' (e.g. Commoner 1972). A more recent rendition of the same idea is the spatial-flow principle: all ecosystems are inter-related, with movement or flow rate dropping sharply with distance, but more gradually between ecosystems of the same type (Forman 1987).

It may be fruitful to regard the ecosphere as a hierarchy of spatial structures and circulations that are constantly readjusting to one another and to changes in the geological and cosmic environments. It is convenient to discuss the flows within the spatial structures according to scale of operation – local, regional, and global.

Local and regional flows

Eight agents, or vectors, effect movements and flows of energy and materials between ecosystems – wind, water, flying animals, terrestrial animals, freshwater aquatic animals, marine animals, humans (which have become flying animals and terrestrial animals rolled into one), and machines (cf. Forman and Godron 1986; Forman 1995, 100). Wind and water transport involve mass flows driven by thermal and gravitational energy gradients. Animal, human, and machine transports involve locomotion of individuals running on their own internal energy.

Mass flows

Wind and water carry heat, sound, gases, aerosols, and particles within and between spatial elements. In meteorology, the vertical movement of heat and energy is called convection, while the horizontal movement is called advection. In a general sense, 'advection' means 'carrying to' and may be applied to the horizontal conveyance of all materials and forms of energy between spatial elements of landscapes and regions (cf. Forman 1995, 333).

Mass flows are often driven by a potential difference of some kind between two adjacent points or regions. An illustrative case is heat advection arising from differential heating within a landscape. Areas that heat more quickly, perhaps owing to smaller albedos (reflectivity), produce heat gradients with areas that heat less quickly. Air will tend to move from the warmer to the cooler places, all other factors being constant. Local temperature differences thus produce a largely horizontal flow of air. The same basic process also produces advection on regional and global scales. The moving air carries with it aerosols and particles. These particles may be removed from one landscape element or ecosystem and deposited in another by fallout or impaction. Advected water vapour may be deposited by condensation or precipitation.

Another example is mass flows induced by the land-surface gradient of potential energy. The difference in height between two adjacent points on the land surface creates a possibility of movement of sediment and water on that surface. If sediment is moved, then some of the potential energy is used (the land surface is lowered a little) and is converted into kinetic energy of the moving sediment. In general, the sediment movement rate will be proportional to the negative gradient of potential energy. With h as the height of the land surface and x as distance, and with the land surface gradient $\Delta h/\Delta x$ equal to the potential energy gradient, and with velocity v, the relationship may be expressed as:

$$v = -\frac{\Delta h}{\Delta x} \qquad (2.11)$$

Mass flows in the ecosphere tend to become channelled along specific paths. In landscapes, the paths traced out by flowing water and sediment define flowlines that run from high areas, down hillsides, into valleys, and along rivers. In air and water bodies, the chief paths of movement are seen in air and water currents.

Locomotive transport

Transport by animals, humans, and machines involves the locomotion of individuals running on their internal energy. The canalisation of the flow of animals, people, and information leads to dispersal routes, migration routes, animal trails, human trails, and transport and communication channels. Interestingly, transport by locomotion and transport by mass flows may lead to opposite results under similar circumstances. A wind blowing from a field to a forest through a narrow gap in the forest edge speeds up, owing to the Venturi effect. Animals moving from a field to a forest often slow down owing to innate caution.

Spatial influences

The spatial arrangement within landscapes is an important functional aspect of ecosystems, as important on a regional scale as the arrangement of continents is on a global scale. Some flows are concentrated (water, silt, pollutant in rivers); other flows are dispersed (erosion, seeds, and atmospheric pollutants). Some move fast; some move slowly. Landscape boundaries act as filters where movement rates change markedly. Different boundaries have different 'permeabilities' that regulate the movement rate between different ecosystems.

Landscape configuration has a major effect upon regional flows, and in turn the flows influence the landscape configuration (Fuentes 1990). A city is a source of people, vehicles, goods, and information. These move out on radiating transport and communications systems. Counterflows of rural people, water supply, and agricultural products come into the city. The routes help to tie landscapes together. The physical flows linking landscapes are eye-catching in mountainous regions (Forman 1995, 25). In mountains, landscapes (alpine, coniferous forest, basin grassland, and so on) are generally small and have sharp boundaries. The landscapes are fine-grained, grading downslope into variable-grained landscapes of flatter land. Gravity carries water and sediments overland, in streams, and in the ground to lower landscapes. Soil creeps overland or dashes down streams. Wind bears seeds and spores to higher landscapes, as well as downwards and along ranges. Animals, often with a cargo of seeds in their fur or feathers, move up and down slopes and along mountains. Water supplies speed downwards in pipes and canals to agricultural and built landscapes.

The landscape configuration in a region influences what happens in neighbouring regions. An illustration of such interregional influence comes from the New Jersey Pine Barrens landscape, USA (Forman 1998). Around the turn of the nineteenth century, anxiety was expressed at possible adverse effects that the Pine Barrens might have on climate in nearby New York and Philadelphia. The cause for concern rested in 250 years of profligate and irresponsible resource use (lumbering, fuel-wood collection, charcoal burning for iron smelting, and sand extraction for glass making) in the 550 000 ha Pine Barren landscape. This 'ruinous exploitation of natural resources' (Wacker 1998, 20) had exposed large areas of soil, caused the spread of shrublands, and led to a halving of average tree height. Even at the end of the nineteenth century, smoke and rampant fires were common in the Pine Barrens. During the next few decades, several events, technological developments, and policy changes led to a material increase in land protection for the Pine Barrens. Fire frequency decreased, forest regeneration occurred, and the climatic threat disappeared (Forman and Boerner 1981).

Global flows

Materials and energy move on a global scale. The chief rapid movements occur in the atmosphere and in the oceans because the Earth's major fluid and gaseous bodies are constantly being stirred. The powerhouse of the stirring is ultimately the Sun. Solar radiation fuels the general circulation of the atmosphere, and, less directly, the general oceanic circulation. These global stirrings ensure that all parts of the ecosphere are linked. Energy and materials are transported within and between the Northern and Southern Hemispheres. Intra-hemispherical flows involve convection (at least for gases, aerosols, and fine particles) and advection. Inter-hemispherical energy and material exchanges are a significant component of the global energy budget and water cycle. In the oceans, global mixing of the southern and northern oceans is effected by deep currents that snake along the ocean bottoms. Atmospheric and oceanic circulation patterns are influenced by topography. The mid-latitude jet stream is commonly anchored to the east of north–south trending mountain ranges. The oceanic circulations are largely dictated by the configuration of the ocean basins. When the ocean basins were in very different positions, the oceanic circulations were correspondingly different.

The global water cycle, in conjunction with local energy budgets, impels global biogeochemical cycles and the denudation–deposition system. Fine dust particles, aerosols, spores, and pollen grains are transported round the world. This transglobal transport is witnessed by the accumulation of polychlorinated biphenyls (PCBs) in arctic environments lying far from the source of such chemicals (e.g. Bright *et al.* 1995). It is also indicated by the movement of fine dust (fine silt and clay) from one continent to another (Simonson 1995). For example, large quantities of dust are carried out of the Sahara every year, some of which may reach the United Kingdom and fall as 'blood' rain (p. 73).

Isolation in a fully connected world

Although it might be argued that no part of the ecosphere is an island, some parts are better connected than others. Connectivity within the ecosphere may be approached through the idea of insularity. The insularity of actual islands can simply be defined by distance from the nearest neighbouring island or distance from the mainland. The insularity of various parts of the ecosphere is not necessarily open to such a ready definition. Patch isolation may be defined as the mean distance between a patch and its neighbouring patches (p. 34). Another measure is patch accessibility. This is defined as the distance along edges (e.g. woodland strips or hedgerows) that connect a patch with its neighbours (p. 34).

The degree of isolation can be surprisingly great, even on continents. The Bitterroot Mountains run north and south for some 80 km on the Montana–Idaho border, USA. They are dissected by deep valleys cutting in from the east. The valleys are very close together, formed in the same rocks, and have almost identical environments. It might be supposed, therefore, that the communities they contain would also be very similar. They are not. Furthermore, the differences cannot be accounted for by any environmental factor, not even a simple latitudinal gradient. So why do they differ? The answer appears to be that the valleys are isolated from one another by the intervening mountain ridges. Propagules from one valley cannot cross a ridge to a neighbouring valley. This means that species cannot colonise new valleys. The constraints are so severe that no unique sort of equilibrium between the vegetation and the environment evolves. Rather, each valley has its own vegetation arising from the happenstance of what was available at crucial times of establishment and a unique disturbance history (McCune and Allen 1985a, 1985b). Such is the power of topographic factors.

References

Adams, L. W. and Geis, A. D. (1983) Effects of roads on small mammals. *Journal of Applied Ecology* **20**, 403–15.

Adkins, K. F. and Merry, C. J. (1994) Accuracy assessment of elevation data sets using the Global Positioning System. *Photogrammetric Engineering and Remote Sensing* **60**, 195–202.

Band, L. E. and Moore, I. D. (1995) Scale: landscape attributes and geographical information systems. *Hydrological Processes* **9**, 401–22.

Bennett, A. F. (1991) Roads, roadsides and wildlife conservation: a review. In D. A. Saunders and R. J. Hobbs (eds) *Nature Conservation 2: The Role of Corridors*, pp. 99–118. Chipping Norton, Australia: Surrey Beatty & Sons.

Blair, D. J. and Biss, T. H. (1967) *The Measurement of Shape in Geography* (Nottingham University, Department of Geography, Quantitative Bulletin 11). Nottingham: Department of Geography, Nottingham University.

Blaszczynski, J. S. (1997) Landform characterization with Geographic Information Systems. *Photogrammetric Engineering and Remote Sensing* **63**, 183–91.

Blöschl, G., Kirnbauer, R., and Gutknecht, D. (1991) Distributed snowmelt simulations in an alpine catchment. I. Model evaluation on the basis of snow cover patterns. *Water Resources Research* **27**, 3171–9.

Bolstad, P. V. and Stowe, T. (1994) An evaluation of DEM accuracy: elevation, slope, and aspect. *Photogrammetric Engineering and Remote Sensing* **60**, 1327–32.

Bosch, W. (1978) A procedure for quantifying certain geomorphological features. *Geographical Analysis* **10**, 241–7.

Boyce, R. B. and Clark, W. A. V. (1964) The concept of shape in geography. *Geographical Review* **54**, 561–72.

Bright, D. A., Dushenko, W. T., Grundy, S. L., and Reimer, K. J. (1995) Effects of local and distant contaminant sources: polychlorinated biphenyls and other organochlorines in bottom-dwelling animals from an Arctic estuary. *Science of the Total Environment* **160–1**: 265–83.

Brown, D. B. and Bara, T. J. (1994) Recognition and reduction of systematic error in elevation and derivative surfaces from $7^1/_2$-minute DEMs. *Photogrammetric Engineering and Remote Sensing* **60**, 189–94.

Burrough, P. A. and McDonnell, R. A. (1998) *Principles of Geographical Information Systems*. Oxford: Oxford University Press.

Carrara, A., Bitelli, G., and Carla, R. (1997) Comparison of techniques for generating digital terrain models from contour lines. *International Journal of Geographical Information Science* **11**, 451–74.

Carter, J. R. (1988) Digital representations of topographic surfaces. *Photogrammetric Engineering and Remote Sensing* **54**, 1577–80.

Carter, J. R. (1992) The effect of data precision on the calculation of slope and aspect using gridded DEMs. *Cartographica* **29**, 22–34.

Ceruti, A. (1980) A method of drawing slope maps from contour maps by automatic data acquisition and processing. *Computers & Geosciences* **6**, 289–97.

Chang, K.-T. and Tsai, B.-W. (1991) The effect of DEM resolution on slope and aspect mapping. *Cartography and Geographical Information Systems* **18**, 69–77.

Chorowicz, J., Kim, J., Manoussis, S., Rudant, J.-P., Foin, P., and Veillet, I. (1989) A new technique for recognition of geological and geomorphological patterns in digital terrain models. *Remote Sensing and Environment* **29**, 229–39.

Clarke, A. L., Grün, A., and Loon, J. C. (1982) The application of contour data for generating high fidelity grid digital elevation models. In *Proceedings of AutoCarto V*, pp. 213–22. Falls Church, VA: American Society of Photogrammetry and Remote Sensing.

Commoner, B. (1972) *The Closing Circle: Confronting the Environmental Crisis*. London: Jonathan Cape.

Cornelius, S. C., Sear, D. A., Carver, S. J., and Heywood, D. I. (1994) GPS, GIS and geomorphological field work. *Earth Surface Processes and Landforms* **19**, 777–87.

Davis, J. C. (1986) *Statistics and Data Analysis in Geology*, 2nd edn. New York: John Wiley & Sons.

Dawson, B. L. (1991) South African road reserves: valuable conservation areas? In D. A. Saunders and R. J. Hobbs (eds) *Nature Conservation 2: The Role of Corridors*, pp. 119–30. Chipping Norton, Australia: Surrey Beatty & Sons.

De Roo, A. P. J., Hazelhoff, L., and Burrough, P. A. (1989) Soil erosion modelling using 'ANSWERS' and Geographical Information Systems. *Earth Surface Processes and Landforms* **14**, 517–32.

DeRose, R. C., Gomez, B., Marden, M., and Trustrum, N. A. (1998) Gully erosion in Mangatu Forest, New Zealand, estimated from digital elevation models. *Earth Surface Processes and Landforms* **23**, 1045–53.

Desmet, P. J. J. and Govers, G. (1997) Comment on 'Modelling topographic potential for erosion and deposition using GIS'. *International Journal of Geographical Information Science* **11**, 603–10.

Dietrich, W. E., Wilson, C. J., Montgomery, D. R., and McKean, J. (1993) Analysis of erosion thresholds, channel networks, and landscape morphology using a digital terrain model. *Journal of Geology* **101**, 259–78.

Dixon, L. F. J., Barker, R., Bray, M., Farres, P., Hooke, J., Inkpen, R., Merel, A., Payne, D., and Shelford, A. (1998) Analytical photogrammetry for geomorphological research. In S. Lane, K. Richards, and J. Chandler (eds) *Landform Monitoring, Modelling and Analysis*. Chichester: John Wiley & Sons.

Downs, P. W. and Priestnall, G. (1999) System design for catchment-scale approaches to studying river channel adjustments using a GIS. *International Journal of Geographical Information Science* **13**, 247–66.

Dreyer, G. D. and Niering, W. A. (1986) Evaluation of two herbicide techniques on electric transmission rights-of-way: development of relatively stable shrublands. *Environmental Management* **10**, 113–18.

Elder, K., Dozier, J., and Michaelsen, J. (1991) Snow accumulation and distribution in an alpine watershed. *Water Resources Research* **27**, 1541–52.

Evans, I. S. (1980) An integrated system of terrain analysis and slope mapping. *Zeitschrift für Geomorphologie*, NF, **36**, 274–95.

Fairfield, J. and Leymaine, P. (1991) Drainage networks from grid digital elevation models. *Water Resources Research* **30**, 1681–92.

Falcidieno, B. and Spagnuolo, M. (1991) A new method for the characterization of topographic surfaces. *International Journal of Geographical Information Systems* **5**, 397–412.

Felicisimo, A. M. (1994) Parametric statistical method for error detection in digital elevation models. *ISPRS Journal of Photogrammetry and Remote Sensing* **49**, 29–33.

Flood, M. and Gutelius, B. (1997) Commercial implications of topographic terrain mapping using scanning airborne laser radar. *Photogrammetric Engineering and Remote Sensing* **63**, 327.

Florinsky, I. V. (1998) Accuracy of local topographic variables derived from digital elevation models. *International Journal of Geographical Information Science* **12**, 47–61.

Forman, R. T. T. (1987) The ethics of isolation, the spread of disturbance, and landscape ecology. In M. G. Turner (ed.) *Landscape Heterogeneity and Disturbance*, pp. 213–29. New York: Springer.

Forman, R. T. T. (1990) Ecologically sustainable landscapes: the role of spatial configuration. In I. S. Zonneveld and R. T. T. Forman (eds) *Changing Landscapes: An Ecological Perspective*, pp. 261–78. New York: Springer.

Forman, R. T. T. (1995) *Land Mosaics: The Ecology of Landscapes and Regions*. Cambridge: Cambridge University Press.

Forman, R. T. T. (ed.) (1998) *Pine Barrens: Ecosystem and Landscape*. New Brunswick, NJ, and London: Rutgers University Press.

Forman, R. T. T. and Boerner, R. E. J. (1981) Fire frequency and the Pine Barrens of New Jersey. *Bulletin of the Torrey Botanical Club* **108**, 34–50.

Forman, R. T. T. and Godron, M. (1986) *Landscape Ecology*. New York: John Wiley & Sons.

Fryer, J. G., Chandler, J. H., and Cooper, M. A. H. (1994) On the accuracy of heighting from aerial photographs and maps: implications to process modellers. *Earth Surface Processes and Landforms* **19**, 577–83.

Fuentes, E. R. (1990) Landscape change in Mediterranean-type habitats of Chile: patterns and processes. In I. S. Zonneveld and R. T. T. Forman (eds) *Changing Landscapes: An Ecological Perspective*, pp. 165–90. New York: Springer.

Garbrecht, J. and Martz, L. W. (2000) Digital elevation model issues in water resources modeling. In D. Maidment and D. Djokic (eds) *Hydrologic and Hydraulic Modeling Support with Geographical Information Systems*, pp. 1–27. Redlands, CA: ESRI Press.

Garbrecht, J. and Starks, P. (1995) Note on the use of USGS level 1 7.5 minute DEM coverages for landscape drainage analyses. *Photogrammetric Engineering and Remote Sensing* **61**, 519–22.

Giles, P. T. (1998) Geomorphological signatures: classification of aggregated slope unit objects from digital elevation and remote sensing data. *Earth Surface Processes and Landforms* **23**, 581–94.

Giles, P. T. and Franklin, S. E. (1996) An automated approach to the classification of the slope units using digital data. *Geomorphology* **21**, 251–64.

Giles, P. T., Chapman, M. A., and Franklin, S. E. (1994) Incorporation of a Digital Elevation Model derived from stereoscopic satellite imagery in automated terrain analysis. *Computers & Geosciences* **20**, 441–60.

Gilvear, D. J., Waters, T. M., and Milner, A. M. (1998) Image analysis of aerial photography to quantify the effect of gold placer mining on channel morphology, interior Alaska. In S. Lane, K. Richards, and J. Chandler (eds) *Landform Monitoring, Modelling and Analysis*, pp. 195–216. Chichester: John Wiley & Sons.

Graff, L. H. and Usery, E. L. (1993) Automated classification of generic terrain features in digital elevation models. *Photogrammetric Engineering and Remote Sensing* **59**, 1409–17.

Griffin, M. L., Beasley, D. B., Fletcher, J. J., and Foster, G. R. (1988) Estimating soil loss on topographically nonuniform field and farm units. *Journal of Soil and Water Conservation* **43**, 326–31.

Griffith, D. A. (1982) Geometry and spatial interaction. *Annals of the Association of American Geographers* **72**, 332–46.

Haggett, P. (1967a) Network models in geography. In R. J. Chorley (ed.) *Models in Geography*, pp. 609–68. London: Methuen.

Haggett, P. (1967b) On the extension of the Horton combinatorial algorithm to regional highway networks. *Journal of Regional Science* **7**, 282–90.

Haggett, P. and Chorley, R. J. (1969) *Network Analysis in Geography*. London: Edward Arnold.

Hannah, M. J. (1981) Error detection and correction in digital elevation models. *Photogrammetric Engineering and Remote Sensing* **61**, 513–17.

Hodgson, M. E. (1995) What cell size does the computed slope/aspect angle represent? *Photogrammetric Engineering and Remote Sensing* **61**, 513–17.

Horn, B. K. P. (1981) Hill shading and the reflectance map. *Proceedings of the Institute of Electrical and Electronics Engineering* **69**, 14–17.

Horton, R. E. (1932) Drainage basin characteristics. *Transactions of the American Geophysical Union* **13**, 350–61.

Huggett, R. J. (1991) *Climate, Earth Processes and Earth History*. Heidelberg: Springer.

Huggett, R. J. (1997) *Environmental Change: The Evolving Ecosphere*. London: Routledge.

Hunter, G. J. and Goodchild, M. F. (1995) Dealing with error in spatial databases: a simple case study. *Photogrammetric Engineering and Remote Sensing* **61**, 529–31.

Hussey, B. M. J. (1991) The flora roads survey – volunteer recording of roadside vegetation in Western Australia. In D. A. Saunders and R. J. Hobbs (eds) *Nature Conservation 2: The Role of Corridors*, pp. 41–8. Chipping Norton, Australia: Surrey Beatty & Sons.

Jensen, J. R. (2000) *Remote Sensing of the Environment: An Earth Resource Perspective*. Upper Saddle River, NJ: Prentice Hall.

Jenson, S. J. and Domingue, J. O. (1988) Extracting topographic structure from digital elevation data for geographic information system analysis. *Photogrammetric Engineering and Remote Sensing* **54**, 1593–600.

Jones, N. L., Wright, S. G., and Maidment, D. R. (1990) Watershed delineation with triangle-based terrain models. *Journal of Hydraulic Engineering* **116**, 1232–51.

Kansky, K. J. (1963) *Structure of Transport Networks: Relationships Between Network Geometry and Regional Characteristics*. University of Chicago, Department of Geography, Research Paper No. 84. Chicago, Illinois: Department of Geography, University of Chicago.

Kidner, D. B. and Smith, D. H. (1992) Compression of digital elevation models by Huffman coding. *Computers & Geosciences* **18**, 1013–34.

Krabill, W. B., Thomas, R. H., Martin, C. F., Swift, R. N., and Frederick, E. B. (1995) Accuracy of airborne laser altimetry over the Greenland ice sheets. *International Journal of Remote Sensing* **16**, 1211–22.

Kraus, K. (1994) Visualization of the quality of surfaces and their derivatives. *Photogrammetric Engineering and Remote Sensing* **60**, 457–62.

Lane, S. N., Richards, K. S., and Chandler, J. H. (eds) (1998) *Landform Monitoring, Modelling and Analysis*. Chichester: John Wiley & Sons.

Lee, J. and Stucky, D. (1998) On applying viewshed analysis for determining least-cost paths on digital elevation models. *International Journal of Geographical Information Science* **12**, 891–905.

Li, Z. (1994) A comparative study of the accuracy of digital terrain models (DTMs) based on various data models. *ISPRS Journal of Photogrammetry and Remote Sensing* **49**, 2–11.

Linton, D. L. (1951) The delimitation of morphological regions. In L. D. Stamp and S. W. Wooldridge (eds) *London Essays in Geography*, pp. 199–218. London: Longman.

López, C. (1997) Locating some types of random errors in digital terrain models. *International Journal of Geographical Information Science* **11**, 677–98.

Ludwig, B., King, J. D. D., and Souchère, V. (1996) Using GIS to predict concentrated flow erosion in cultivated catchments. In *HydroGIS 96: Applications of Geographic Information Systems in Hydrology and Water Resources Management (Proceedings of the Vienna Conference, April 1996)* (IAHS Publication No. 235), pp. 429–36. Rozendaalselaan, The Netherlands: International Association of Hydrological Sciences.

Mark, D. M. (1975) Computer analysis of topography: a comparison of terrain storage methods. *Geografiska Annaler* **57A**, 179–88.

McCune, B. and Allen, T. F. H. (1985a) Will similar forests develop on similar sites? *Canadian Journal of Botany* **63**, 367–76.

McCune, B. and Allen, T. F. H. (1985b) Forest dynamics in Bitterroot Canyons, Montana. *Canadian Journal of Botany* **63**, 377–83.

McGarigal, K. and Marks, B. J. (1995) *FRAGSTATS: Spatial Pattern Analysis Program for Quantifying Landscape Structure* (US Forest Service General Technical Report PNW-GTR-351). Portland, OR: Pacific Northwest Research Station, USDA Forest Service.

McKenney, D. W., Mackey, B. G., and Zavitz, B. L. (1999) Calibration and sensitivity analysis of a spatially distributed solar radiation model. *International Journal of Geographical Information Science* **13**, 49–65.

Miller, V. C. (1953) *A Quantitative Geomorphic Study of Drainage Basin Characteristics in the Clinch Mountain Area, Virginia and Tennessee* (Office of Naval Research, Geography Branch, Project NR 389–042, Technical Report 3). Arlington, VA: Office of Naval Research.

Milne, G. (1935) Some suggested units of classification and mapping, particularly for East African soils. *Soil Research* **4**, 183–98.

Milne, J. A. and Sear, D. A. (1997) Modelling river channel topography using GIS. *International Journal of Geographical Information Science* **11**, 499–519.

Mitasova, H., Hofierka, J., Zlocha, M., and Iverson, L. (1996) Modelling topographic potential for erosion and deposition using GIS. *International Journal of Geographical Information Systems* **10**, 629–41.

Mitasova, H., Hofierka, J., Zlocha, M., and Iverson, L. (1997) Reply to Desmet and Govers. *International Journal of Geographical Information Science* **11**, 611–18.

Mitchell, C. W. (1991) *Terrain Evaluation: An Introductory Handbook to the History, Principles, and Methods of Practical Terrain Assessment*, 2nd edn. Harlow: Longman.

Monckton, C. G. (1994) An investigation into the spatial structure of error in digital elevation data. In M. F. Worboys (ed.) *Innovations in GIS 1*, pp. 201–11. London: Taylor & Francis.

Montgomery, D. R. and Dietrich, W. E. (1989) Source areas, drainage density and channel initiation. *Water Resources Research* **25**, 1907–18.

Moore, I. D. (1988) A contour-based terrain analysis program for the environmental sciences (TAPES). *Transactions of the American Geophysical Union* **69**, 345.

Moore, I. D. and Grayson, R. B. (1989) Hydrologic and digital terrain modelling using vector elevation data. *Transactions of the American Geophysical Union* **70**, 1091.

Moore, I. D. and Grayson, R. B. (1990) Terrain-based catchment partitioning and runoff prediction using vector elevation data. *Water Resources Research* **27**, 1177–91.

Moore, I. D., O'Loughlin, E. M., and Burch, G. J. (1988) A contour-based topographic model for hydrological and ecological applications. *Earth Surface Processes and Landforms* **13**, 305–20.

Moore, I. D., Grayson, R. B., and Ladson, A. R. (1991) Digital terrain modeling: a review of hydrological, geomorphological and biological applications. *Hydrological Processes* **5**, 3–30.

Moore, I. D., Turner, A. K., Wilson, J. P., Jenson, S., and Band, L. (1993) GIS and land-surface–subsurface process modelling. In M. F. Goodchild, B. O. Parks, and L. T. Steyaert (eds) *Environmental Modeling with GIS*, pp. 196–230. New York: Oxford University Press.

Morgan, R. P. C. (1986) *Soil Erosion and Conservation*, revised and enlarged edn. Harlow: Longman Scientific and Technical.

Morison, C. G. T., Hoyle, A. C., and Hope-Simpson, J. F. (1948) Tropical soil–vegetation catenas and mosaics: a study in the south-western part of Anglo-Egyptian Sudan. *Journal of Ecology* **36**, 1–84.

Nelson, R., Krabill, W., and Tonelli, J. (1988) Estimating forest biomass and volume using airborne laser data. *Remote Sensing of the Environment* **24**, 247–67.

Nilsson, M. (1996) Estimation of tree heights and stand volume using an airborne LIDAR system. *Remote Sensing of the Environment* **56**, 1–7.

O'Callaghan, J. F. and Mark, D. M. (1984) The extraction of drainage networks from digital elevation data. *Computer Vision, Graphics and Image Processing* **28**, 323–44.

Oswald, H. and Raetzsch, H. (1984) A system for generation and display of digital elevation models. *Geo-Processing* **2**, 197–218.

Peuker, T. K. (now Poiker), Fowler, R. J., Little, J. J., and Mark, D. M. (1978) The triangulated irregular network. In *Proceedings of the DTM Symposium*, American Society of Photogrammetry – American Congress on Survey and Mapping, pp. 24–31. St Louis, MO: American Society of Photogrammetry.

Polidori, L., Chorowicz, J., and Guillande, R. (1991) Description of terrain as a fractal surface and application to digital elevation model quality assessment. *Photogrammetric Engineering and Remote Sensing* **57**, 1329–62.

Prosser, I. P. and Abernethy, B. (1996) Predicting the topographic limits to a gully network using a digital terrain model and process thresholds. *Water Resources Research* **32**, 2289–98.

Ridgway, J. R., Minster, J. B., Williams, N., Bufton, J. L., and Krabill, W. B. (1997) Airborne laser altimeter survey of Long Valley, California. *Geophysical Journal International* **131**, 267–80.

Ritter, P. (1987) A vector-based slope and aspect generation algorithm. *Photogrammetric Engineering and Remote Sensing* **53**, 1109–11.

Scheidegger, A. E. (1986) The catena principle in geomorphology. *Zeitschrift für Geomorphologie*, NF **30**, 257–73.

Schumm, S. A. (1956) The evolution of drainage systems and slopes in badlands at Perth Amboy, New Jersey. *Bulletin of the Geological Society of America* **67**, 597–646.

Simonson, R. W. (1995) Airborne dust and its significance to soils. *Geoderma* **65**, 1–43.

Skidmore, A. K. (1990) Terrain position as mapped from a gridded digital elevation model. *International Journal of Geographical Information Systems* **4**, 33–49.

Skidmore, D. E. (1989) A comparison of techniques for calculating gradient and aspect from a gridded digital elevation model. *International Journal of Geographical Information Systems* **3**, 323–34.

Smith, D. H. and Lewis, M. (1994) Optimal predictors for compression of digital elevation models. *Computers & Geosciences* **20**, 1815–22.

Speight, J. G. (1974) A parametric approach to landform regions. In E. H. Brown and R. S. Waters (eds) *Progress in Geomorphology: Papers in Honour of David L. Linton* (Institute of British Geographers Special Publication No. 7), pp. 213–30. London: Institute of British Geographers.

Srinivasan, R. and Engel, B. A. (1991) Effect of slope prediction methods on slope and erosion methods. *Applied Engineering in Agriculture* **7**, 779–83.

Stoddart, D. R. (1965) The shape of atolls. *Marine Geology* **3**, 369–83.

Tajchman, S. J. (1981) On computing topographic characteristics of a mountainous catchment. *Canadian Journal of Forest Research* **11**, 768–74.

Theakstone, W. H. and Jacobsen, F. M. (1997) Digital terrain modelling of the surface and bed topography of the glacier Austre Okstindbreen, Okstindan, Norway. *Geografiska Annaler* **79A**, 201–14.

Turner, M. G. (1989) Landscape ecology: the effect of pattern on process. *Annual Review of Ecology and Systematics* **20**, 171–97.

Turner, M. G. (1990) Spatial and temporal analysis of landscape patterns. *Landscape Ecology* **4**, 21–30.

Twigg, D. R. (1998) The Global Positioning System and its use for terrain mapping and monitoring. In S. Lane, K. Richard, and J. Chandler (eds) *Landform Monitoring, Modelling and Analysis*, pp. 37–61. Chichester: John Wiley & Sons.

Unwin, D. (1981) *Introductory Spatial Analysis*. London and New York: Methuen.

Vaughn, C. R., Bufton, J. L., and Rabine, D. (1996) Georeferencing of airborne laser altimeter measurements. *International Journal of Remote Sensing* **17**, 2185–200.

Vertessy, R. A., Wilson, C. J., Silburn, D. M., Connolloy, R. D., and Ciesiolka, C. A. (1990) Predicting erosion hazard areas using digital terrain analysis. In *Research Needs and Applications to Reduce Erosion and Sedimentation in Tropical Steeplands (Proceedings of the Fiji Symposium, June 1990)* (IAHS–AISH Publication No. 192), pp. 298–308. Rozendaalselaan, The Netherlands: International Association of Hydrological Sciences.

Voorhees, L. D. and Cassell, J. F. (1980) Highway right-of-way: mowing versus succession as related to duck nesting. *Journal of Wildlife Management* **44**, 155–63.

Wacker, P. O. (1998) Human exploitation of the New Jersey Pine Barrens before 1900. In R. T. T. Forman (ed.) *Pine Barrens: Ecosystem and Landscape*, pp. 3–23. New Brunswick, NJ, and London: Rutgers University Press.

Way, J. M. (1977) Roadside verges and conservation in Britain: a review. *Biological Conservation* **12**, 65–74.

Weibel, R. and Heller, M. (1991) Digital terrain modelling. In D. J. Maguire, M. F. Goodchild, and D. W. Rhind (eds) *Geographical Information Systems, Vol 1: Principles*, pp. 269–97. Harlow: Longman.

Weltz, M. A., Ritchie, J. C., and Fox, H. D. (1994) Comparison of laser and field measurements of vegetation height and canopy cover. *Water Resources Research* **30**, 1311–19.

Wilson, J. P. and Gallant, J. C. (2000) Digital terrain analysis. In J. P. Wilson and J. C. Gallant (eds) *Terrain Analysis: Principles and Applications*, pp. 1–27. New York: John Wiley & Sons.

Wise, S. M. (1998) The effect of GIS interpolation errors on the use of Digital Elevation Models in geomorphology. In S. Lane, K. Richards, and J. Chandler (eds) *Landform Monitoring, Modelling and Analysis*, pp. 139–64. Chichester: John Wiley & Sons.

Wise, S. (2000) Assessing the quality for hydrological applications of digital elevation models derived from contours. *Hydrological Processes* **14**, 1909–29.

Wood, J. D. and Fisher, P. F. (1993) Assessing interpolation accuracy in elevation models. *IEEE Computer Graphics and Applications* **13**, 48–56.

Wooldridge, S. W. (1932) The cycle of erosion and the representation of relief. *Scottish Geographical Magazine* **48**, 30–6.

Zevenbergen, L. W. and Thorne, C. R. (1987) Quantitative analysis of land surface topography. *Earth Surface Processes and Landforms* **12**: 47–56.

Zhang, W. H. and Montgomery, D. R. (1994) Digital elevation model grid size, landscape representation, and hydrologic simulations. *Water Resources Research* **30**, 1019–28.

PART II Topographic influences

CHAPTER 3

Climate

Meteorologists and climatologists classify atmospheric systems, as landscape ecologists classify landscapes, according to their scale. The favoured terms are microscale, local, mesoscale, and macroscale. Microscale atmospheric systems have a maximum area of about 1 km^2; they include small-scale turbulence, dustdevils, and small cumulus clouds. Local systems, such as tornadoes and large cumulus clouds, have areas in the range 0.1–2500 km^2, so overlapping the top end of microscale systems and the bottom end of mesoscale systems. Mesoscale systems fall in the range 100–40 000 km^2 and include sea breezes, cyclonic eddies, dry lake circulations, thunderstorms, and squall lines. Macroscale systems range from about 10 000 km^2 to the global circulation and include anticyclones, cyclones, fronts, hurricanes, and jet streams.

The scales of atmospheric systems and the terms applied to climates at the same scales – microclimate, local climate, mesoclimate, and macroclimate – do not always correspond. The microclimate is the climate at or near the ground, in the lowest layer of the atmosphere, but microscale atmospheric systems occur everywhere in the atmosphere. Microclimates are also associated with buildings, other human-made surfaces and features, and with such distinctive natural environments as caves and animal burrows. They occupy the vegetation canopy, the urban canopy, the soil layer, and buildings. They display horizontal differences over a few centimetres to a hundred metres or so (in clearings). Their vertical dimension is about 1 m in grassland and low crops, to 30 m in forests, and hundreds of metres in some parts of cities.

Topographic differences over tens of metres to about 100 km produce a mosaic of microclimates that has much in common with the mosaic of land surface units and landscape mosaic. At this scale, topography produces distinctive topoclimates (local climates). Topoclimates are the climates of the lower troposphere that is influenced by topography, that is the climate of air in contact with the natural and human-made planetary cover (e.g. Yoshino 1975; Obrębska-Starklowa 1995). Some authors equate local climates with mesoclimates (e.g. Bogdan 1988; McKendry 1993), but others prefer to limit mesoclimatology to mesoscale weather systems (e.g. Barry 1992, 12). Macroclimates are regional, continental, and global climates. They are normally associated with synoptic-scale and planetary circulation systems, but some mountains generate their own distinct regional weather systems. A case may be made for recognising three levels of climate – microclimates, topoclimates (local climates), and macroclimates (Bogdan 1988). This scheme gives a prominent place to topoclimates, so emphasising the influence of all topographic factors (relief, altitude, aspect, water bodies, human features, vegetation, soils, and so on) on the lower troposphere.

Topography affects climates through altitude, which has a similar effect to increasing latitude, though there are important differences. It also affects air movement at all scales to create a hierarchy of circulation systems. The mechanisms involved are modifications of airflow (aerodynamic effects) and modifications of surface heat balances (thermal effects). The circulation systems across the scales are interdependent and not easy to tease apart. Nonetheless, three scales of topographically modified circulation system stand out – local motions, synoptic-scale weather systems (especially fronts), and planetary-scale wave motion (cf. Barry 1992, 108–57). This chapter will explore the effects of altitude on climate, and the topographic modification of boundary layer climates and circulation systems, synoptic-scale atmospheric systems, and global climates. It will start by examining how climatic factors related to topography are modelled.

Modelling climatic phenomena

Many climatic modelling applications require a surface or areal representation of different climatic phenomena. A problem is that climate variables are usually measured using a network of gauging stations that provide point-based estimates. Many studies have attempted to transform these point data to a surface representation, usually using an interpolation algorithm. However, the spatial pattern of most climate variables is very complex and they often display non-stationarity, that is, they may exhibit systematic spatial trends or 'drift' in the mean or variance of the process. Often, such a trend is a result of the underlying topographic influence and hence much work has been carried out to identify the relationship between elevation and the climatic parameter and thus remove the drift before application of the data to an interpolation algorithm. Another problem is that climate stations are usually located in low-elevation easily accessible sites, which means that the data collected do not necessarily well represent the climatic conditions experienced across the site. For instance, all 149 meteorological stations in Scotland that measure daily temperature are located below 400 m altitude and mainly in agricultural areas (Hudson and Wackernagel 1994). In Scotland, 21 per cent of the land lies above 400 m and the highest point – Ben Nevis – is 1344 m. Significant work has been carried out to overcome such difficulties, particularly for precipitation, but also for temperature and solar radiation.

Modelling precipitation

Areal mean-annual precipitation is one of the most important climatic parameters for any region or catchment for hydrologists. Mean annual precipitation estimates are employed in many critical calculations, such as for water and chemical balances of catchments and lakes (Dingman and Johnson 1971; LaBaugh 1985), for site assessment for reservoir installation, or even for nuclear waste repository plants (Hevesi et al. 1992b). Areal precipitation measurements are usually produced using a set of point precipitation measurements taken from the gauging network and the application of some interpolation technique to convert these point values to a surface representation. Uncertainties in the estimates are propagated through many calculations, so clearly the most accurate estimates possible are desirable from the outset. Uncertainties in the estimates of mean annual precipitation are a result of (1) failure of individual gauges to catch the volume of precipitation that actually falls at the gauge site; (2) failure of the gauge network to sample the area of interest adequately (Dingman et al. 1988); and (3) the accuracy of the interpolation algorithm.

Measurement errors at individual gauges can usually be attributed to: (1) observer error; (2) location of the gauge with respect to obstruction; (3) wind effects; (4) failure to collect inputs from horizontal interception; and (5) topographic position. Out of these sources of errors, wind effects are the only ones to have been reasonably well investigated (Black 1954; Wilson 1954; Larson and Peck 1974; Neff 1977; Helvey and Patric 1983). Wind often causes precipitation to fall at a considerable inclination. The precipitation incident on a hillslope may therefore deviate widely from measurements made in conventional rain-gauges with horizontal orifices (Sharon 1980; Sevruk 1982, 1986; Folland 1988; Sharon et al. 1988). Accurate estimates of rainfall intensity are therefore necessary if accurate estimates of rain-dependent processes are to be made. A numerical model of the rain-gauge exposure problem that accounts for the interaction between wind speed and direction, rain-drop size, and the physical characteristics of the collector was developed by Folland (1988). From this model a 'flat champagne-glass'-shaped collector that minimised errors was designed.

Topographic position also introduces precipitation measurement errors. Sharon (1980) developed a crude but simple method of accounting for these and wind-induced errors as follows. Sharon used Fourcade's (1942) equation to develop a correction factor, C_f, to adjust rainfall measurements made using a gauge with a horizontal orifice, P_o, of rainfall on an inclined surface, defined in terms of per unit of projected horizontal area, P_a, on the basis of the topographic attributes of slope, β, and aspect, ψ, and the inclination of the rainfall vector, ε, measured from the zenith in the direction from which rain is falling, ϕ:

$$C_f = \frac{P_a}{P_o} = 1 + \tan\beta \, \tan\varepsilon \, (\psi - \phi) \qquad (3.1)$$

In practice ε is dependent on raindrop size and windspeed and ϕ will vary during a storm. The values of these variables can be inferred from wind speed and direction measurements and effective drop sizes that are related to rainfall intensity. Sharon showed that for rainfall inclined at 40°, C_f varies from 0.52 on the lee side to 1.48 on the windward side of a 30° slope, and concluded that the above equation is accurate providing the rainfall vector can be specified within narrow limits. These effects even occur at the micro-topographical scale on cultivated fields with ridges (Sharon *et al.* 1988). Sharon (1980) also found that the variations of hydrological rainfall between windward and lee slopes are 'reduced near wind-exposed ridges, owing to systematic small-scale variations of meteorological rainfall and of rain inclination'.

The introduction of errors caused by the inadequacy of the representative gauge network is usually more significant in mountainous or topographically varied regions, where orographic effects and aspect, and other local conditions, induce a high degree of spatial variability in precipitation. In mountainous regions, most rain gauges are located in valleys and thus do not represent the precipitation that falls on an entire catchment. This means developing procedures for estimating precipitation amounts and distribution on mountainous catchments. Molnau *et al.* (1980) found an average of 8 per cent error with a density of 46 km^2 per gauge in a mountainous research catchment in Idaho, USA.

Orographic effects explain many spatial features of the regional distribution of precipitation. The influence of topographic barriers on precipitation distribution and amount has been recognised for many years (e.g. Salter 1921). Although reasonably well understood (Browning and Hill 1981; Sumner 1988; Barros and Lettenmaier 1994), attempts to incorporate topographic influences into areal models of precipitation have yet to be fully resolved; this is a result of the poor instrumentation of most upland and mountainous areas (see p. 56). Adaptations of standard interpolation techniques have been the most popular approach to this problem.

A number of studies and reviews have been carried out to evaluate the performance of interpolation techniques generally, and more specifically to model precipitation spatially (Court and Bare 1984; Creutin and Obled 1982; Dingman 1994; Hutchinson 1995; Hutchinson and Gessler 1994; Lebel *et al.* 1987; Mandeville and Rodda 1970; Shaw and Lynn 1972; Tabios and Salas 1985). In general it is considered that optimal interpolation/kriging methods provide the best estimates of regional precipitation in a variety of situations. Kriging is a method developed by Matheron (1963) and Krige (1966) and was originally used to map ore concentrations (Journel and Huijbregts 1978). These methods perform more accurately because they are based on the spatial correlation structure of pre-

cipitation in the region of application, whereas other methods impose essentially arbitrary spatial structures. Kriging also provides surface estimates of uncertainty.

Early work into the incorporation of topography into the model of areal precipitation involved regression or correlation to obtain some objective quantitative relationship between precipitation and topographic variables, usually altitude (Bleasdale and Chan 1972; Chuan and Lockwood 1974; Rodda 1962; Spreen 1947; Wigley *et al.* 1984). Each of these found a strong relationship between precipitation and altitude. Most of these studies involved much laborious manual work. Early computer developments have included the use of a regular 5-km rainfall grid covering the whole of the UK (Shearman and Salter 1975). More recently, much work has been carried out using automated cartography techniques or GIS. Bleasdale and Chan (1972) suggested two possible approaches, one through the use of digitised ground contours, the other through a sufficiently fine regular grid of spot heights. Today, it is possible to obtain this information from a digital elevation model (DEM), and using algorithms provided within many GIS's one can obtain topographical information from the DEM such as aspect and slope (p. 21).

Kriging has led to significant progress in characterising the spatial relationship between precipitation and orography. Kriging requires a stationary field for estimation; that is, there must be no systematic spatial trend or 'drift' in the mean or variance of the process. This condition is not met in mountainous areas where precipitation usually increases with elevation and is influenced by topographic aspect and storm direction (Garen *et al.* 1994). The problem of kriging in mountainous terrain has been tackled in a variety of ways.

Dingman *et al.* (1988) attempted to evaluate errors introduced by network coverage inadequacy by applying kriging in two different approaches to estimate mean areal precipitation in New Hampshire and Vermont, USA, a mountainous area with known orographic influences. The data used were 1951–1980 normal precipitation at 120 gauges in two states and in adjacent portions of bordering states and provinces. In the first approach, kriging is directly applied to the mean areal precipitation values, and in the second, kriging is applied to a 'precipitation delivery factor', which is a linear function of station elevation and represents the mean areal precipitation with the orographic effect removed. PDF is an attempt to 'detrend' the data using information from outside the data set. Seven validation stations were not included in the original analysis, and when the values of these were tested against the estimates (cross-validation) from the different approaches it was found that the first method gave slightly better kriged estimates but resulted in an error surface that was highly contorted and with larger maximum errors over most of the region. The second method had a considerably smoother error surface and was considered preferable as a basis for point and areal estimates of mean areal precipitation.

Hevesi *et al.* (1992a, 1992b) required areal mean annual precipitation values for hydrological characterisation of a potential high-level nuclear waste repository site at Yucca Mountain, Nevada, USA, for a 2600-square-mile (6730 km^2) catchment containing Yucca Mountain. Kriging and co-kriging were both utilised, with co-kriging (with elevation as the covariate) defining an irregular surface that more accurately represented expected local orographic influences on the mean areal precipitation. Kriging estimates tended to be lower than co-kriging estimates because the increased mean areal precipitation expected for remote mountainous topography was not adequately represented by the available sample of gauges.

Chua and Bras (1982) considered the application of kriging to estimate mean areal precipitation of mountainous regions under two different assumptions of the drift and spatial dependency of precipitation influenced by orographic effects. They used the 'detrending' method which assumes a linear relationship between the drift and ground elevation. The variogram is calculated from the residuals resulting from subtraction of the linear relationship from the original data. The second procedure used – generalised covariance (GC) – assumes a polynomial drift with unknown coefficients and a generalised covariance estimated directly from the raw data. The study site was San Juan Mountains, northwestern Colorado, USA. Kriging with the GCs identified produced mean areal precipitation estimates that were consistently lower than the 'detrending' estimates. Point suppression tests indicated that the 'detrending' procedure performed better than the GC method in terms of providing accurate kriged estimates but was less successful in representing the kriging estimation error variance.

Another popular interpolation method is splines. Hutchinson (1995) interpolated mean rainfall using thin plate smoothing splines for a region in southeast Australia. In order to determine the influence of topography, partial thin plate splines with increasingly

complex dependencies on topography (elevation and aspect) were fitted. Hutchinson (1995) found that incorporating any topography parameter other than actual station elevation does not necessarily contribute significantly to interpolation accuracy, especially when data sets are reasonably dense. However, in a different study of annual and monthly areal precipitation estimates, aspect was found to be an important parameter (Hanson 1982).

Hanson (1982) carried out a study to investigate the distribution of precipitation on a mountainous watershed that lies within a mountain range in southwest Idaho, USA, and also developed procedures to generate annual and monthly precipitation series that simulate the precipitation input to the watershed. Analysis of the data for Reynolds Creek Experimental Watershed showed a linear relationship between annual amounts and elevation. This relationship was strongest when the gauges were grouped into downwind and upwind sites. This grouping was considered appropriate because most of the winter storms moved over the watershed from the west and southwest, and the heaviest precipitation was on the west (downwind) side of the watershed. Gauge sites along the western and southern watershed borders were found to be most representative of the upwind gauges on the east side, because they measured the precipitation from the air moving upwind onto the watershed. The maximum annual precipitation on the watershed was just leeward of the western watershed boundary. Hanson also discovered that the monthly precipitation and elevation relationship was best represented by grouping the gauge sites into upwind and downwind sites. In the summer, however, when there are only small amounts of precipitation, one equation could be used to present the elevation relationship. Precipitation differences across the catchment were found to be associated with elevation and storm patterns. The major winter storms moved on to the watershed from the west and south-west, and caused a zone of high precipitation on the leeward slopes of the watershed, and less precipitation on the north and north-east sections.

It is acknowledged that there are significant effects on rainfall related to elevation and aspect, although incorporation of the latter into areal precipitation models has not proved as easy to define as precipitation. It is also generally accepted that the effects of topography on received rainfall are spatially coarser than the finest descriptions of topography (Hutchinson 1995), although this phenomenon has rarely been addressed. Both of these effects have been examined using local regression methods in the PRISM model (Daly *et al.* 1994). The PRISM model (Precipitation-elevation Regressions on Independent Slopes Model) uses spatially smoothed elevations ('orographic elevation') instead of actual (point) elevations, and also groups rain gauges according to aspect similarity. When compared to commonly used geostatistical techniques, PRISM exhibited a superior performance. However, Hutchinson (1995) found that incorporating dependence on aspect, determined from a digital elevation model, made only a marginal further improvement upon the results derived when using splines.

Modelling temperature

With temperature there has been some controversy over the most appropriate method for producing a surface from a limited set of stations. White and Smith (White 1979; White and Smith 1982) used a regression model to predict a number of climatic variables from a set of topographic and locational variables, but their efforts were strongly criticised by Gregory (1982). Despite the condemnation, research has continued down this road.

An early attempt was carried out by Tabony (1985) who tried to quantify the relationship between minimum temperature and topography in order to allow minima to be 'reduced' to a standard topography, so enabling the spatial interpolation of minimum temperature. Temperature data were obtained from the climatological network in the UK, and a data set containing spot heights on a 0.5-km grid was used as the principal source of topographic information. The differing roles of 'local' and 'large-scale' shelter were recognised, and the topographic data were used to identify parameters that would provide objective representations of these variables. The topographic variables were then regressed against measures of minimum temperature associated with given return periods.

Lennon and Turner (1995) attempted to model temperature in Great Britain using four different models: simple interpolation, thin plate splines, multiple linear regression, and mixed spline-regression. The British distribution of mean daily temperature was predicted with the greatest accuracy by using a mixed model that involved fitting a thin-plate spline to the surface of the country, after correction of the data by a selection from independent topographical variables (e.g. altitude, distance from the sea, slope, and topographic roughness) chosen using multiple regression

from a digital elevation model of the country. Prediction of temperature by this method was more than 95 per cent accurate in all months of the year when tested using cross-validation. The next most accurate method was a pure multiple-regression model using the DEM. Both regression and thin-plate spline models based on a few topographic variables (latitude, longitude, and altitude) were only comparatively unsatisfactory. Differences between the methods were considered to be dependent largely on their ability to model the effect of the sea on land temperatures.

Hudson and Wackernagel (1994) also used elevation for interpolation of point-based temperature data. They investigated the use of elevation data to help map January mean temperatures for the period 1961–1980 in Scotland. Elevation was used as an external drift variable for kriging and co-kriging of temperatures. They found that for each of the 12 months, the average temperature was negatively correlated with elevation more strongly than with latitude or longitude, and there were distinctly seasonal changes in this correlation. The seasonal effects were due to the deterministic period variations in solar radiation, which in the winter were overlain by random air movements. They found that their correlations with elevation changed with time and space. The correlations between July temperatures and elevation were weakest because of the stronger correlations with latitude in summer.

Hudson and Wackernagel (1994) found that there was a strong linear relationship between January mean temperatures and elevation for elevations above 100 m. They made the assumption that this linear behaviour extended to areas about 400 m and carried out kriging using elevation as an external drift variable. This improved the estimates of temperature when compared to kriging based on temperature data alone, although some exaggerated extrapolation effects were observed in a region in the Highlands, where stations are scarce and located at low altitudes. They felt that if a non-linear function, describing more precisely the average relationship between temperature and elevation, were available for Scotland, the results could have been further improved.

Ishida and Kawashima (1993) predicted surface air temperature in central Japan from temperature recordings using seven different procedures: simple and universal kriging, co-kriging estimators, regression analysis and inverse distance weighting. The co-kriging incorporated DEM as well as the air temperature readings. From cross-validation the kriging estimator provided better estimates than regression (which treats the data as spatially independent observations) and the inverse distance-weighted method. Co-kriging provided further improvements because of the strong correlation between air temperature and elevation. The accuracy of spatial prediction decreased due to nocturnal cooling in winter and summer daytime heating. This decrease implies that a strong radiation balance at the surface, positive or negative, causes a relatively short-range variation in surface air temperature through the effects of local environment.

Solar radiation modelling

Solar radiation powers or influences many biophysical and meteorological processes both indirectly and directly, and is of importance within hydrology, ecology, agriculture, forestry, engineering, and many other fields. More specifically, it influences processes such as soil temperature and soil heat flux, sensible heat flux, surface and air temperatures, wind and turbulent transport, evapotranspiration, and the growth and activity of fauna and flora. The Sun supplies 99.8 per cent of energy at the Earth's surface (Kumar *et al.* 1997) and topography is one of the most important factors controlling the amount of solar energy incident at any location on the Earth's surface. Variability in topographic variables, including elevation, slope, slope orientation (aspect or azimuth), and shadowing, can create steep local gradients in solar radiation receipt. The importance of topographic effects has been recognised for some time; however, only recently have efforts been made to integrate topographic effects quantitatively in a systematic manner through environmental modelling (e.g. Dubayah and Rich 1995; Kumar *et al.* 1997; McKenney *et al.* 1999). Dubayah and Rich (1995) recognise several factors that have hindered progress on topographic solar radiation modelling: (1) the complexity of physically based solar radiation formulations for topography; (2) a lack of representative data to process these formulations; and (3) a lack of suitable modelling tools. GISs and DEMs have provided solutions in part to problems (2) and (3).

Global radiation is made up of shortwave and longwave radiation. Shortwave radiation can be absorbed by terrestrial bodies and cloud cover and re-emitted as longwave radiation. Shortwave radiation actually reaching the Earth's surface may be direct, diffuse, or reflected (Figure 3.1). Direct radiation strikes the Earth's surface from the solar beam without any interactions with atmospheric particles. Diffuse radiation is scattered out of the solar beam by

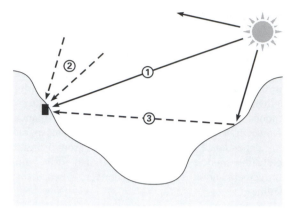

Figure 3.1 Downward irradiance received on a slope: (1) direct irradiance; (2) diffuse irradiance from the sky; (3) terrain-reflected irradiance.

gases (a phenomenon known as Rayleigh scattering) and by aerosols (including dust particles, sulphate particles, soot, sea-salt particles, pollen, etc.). The amount of direct and diffuse radiation received at a point is influenced by topography, owing to self-shadowing by the slope itself and shadows cast by neighbouring terrain. In addition, reflected radiation is predominantly reflected from nearby terrain towards a point of interest and is therefore more important in mountainous regions.

Direct shortwave radiation is the most important component of global radiation, since it contributes the most to the energy balance and the other components depend on it, directly and indirectly (Kondratyev 1965). Global radiation at any point is roughly proportional to direct solar radiation and varies with the geometry of the surface (Kumar *et al.* 1997). The other components, including diffuse radiation, vary only slightly from slope to slope within a small area and these variations can be linked to slope gradient (Williams *et al.* 1972). The flux of clear-sky diffuse radiation varies with slope orientation in a similar way to the flux of direct solar radiation, hence preserving the spatial variability in total radiation (Dubayah *et al.* 1989). The overall influence of terrain upon the actual radiation received at any point on the Earth's surface is sometimes referred to as a toposcale process (McKenney *et al.* 1999).

An areal representation of solar radiation is necessary for many applications but extrapolating vertical atmospheric values is very difficult. It is not feasible to record solar radiation at one site and extrapolate it to another, because it is highly variable from site to site due to changing slope and aspect (Kumar *et al.* 1997). In flat regions with clear-sky conditions, the downwelling shortwave radiation is nearly the same from one point to another over relatively large areas, so it is feasible to take one point measurement as representative of the entire area. But in mountainous terrain, localised weather conditions are set up (Barry 1981) and a single point measurement would not be sufficient to represent the region. In such regions point measurements are only representative of the locality from which they are collected. To obtain a reasonable accuracy in the measurement of incoming fluxes in mountainous terrain, either a very dense network of measuring stations must be established or radiation modelling approaches must be used (Duguay 1993). A number of solar radiation models have been built for different purposes at different scales (Kumar *et al.* 1997; Dubayah and Rich 1995; Swift 1976; Hutchinson *et al.* 1982; Bland and Clayton 1994; Lourens *et al.* 1995; Cooter and Dhakhwa 1995), and most have incorporated topographic variables as model components.

An early example is that of Kondratyev (1969), who developed empirical relationships between slope exposure and clear-sky radiation with results in graphical and tabular form. This information is of limited use for studies that require spatial variations in solar radiation. Swift (1976) modelled total shortwave radiation over a large area from measurements of solar radiation on a nearby site, in addition to slope and aspect information. The drawback to this model was that the shading effect by adjacent features had to be visually estimated and manually calculated. More recent models have employed DEMs and DEM derivatives to model the different components of solar radiation: direct, diffuse and reflected irradiance.

Topographic radiation models require data about the terrain of the area of interest. In particular, they require digital elevation, surface reflectance, and ground-based radiation data. DEMs are used to produce slope, aspect, sky-view factor, and terrain configuration maps that are employed in the modelling process. Dubayah and Rich (1995) present a series of equations to calculate total irradiance on a slope using a DEM and ground-based radiation data:

$$E_\downarrow(\text{slope}) = [V_d \bar{F}_\downarrow(\tau_0) + C_t \bar{F}_\uparrow(\tau_0) \\ + (\cos i) S_0 \exp(-\tau_0 \cos\theta_0)] \quad (3.2)$$

where V_d is the sky view factor, $\bar{F}_\downarrow(\tau_0)$ is the average diffuse radiation on a level surface at the elevation,

and τ_0 is optical depth, so that $V_d \bar{F}_\downarrow(\tau_0)$ is the diffuse radiation (calculated with an isotropic assumption). C_t is a terrain configuration factor used to adjust an average reflected radiation term (of radiation reflected from nearby terrain) and assumes (unrealistically) that radiation reflected off the terrain is isotropic, and $\bar{F}_\uparrow(\tau_0)$ is the amount of radiation reflected off the surface with an average reflectance, so that $C_t \bar{F}_\uparrow(\tau_0)$ is the reflected radiation from surrounding terrain. Finally, $(\cos i) S_0 \exp(-\tau_0 \cos\theta_0)$ is the direct irradiance on a slope in which $\cos i$ is the cosine of the solar illumination angle of the slope, S_0 is the flux at the top of the atmosphere, and θ_0 is the solar zenith angle. V_d, C_t, and $\cos i$ (including slope azimuth and slope angle) are all calculated using a DEM and all vary spatially. Diffuse radiation and direct irradiance will vary spatially with elevation because τ_0 is a function of pressure. In practice, the optical depth at a certain elevation is specified and then the changes in irradiance caused only by changes in elevation are computed, before taking account of other terrain factors (Dubayah and Rich 1995). The Dubayah–Rich equation is a function of wavelength and to calculate total irradiance it is integrated with respect to wavelength over the required spectrum interval. This would be very time-consuming to carry out over many spectral intervals. An approximation can be made by dividing the solar spectrum into two broad bands, one mainly scattering and the other absorbing, corresponding to the visible and near-infrared, and by using the equation in each wavelength region.

Net solar radiation is calculated by multiplying the Dubayah and Rich equation by $(1 - R)$ where R is the spatially varying solar reflectance map, which is distinct from R_0, the locally averaged solar reflectance used to approximate reflected irradiance. Dubayah and Rich (1995) present two examples which utilise the above approach to calculating total solar radiance: ATM (Atmospheric and Topographic Model) (Dubayah 1992) and SOLARFLUX (Hetrick *et al.* 1993a, 1993b). ATM is a collection of separate programs, while SOLARFLUX is implemented on a GIS.

The main objective of ATM was to provide inputs for hydrological and snowmelt models in mountainous terrain. It is, therefore, explicitly concerned with atmospheric radiation problems and allows different radiative drivers and interfaces. ATM is able to produce detailed topoclimatologies for large areas at arbitrary time intervals, using existing data. An example application was in the Rio Grande River basin of Colorado, United States. Thirty-nine DEMs at 30 m resolution were laid out in a mosaic to cover the upper section of the basin, above Del Norte, Colorado, and west to the continental divide. A four-year time series of hourly pyranometer measurements of direct and diffuse fluxes from 1987 was used with Landsat TM satellite estimates of reflectance and NOAA estimates of snow cover to produce a four-year monthly climatology of incoming radiation for the entire basin.

SOLARFLUX uses a DEM, latitude, a time interval for calculation, and atmospheric conditions to produce output of direct radiation flux, duration of direct radiation, sky view factor, hemispherical projections of horizontal angles, and diffuse radiation flux for each surface location. SOLARFLUX has been used for a range of temporal and spatial scales. At a landscape scale it has been employed with landscape-level microclimate-based habitat models for diverse topographic regions (Hetrick *et al.* 1993a, 1993b; Saving *et al.* 1993). It has also been used to examine microclimate heterogeneity as it affects sites where young shrubs and trees can become established (Rich *et al.* 1993).

Problems are associated with the above approaches. For instance, close to the ground, sky obstruction results from local features such as plant canopies, nearby terrain, or human constructions. These factors may act together to provide a complex pattern of sky obstruction. Also, there are many problems with modelling and measuring incident solar radiation with such obstructions, including the high temporal variability of sky conditions, anisotropic irradiance distributions, the geometric complexity of plant canopies, and the resulting complex patterns of reflectance off the many surfaces. Furthermore, in the method used by Dubayah and Rich (1995) to calculate actual solar flux, field data such as pyranometer data (which measures actual incoming solar flux at a station), atmospheric optical data, or atmospheric profiling must be used and these may not be available.

Kumar *et al.* (1997) developed a model to compute potential solar radiation (the amount of shortwave radiation received under clear-sky conditions) over a large area using only digital elevation and latitude data and incorporating the variation in radiation at different aspects and slopes. The model is able to handle flat or mountainous terrain, as well as shading by adjacent features. The model calculates only the potential beam solar radiation and a simplified diffuse radiation, but it can be modified to include other parameters such as effects of cloud cover and precipitable water content of the atmosphere, if the data should be available. The emphasis of their work was to describe relative spatial variation in solar radiation, with the modelled values as close to the actual values as possible, but not necessarily exact values for validation and calibration purposes. Relative solar radiation is still useful for many applications; for instance, it can be linked to the distribution of flora and fauna and productivity. Kumar *et al.* (1997) also used widely accepted empirical relations rather than rigorous computational methods developed for specific sites and requiring much site-specific data. The model was used to compute the potential shortwave radiation at Nullica State Forest near Eden, New South Wales, Australia, at a latitude of 36.5°S. The terrain at the site was considered to be fairly rugged with elevation ranging from 9 to 880 m. The total area of the study site was 1640 ha and the resolution of the model grid was 30 m.

In another study, McKenney *et al.* (1999) examined the application of SRAD, a grid-based spatially distributed radiation model, to a boreal environment, a major forest biome in Canada. They have found extra difficulties in developing surface radiation models for forested environments as these often have high relief or rugged terrain and are usually poorly covered by meteorological data collection networks. Once again, the problem of incorporating terrain effects is tackled by the development of modelling strategies that use DEMs. The SRAD model also overcomes the problem of lack of actual on-site meteorological measurements by using general atmospheric information about the area of interest. SRAD was developed by Ian Moore (1992) and calculates spatially distributed estimates of a full radiation budget using as inputs (1) a user-supplied parameter file which calibrates the mode for local meso-scaled atmospheric conditions, and (2) a DEM. The SRAD approach computes the Sun's position for each time-step using latitude, date, and time; determines ground-level clear-sky direct and diffuse irradiance components using the transmittance and circumsolar coefficient; compiles a set of values for the day (some including topographic effects) and, using cloudiness, sky view, and albedo, calculates the horizontal and inclined surface irradiance values; and computes temperatures and outgoing longwave irradiance using the radiation ratio (McKenney *et al.* 1999).

McKenney *et al.* (1999) found the SRAD-generated irradiance estimates to be consistent with irradiance data from other sources – that is, radiation as (1) measured at nearby radiation stations, (2) estimated from interpolated radiation surfaces based on radiation and sunshine hour data, and (3) estimated from a published map of national radiation isolines. They found that the estimates of irradiance were most sensitive to the parameters of sunshine fraction and cloudiness. They produced and tested radiation estimates at both 20 m and 100 m resolution. Very low irradiance estimates that were produced by the fine scale were absent at the coarser scale, although the mean value of irradiance was estimated as 12.4 MJ/m^2/day annually by both scales. They also determined radiation estimates at the two scales using a series of forest research plots. They found that for 91 per cent of the plots, the irradiance estimates throughout the growing season differed by less than 0.5 MJ/m^2/day between the coarse and fine scales. This concurs with findings from other studies (Dubayah *et al.* 1989; Dubayah and Van Katwijk 1992) that regional radiation means do not change noticeably with resolution generalisations. However, it has been found that regional variances decrease significantly for different types of terrain.

There are errors associated with every step of solar radiation modelling: (1) with the radiative transfer calculations; (2) with interpolating and extrapolating empirical measurements over a landscape; (3) with registration (between reflectance and DEM data); and (4) with approximations, particularly to a physical model (Dubayah and Rich 1995). However, the most serious source of error is considered to be the poor quality of most digital elevation data. Noise in the DEM will produce inaccuracies in slope and aspect. This means that, as well as gradient error, error will also propagate into the horizon angles, sky-view factors and terrain configuration factors. This will affect the direct and diffuse irradiance calculations.

Solar radiation models can become highly elaborate, due to the complex interactions between the atmosphere, topography, and plant canopies. Dubayah and Rich (1995) recommend that research should focus on the ability to calculate either potential radiation or some type of relative radiation (as with the approaches of Kumar *et al.* 1997 and McKenney *et al.* 1999), rather than attempting to obtain increasingly better estimates of actual solar radiation. If a model becomes too complex, it becomes too difficult to use, due to input data requirements. Future research is likely to include adding anisotropy to the diffuse and reflected terrain calculations and incorporating more canopy effects (Dubayah and Rich 1995). Clouds should also be included, which is critical since the spatial variability of clouds can dominate the topographic variability of even rugged terrain.

Multi-climate parameter models

Running *et al.* (1987) developed a model, MT-CLIM, for calculating daily microclimate conditions in mountainous terrain. Maximum–minimum daily air temperatures, humidity, incoming shortwave radiation, and precipitation are extrapolated from the point of measurement (station) to the study site of interest, making corrections for differences in elevation, slope, and aspect between the two sites.

Hutchinson *et al.* (1996) developed a gridded topographic and monthly mean climate database for the entire African continent. The gridded climate data was obtained by fitting thin-plate smoothing spline functions of longitude, latitude, and elevation (DEM) to point values of monthly mean daily minimum temperature, daily maximum temperature, and precipitation. Incorporation of a continuously spatially varying dependence on elevation was a critical factor in the accuracy of the surfaces. Validation showed that the gridded monthly mean temperatures were interpolated to within standard errors of about ±0.5°C and monthly mean precipitation to within errors of about ±10–30 per cent.

Altitude and climate

Mountains are often forceful agents in their fluid environment, actively creating patterns of weather and climate (cf. Washington 1996). This section will explore the effects of altitude on selected climatic variables – air pressure, temperature, rainfall, and snowfall.

Pressure and temperature

Air pressure

It is well known that air pressure decreases with increasing altitude. The relationship was first discovered in September 1648, when Florin Périer, at the request of his brother-in-law, Blaise Pascal, worked a simple barometer (a Torricellian mercury tube) at the base and summit of Puy de Dôme, France. For a Standard Atmosphere (which approximates annual conditions in mid-latitudes), air pressure at mean sea-level is 1013.25 mb and declines with geopotential altitude to 540.5 mb at 5000 m. Values are a little higher at equivalent altitudes in the tropics. Air density and vapour pressure also decrease with altitude, although mountain air may contain more water than free air at the same altitude. Lower air pressure, low vapour pressure, and low density at high elevation have a marked effect on radiation and bioclimatology. This is partly because oxygen pressure, which is a constant fraction of barometric pressure, decreases at the same rate. In consequence, oxygen pressure at 2000 m in tropical latitudes is 20 per cent less than at sea level; and by about 5400 m, it is 50 per cent less.

Air temperature

Temperature decreases with altitude. In the free atmosphere, the average rate at which it does so (the environmental lapse rate) is about 6°C/km. A study on Mount Kinabalu, in north Borneo, used thermohydrographs placed at five altitudes – 10, 500, 1680, 3200, and 3780 m (Figure 3.2) – to measure altitudinal changes in air temperature and humidity (Kitayama 1992). The temperature lapse rate was 5.5°C/km (Figure 3.2). Mean daily maximum and minimum temperatures also decreased with altitude, the mean daily maximum temperature having a higher lapse rate (6.2°C/km) than the mean daily minimum temperature (4.9°C/km). This difference in lapse rate for maximum and minimum temperatures means that the daily temperature range also decreased with altitude.

Under some conditions, mainly during night and during winter, lapse rates may be reversed near the ground in a layer of temperature inversion (a layer in which temperature increases with height). Lapse rates are modified by mountains because slope air over a mountain is influenced by radiative and turbulent heat exchanges, which are also affected by snow cover. In the Austrian Alps, the lapse rates between Kolm Saigurn (1600 m) and Sonnblick (3106 m) were, in

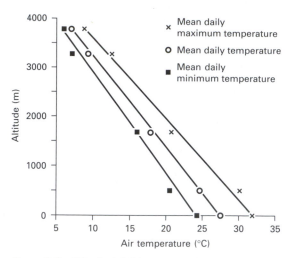

Figure 3.2 Altitudinal fall in measured air temperatures on Mount Kinabalu, north Borneo. The mean daily temperature lapse rate is 5.5°C/km. (Adapted from Kitayama, 1992.)

winter, 4.9°C/km at 2 a.m. and 6.6°C/km at noon; and, in summer, 6.0°C/km at 2 a.m. and 8.9°C/km at noon (Hahn 1906, 102). Such diurnal changes in the lapse rate are partly caused by local topographic effects.

Topography may alter the structure of the atmosphere at a much larger scale and materially affect temperature lapse rates. These larger-scale effects may be seen by considering three idealised topographic cases – an isolated mountain, a small plateau, and a large plateau (Tabony 1985). The temperature curve in all cases represents a surface inversion on a clear winter's night: air temperature increases with height, then stays constant, then decreases (Figure 3.3). In the isolated mountain case, the mountain temperatures roughly match temperatures in the free atmosphere.

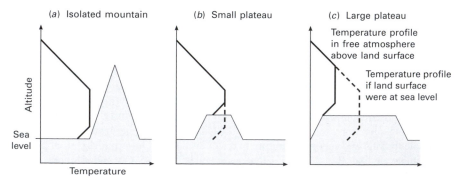

Figure 3.3 Schematic, clear-winter-night vertical temperature profiles for three topographic cases: (a) isolated mountain; (b) small plateau; (c) large plateau. (Adapted from Tabony, 1985.)

In the large plateau case, the entire temperature curve is displaced upwards, so producing a modified atmosphere (in winter) and lower surface temperatures over the plateau. In the small plateau case, the lapse rates are similar to those in the seasonally modified atmosphere, but the terrain does not force the temperature curve upwards and temperatures are independent of altitude.

Altitudinal thermal zones

In the same way that the general decrease of temperature with increasing latitude produces latitudinal climatic zones (torrid, temperate, and frigid), so decreasing temperature with increasing elevation produces altitudinal climatic zones. The climatic zones met with on ascending a mountain are termed submontane, montane, subalpine, alpine, and nival. Variations on this basic scheme are common. Some systems recognise an upper montane belt, some a subnival zone (Figure 3.4). Many South American countries employ a highly logical scheme that recognises four temperature zones – *tierra caliente* (hot or warm land), *tierra templada* (temperate land), *tierra fria* (cool land), and *tierra helada* (land of frost). By whatever name they go, altitudinal climatic zones affect soils and life, as will be seen in later chapters.

Altitudinal climatic zones on Mount Kinabalu range from tropical to polar. Kanehiro Kitayama (1992) modified Köppen's (1936) thermal climate system and Kira's (1976) warmth index to work with altitude, rather than latitude, and with diurnal, rather than annual, temperature regimes. In the Köppen system, the mean daily maximum and mean daily minimum air temperatures were used in place of mean monthly temperatures of the warmest and coldest months. In Kira's system, mean daily temperatures were substituted for monthly means in the equation:

$$\text{warmth index} = \sum^{n}(t - 5) \qquad (3.3)$$

where n is the number of months with a mean air temperature (t) over 5°C, a temperature deemed critical for plant growth. Transcribing Köppen's climatic classification to Mount Kinabalu sets tropical climates below 1200 m, temperate climates between 1200 and 3610 m, and polar climate above 3610 m. Similarly, transcribing Kira's warmth index to Mount Kinabalu gives all climatic zones from tropical to polar, all much compressed compared to their latitudinal counterparts (Figure 3.5).

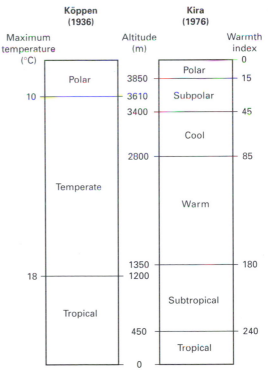

Figure 3.5 Altitudinal climatic zones on Mount Kinabalu, north Borneo. Köppen's and Kira's systems are adapted so that mean daily maximum temperatures are used in place of mean monthly maximum temperatures. (Adapted from Kitayama, 1992.)

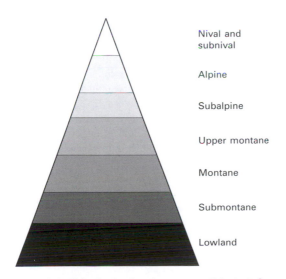

Figure 3.4 Altitudinal climatic zones. (Adapted from Leuschner, 1998.)

Rain and snow

The effect of mountains on the distribution and amount of precipitation is a somewhat complex subject. Orographic features are capable of exerting a significant influence upon the development and distribution of precipitation (Smith 1979; Barros and Lettenmaier 1994). Orographic influences on precipitation include the forced uplifting of air parcels, lee-wave formation, enhanced cyclonic convergence, convective currents triggered by the elevated heating of air surrounding mountain peaks, cooling of warm moist winds over a snowpack, and the seeder–feeder mechanism in which precipitation from higher clouds scours out droplets from orographically induced cap clouds over terrain peaks. In essence, precipitation falling over mountains has two components, one that would fall even without the mountains being there and one that is directly attributable to the intensifying precipitation processes of the mountains. Over high mountains and isolated peaks with steep slopes, a single physical mechanism – forced lifting of moisture-laden air under unstable conditions – commonly produces heavy precipitation. This process occurs in the Cascades and Rockies in North America. Over extensive mountain masses with less steep slopes and moderate elevation (hundreds of metres), the orographic enhancement of precipitation results from several physical mechanisms. This kind of enhancement characterises the Appalachian Mountains. The term 'orographic rainfall' is applied to rainfall enhancement resulting solely from forced uplift of air, while the term 'orographic component of rainfall' covers precipitation produced by the several physical precipitation-inducing mechanisms that exist in mountainous terrain (Bonacina 1945).

In the British Isles, data from over 6500 stations gave the following relationship between average annual precipitation, R (mm), and altitude, H (m) (Bleasdale and Chan 1972):

$$R = 714 + 2.42H \qquad (3.4)$$

Even fairly low hills, such as the Chilterns and South Downs of southern England, produce an orographic component of rainfall – they receive some 120–130 cm more rain per year than the surrounding lowlands. In northern Italy, a study showed that orographic enhancement of precipitation in cold fronts varied strongly with height (Corradini and Melone 1989). The enhancement for storms attaining maturity before the passage of the front was 50 per cent, and for storms maturing after the passage of the front it was 150 per cent. However, the effect of altitude on the vertical distribution of precipitation seems highly variable. The question is difficult to address in detail owing to a relative dearth of high-altitude stations. In some areas, including the European Alps, high-resolution rain-gauge observations can be called into service. In addition, it is now possible to use models that have a high enough resolution to simulate the precipitation processes involved (e.g. Barros and Lettenmaier 1993; Kuligowski and Barros 1999).

Precipitation in the European Alps

A new precipitation climatology of the European Alps was based on a high-resolution data set (Frei and Schär 1998). It covered the entire Alpine range, including a broad ribbon of its adjacent foreland, with a resolution of about 25 km. The observational database comprised bias-uncorrected, quality-controlled daily rain-gauge totals for the 20-year period 1971–1990. It was assembled from one of the densest rain-gauge networks over a mountain range in the world, and used more than 6600 stations from the high-resolution networks of the Alpine countries. Daily precipitation fields were produced by an advanced distance-weighting scheme commonly employed for the analysis of precipitation on a global scale. Annual, seasonal, and monthly means were derived from these daily fields.

The results of the study depicted the mesoscale distribution of the Alpine precipitation climate, its relation to the topography, and its seasonal cycle. Gridded analysis results were also provided in digital form. The most prominent Alpine effects include the enhancement of precipitation along the Alpine foothills, and the shielding of the inner-Alpine valleys (Figure 3.6). A detailed analysis along a section across the Alps also demonstrated that a simple precipitation–height relationship does not exist on the Alpine scale, because the topographic signal is largely associated with slope gradient and shielding rather than height effects. Although systematic biases associated with the rain-gauge measurement and the topographic clustering of the stations were not corrected for, a qualitative validation of the results using existing national climatologies showed good agreement on the mesoscale. Results indicated that the pattern and magnitude of analysed Alpine precipitation critically depend upon the density of available observations and the analysis procedure adopted.

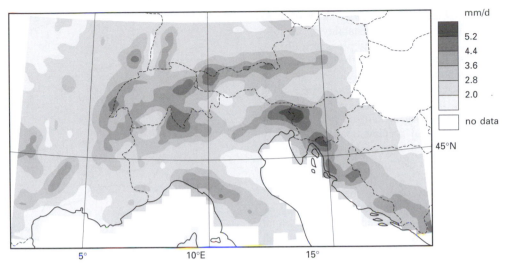

Figure 3.6 Mean annual precipitation (mm/day) in the Alpine region, 1971–1990. (Adapted from Frei and Schär, 1998.)

Rain on windward and leeward slopes

Differences between the climate of windward and leeward slopes may be considerable: large mountain ranges, and even some hills, cast a rain shadow in their lee that influences vegetation and soils. The Scandinavian mountains lie across the prevailing westerly airflow. Maritime air is forced to rise over the mountains and annual precipitation is high – in western Norway, over 2500 mm falls on the mountains. In the lee of the mountains, annual precipitation totals are much lower. The upper Gudbrandsdalen and Osterdalen, sheltered in the lee of the Jotunheim and Dovre Mountains, receive less than 500 mm per year. In Scotland, anomalies from Bleasdale and Chan's (1972) regression of rainfall versus altitude exceed 600 mm on the western side, from Cape Wrath to Glasgow; to the east, the Cairngorms have a negative anomaly greater than 600 mm, showing a prominent west to east rain-shadow effect. In north Wales, sites in Snowdonia receive over 2500 mm of precipitation a year, while the lower Dee valley, which lies in the lee of the North Welsh mountains, receives less than 750 mm.

The forced ascent of air by hills and mountains is in places intensified by topographic constrictions. Examples are found in the English Lake District and in North Wales. In Snowdonia, convergence of air up the Glaslyn Valley leads to even higher rainfall than in the surrounding mountains, as does the convergence of airstreams entering several radiating valleys on the knot of mountains at the head of Borrowdale and Langdale in the Lake District (Manley 1952, 136).

The mechanisms of increased precipitation on the windward side of mountains and hills are complex. A study of heavy-rain events and associated flash-flood episodes occurring from 1978 to 1992, with daily falls of more than 100 mm at one or more rain-gauge sites, over the south-east flank of Mauna Loa volcano, Hawaii, revealed something of the complexity involved (Kodama and Barnes 1997). The study used surface, rawinsonde (radar wind-sounding), rain-gauge, and satellite data. The events were associated with four kinds of synoptic-scale disturbances – Kona storms, cold fronts, upper-tropospheric troughs, and tropical systems, each of which can induce heavy-rain events. Kona storms, which are associated with Kona cyclones, occur ordinarily in the trade-wind lee on mountain ridges exposed to onshore winds, and bring spectacular rainfalls two or three times in winter to areas of Hawaii (Simpson 1952). In the study, Kona storms, and not the ubiquitous trade winds, were the dominant synoptic factor causing the heavy-rain events on the south-east flank of Mauna Loa. The heaviest rains fell at elevations above 500 m over slopes lying athwart the prevailing low-level airflow. Far less rain fell on lee slopes. The duration and pattern of rainfall indicated that more than one convective cell was causing the storms. The air was only a little unstable before the heavy-rain events, and of all the standard stability indices, the K index (George 1960), which measures the static stability of the 805–500-mb layer, was the only useful predictor of heavy rain. It was also found that, before the heavy rain fell, the moisture content of the mid-levels (750–450 mb) in the atmosphere and the onshore flow normal to the volcanic slopes increased. Mid-level parcels of air are not conditionally unstable and are unlikely to contribute directly to total precipitation; they possibly indicate the presence of large-scale rising air motion in the absence of a trade-wind inversion. However, they may help to produce a heavy-rain event in two ways. First, they may result in the entrainment of air with a higher virtual temperature into the rising air parcel, resulting in a smaller reduction of cloud parcel buoyancy and, therefore, a deeper cloud with higher storage of liquid water for precipitation processes. Second, moist mid-level air may suppress the formation of cold downdraughts, which result in significant cold outflows that tend to cause stability and prevent prolonged precipitation.

Smaller-scale rainfall and relief

A link exists between heavy precipitation cells and orography. On 19 January 1996, a cold front crossed the central Appalachian region of the eastern United States. Several heavy precipitation cells evolved causing severe rainfall and, in conjunction with rapid snowmelt, flooding of much of the region. Six mid-Atlantic states were declared federal disaster areas. Analysis of radiosonde, rainfall, streamflow gauge data, and WSR-88D images showed a connection between heavy rainfall and orography for the 19 January event (Barros and Kuligowski 1998). In particular, tracks of the centre-of-mass of storm precipitation were found closely to follow the contours of regional orographic features. Higher-intensity precipitation cells were ordinarily located to windward of the orographic crest. This windward effect operated at the synoptic scale and involved topography modifying large-scale airflow. Leeward effects operated on the mesoscale. They involved the growth of orographically induced regional circulation and cloudiness features that led to a strong, localised augmentation of precipitation on the eastern slopes of the Appalachians and the Poconos in Pennsylvania. These leeside storms produced the strongest runoff responses in the Susquehanna River basin. It seems likely that they were produced by a mixture of mesoscale mechanisms, mainly forced ascent at the foot of a low-level cold air pool in southern and south-eastern Pennsylvania; forced orographic

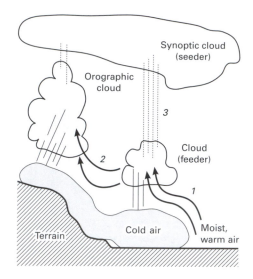

Figure 3.7 Schematic representation of leeside mesoscale mechanisms of rainfall enhancement triggered by regional orography in Pennsylvania. The numbers **1**, **2**, and **3** mark sites where specific mechanisms of rainfall enhancement may be dominant. **1** Forced lifting of air at the head of a low-level cold-air pool, southern and south-eastern Pennsylvania. **2** Forced orographic uplift on the leeside of the Allegheny Front and southern slope of the Poconos. **3** Seeder–feeder mechanism. (Adapted from Barros and Kuligowski, 1998.)

uplift on the leeside of the Allegheny Front and on the southern slopes of the Poconos; and seeding by prefrontal precipitation (precipitation falling ahead of a front) from high-level clouds – the seeder–feeder mechanism (Figure 3.7).

Snowfall in mountains

Snowfall is more likely to occur in mountains, partly because precipitation is higher and partly because atmospheric and ground-surface temperatures are lower than in surrounding lowlands. The chances of snow falling and the likelihood of its accumulation and persistence as snowbeds broadly depend upon latitude, altitude, and, to a lesser degree, aspect. The pattern of mountain snowfall is complicated and depends on the frequency of different snow-bearing airstreams. In the Scottish mountains, high altitude and northerly latitude combine to give many winter days with snow lying – hence the expanding Scottish skiing industry. However, easterly continental and westerly maritime snow-bearing airstreams tend to favour different types of snowfall with contrasting geographical distributions (Harrison 1993). The winter of 1985–1986, one of the severest in Scotland in recent years, saw heavy snowfalls associated with south-east and north-east winds. February was the snowiest month, with snow falling from a vigorous easterly airstream that lasted while high pressure was established to the north of Scotland. Snow lay for the entire month above altitudes of 200 m. In contrast, the winter of 1988–1989 was then the mildest on record in the British Isles, with easterly winds rarely blowing. A strong and relentless westerly circulation brought heavy precipitation and strong winds. Most of the winter snow fell in February and March in strong westerly to north-westerly winds. A late snowfall occurred from 23 to 25 April.

To compare snowfall in the two winters, regression equations were used to relate days with snow lying at 09.00 GMT to altitude, latitude, and longitude. The total number of 'snow days' for the winter season (October to April), as well as for February, March, and April, was computed from the *Monthly Weather Report* at 39 Scottish climatological stations. The regression equations are shown in Table 3.1. Differences in the longitude terms are apparent, which confirm the increasing south-west to north-east snow cover in cold winters, and the increasing south-east to north-west snow cover in mild winters, in accordance with the source of the snow. Further analysis considered the effects of a change from cold winters dominated by easterly airstreams to a run of warm winters dominated by maritime westerly airstreams. The results showed that such a shift would lead to an overall reduction in the number of days with snow lying. This reduction would be greater in the eastern Grampians than in the western Grampians. It would

improve winter access to skiing sites, in the Cairngorm area for example, but would have a greater effect in reducing the availability of ski slopes in the east of the region. At high altitudes, these east–west differences are fairly small and may be masked by local topographic controls on snowfall – winds coming from all directions between west and north may produce greater accumulations of snow in corries and sheltered hollows facing those directions than would winds from the east. However, temperatures would be higher and the snow less persistent.

Local climates (topoclimates)

All topographic features potentially modify radiation fluxes, heat balances, moisture levels, and aerodynamics in the local environment. Radiation modification depends largely on the aspect and inclination of ground surfaces and the walls and the roofs of buildings, on the albedo (the reflectivity of topographic features – different vegetation types, bare soil, human-made surfaces, water bodies), on shading effects, and in some cases, on local energy sources (domestic fires, industrial plants, and so on). Some parts of landscapes receive more sunlight than others, while emitting and absorbing different amounts of longwave radiation. In turn, the altered radiation balances produce hotter and cooler areas within landscapes by modifying local heat balances. Moisture levels vary owing to spatial variability in precipitation receipt (caused by differing interception rates and shelter effects), in evaporation rates, and in soil drainage. Aerodynamic modifications concern the physical effects of small-scale topographic features on airflow.

Table 3.1 Regression equations for total number of days with snow lying, y, versus altitude, h, latitude, x_1, and longitude, x_2, for mild and cold winters for elevations between 75 and 400 m in Scotland

Cold winter, 1985–1986		Warm winter, 1988–1989	
Regression equation	R^2 (%)	Regression equation	R^2 (%)
Winter season			
$y = -552 + 93.4 \ln(h + 375) + 16.60x_1 - 4.14x_2$	79.5	$y = -659 + 91.2 \ln(h + 1250) + 3.80x_1 + 3.09x_2$	69.0
February, March, April			
$y = -309 + 53.8 \ln(h + 375) + 3.88x_1 - 2.90x_2$	74.6	$y = -506 + 69.7 \ln(h + 1250) + 3.53x_1 + 2.97x_2$	69.1

Source: Adapted from Harrison (1993).

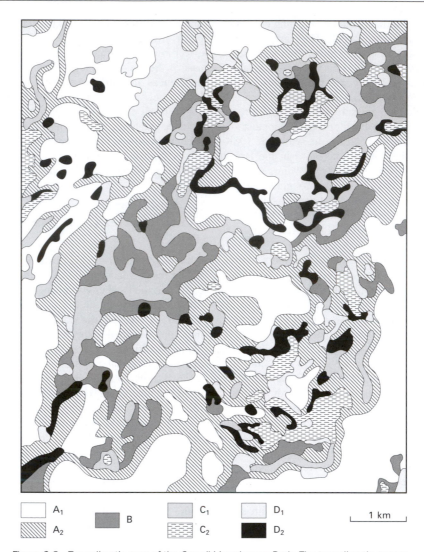

Figure 3.8   Topoclimatic map of the Suwałki Landscape Park. The topoclimatic groups are explained in Table 3.2. (Adapted from Błażejczyk and Grzybowski, 1993.)

Table 3.2   A topoclimatic classification

Topoclimatic group	Topoclimatic subgroup	Occurrence in Suwałki Landscape Park, north-east Poland
A Convective	A_1 Average solar radiation influx to the substrate	Plains
	A_2 Variable solar radiation flux	Slopes
B Evaporative–convective		Higher levels of valleys and basins
C Evaporative	C_1 High evaporation	Moist meadows
	C_2 Very high evaporation	Lakes
D Radiative	D_1 Convection prevalent	Dry forests in plains and slopes
	D_2 Evaporation prevalent	Marshy meadows and woods

Source: Adapted from Błażejczyk and Grzybowski (1993).

Topoclimatic features

Topoclimatic mapping

Detailed topoclimatic maps indicate the most important processes in the active ground surface and the atmospheric boundary layer (Vysoudil 1996a). They are constructed using ground-surface attributes (slope, aspect, local relief, slope curvature, contour curvatures, and so on) and land-cover and land-use information. GIS techniques and remotely sensed databases are very helpful in the construction process (e.g. Vysoudil 1996b, 1998).

Topoclimates may be classified according to the relative importance of the surface heat-balance components. The heat balance is written:

$$R = H + LE + G \qquad (3.5)$$

It states that the net surface radiation, R, is used to heat the air by turbulent transfer of sensible heat in convection, H; to evaporate water by turbulent transfer of latent heat, E (L is the latent heat of evaporation); or to heat the soil by conduction, G. The relative importance of the components H, E, and G depends on landform and landscape properties – slope, aspect, albedo, soil type, vegetation height, and so on. One classification recognises four topoclimatic groups, three of which contain two subgroups (Figure 3.8; Table 3.2) (Błażejczyk and Grzybowski 1993). Convective topoclimates have heat exchange dominated by sensible heat transfer, and the two subgroups are designated according to relief (plains and slopes). Evaporative–convective topoclimates have sensible and latent heat fluxes as the co-dominant modes of turbulent heat transfer. Evaporative topoclimates have heat exchange chiefly through evaporation, and the subgroups depend upon the level of evaporation – high or very high. Radiative topoclimates have relatively low levels of longwave radiation emission from the active surface and occur mainly in forests. They are subdivided according to the prevalence of convection and evaporation.

The significance of topoclimatic differences to near-surface temperatures is clear from a study of a summer temperature transect through woods and clearings near Leipzig, Germany (Figure 3.9) (Koch 1934). In the afternoon, the clearings heat up to above 25°C, while the forests are a little cooler. By late afternoon, temperatures along the transect have evened out and fallen to around 21°C. After sunset, radiative heat loss under clear skies in the clearings leads to their cooling to 11–12°C, well below the woods that retain heat to sunrise and maintain a temperature of around 15°C. After sunrise, the clearings warm fast and temperatures even out again to 16–17°C. As the morning progresses, the clearings heat faster than the woods and they become hotter.

Slope inclination as a topoclimatic factor

Slope exposure and slope inclination (gradient) substantially influence topoclimates. Slope inclination varies at all latitudes, both tropical and extratropical. It has intriguing effects on radiation receipt and heat balances. Very steep slopes receive maximum radiation intensity at low angles of the Sun, which occur during the winter and in the early morning and the evening. This radiation can produce high surface temperatures on vertical objects, even in winter. The bark of tree boles and the sides of grass tussocks display such wintertime warming at low angles of the Sun. Scots pine (*Pinus sylvestris*) bark measured 30°C above air temperature in December and 40°C above air temperature in March (Stoutesdijk 1977). The bole temperatures did not attain such high values on trees with thin bark, such as beech (*Fagus sylvatica*). On the south side of a large green tussock of wavy hairgrass (*Deschampsia flexuosa*) the temperature was 19°C, on the north side −2°C, and the air temperature 1°C on 24 December (Stoutesdijk and Barkman 1992, 77).

Slope exposure as a topoclimatic factor

Slope exposure, or aspect, is different in the tropics than in temperate and frigid (extratropical) zones. Outside the tropics, slope exposure is the same at all seasons, the Sun always lying to the south in the Northern Hemisphere and to the north in the Southern Hemisphere. This means that slopes aligned east-to-west will always be sunny or shady, depending on whether or not they face the Sun. Within the tropics, exposure north or south depends on the season. A south-facing slope lying on the equator will be exposed to the Sun during the Southern Hemisphere summer, but not during the Southern Hemisphere winter. In consequence, all east–west aligned slopes within the tropics are sunny for part of a year and shady for the rest.

Extreme topoclimatic differences occur between extratropical north-facing (distal) and south-facing (proximal) slopes. There is often a stark difference between the sunny south-facing slope and the shady north-facing slope in Alpine mountain valleys (Garnett

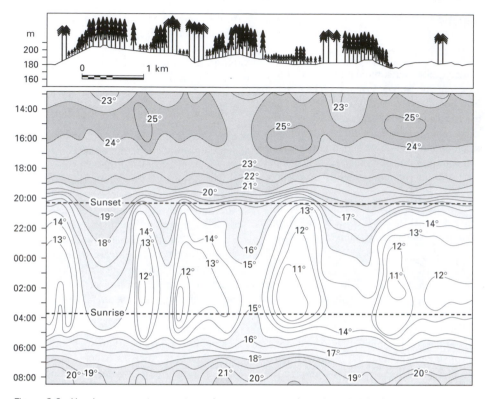

Figure 3.9 Hourly summer temperature changes, measured at chest height in woods and clearings, during the course of a day. Temperatures are in °C. The maximum relief is 20 m. (Adapted from Geiger, 1965.)

1937). In many areas, wooded north-facing slopes contrast totally with alpine meadows on south-facing slopes. The more or less southwards-facing sunny slope goes by various names: *adret* (French), *adretto* (Italian), and *Sonnenseite* (German). It receives more light and warmth than the shaded slope owing to its exposure for a longer period to the Sun's rays. The more or less northwards-facing shady slope is called the *ubac* (French), *opacco* (Italian), and *Schattenseite* (German). At Kinlochleven, Scotland, the potential insolation received at noon in summer by the *adret* slope (about 25°) at 457 m is nearly three times the amount received at noon in summer by the *ubac* slope (30°) at 396 m (Garnett 1939). In terms of human occupancy, a southerly exposure has some of the advantages of a more southerly latitude and a northerly exposure carries some of the risks of a more northerly latitude. Indeed, the climatic effect of aspect may be expressed as an 'equivalent latitude'. On average, in middle latitudes, a south-facing slope of 20° is equivalent to a southward climatic shift of 8° to 9° of

latitude; a 20° north-facing slope is equivalent to a northward climatic shift of 12° to 15° of latitude (Crowe 1971, 28–9). Table 3.3 shows the solar income on north-facing and south-facing slopes at 45°N at three times of year. Such variations in radiation receipt lead to substantial temperature differences. In a Derbyshire dale, England, the summer mean temperature was 3°C higher on a south-facing slope than on a north-facing slope (Rorison *et al.* 1986), a difference equivalent to a latitudinal shift of hundreds of kilometres!

Urban heat islands

The air in the urban canopy is often warmer than air in the surrounding rural areas, so creating an urban heat island. The precise form and size of urban heat islands are variable and depend upon meteorological, locational, and urban factors. Heat islands occur in both tropical and extratropical cities (e.g. Goldreich 1995).

Table 3.3   Daily solar income on north-facing and south-facing slopes of 20° at latitude 45°N

Date	Solar income on a horizontal surface (cal/cm²/day)	Solar income on a south-facing slope		Solar income on a north-facing slope	
		(cal/cm²/day)	(% of horizontal surface)	(cal/cm²/day)	(% of horizontal surface)
22 June	577	590	102	495	86
21 March	315	408	129	191	61
22 December	68	131	193	2	3

Source: Adapted from Crowe (1971, 28).

Heat islands are best developed in large cities under clear skies and light winds, just after sunset. Along a rural–city-centre transect, the general pattern of temperature change is from a cool rural 'sea', to a rural–suburban 'cliff', to a suburban 'plateau', and eventually to a city centre 'peak' (Figure 3.10). The 'cliff' is a steep temperature gradient that may reach 4°C/km. The 'plateau' of warm air has cooler and warmer spots within it corresponding to differing land-uses – parks, lakes, and open areas are cooler, while commercial centres, industrial areas, and regions of dense building are warmer and form micro-urban heat islands. Micro-urban heat islands in a portion of Dallas, Texas, were mapped using LANDSAT TM satellite data and GIS software (Aniello *et al.* 1995). They correlated with areas devoid of tree cover – newly developed residential neighbourhoods, parking lots, business districts, apartment complexes, and shopping centres. The 'peak' is especially common in North American cities and is the site of the hottest urban temperatures. The differences between the peak temperature and the background rural temperature is the urban heat-island intensity. Urban heat-island intensity depends on many factors, of which city size is one of the most important (e.g. Ono *et al.* 1994). The maximum heat island intensity of Dublin, Ireland, is about 8°C (Graham 1993); it is also about 8°C for Barcelona, Spain (Moreno-Garcia 1994); it is about 10°C for metropolitan Washington, DC (Kim 1992). New York City has satellite-detected urban heat-island extremes of up to 17°C (Price 1979).

At least seven mechanisms may explain the occurrence of urban heat islands (Oke 1982a, 1982b):

1. Counter-radiation in cities is increased because outgoing longwave radiation is absorbed and re-emitted by the polluted urban atmosphere.
2. Net longwave radiation from urban canyons (streets) is decreased due to a reduction of the sky-view factor by buildings.

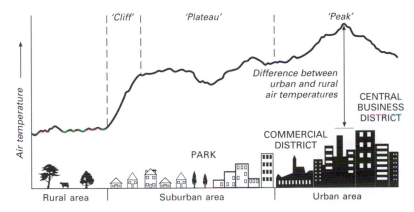

Figure 3.10   Schematic cross-section through a typical urban heat island. (Adapted from Oke, 1978, 288.)

3. Shortwave radiation absorption is enhanced by the effect of canyon geometry on the albedo.
4. Extra heat is stored during the day, owing to the thermal properties of the urban fabric, and then released at night.
5. Heat from human sources is released from the sides of buildings.
6. Evaporation is decreased because vegetation is removed and the surface of the city 'waterproofed'.
7. Loss of sensible heat is lessened because wind speed is reduced within the urban canopy.

All or any of these mechanisms may operate in a particular town or city, and their relative roles are unclear. A simple energy-balance model suggested that two factors are the chief, and almost equal, mechanisms on most occasions: (1) the effect of street canyon geometry on radiation; and (2) the effect of thermal properties on heat storage release (Oke *et al.* 1991). In very cold conditions, space heating of buildings can become a dominant mechanism, depending on the effectiveness of wall insulation. The effects of the urban 'greenhouse' and surface emissivity appeared to be relatively minor.

Heat islands in non-urban areas also exist. Winter morning temperatures in a very small 'downtown' area of Hanover, New Hampshire, USA, were found to be 1–2°C warmer than nearby open areas at the same elevation, but winter morning air temperatures in Lyme, a hamlet consisting of about 60 wooden buildings (including large houses, two inns, and a multi-storey church) with an area of 0.3 km² and lying a few kilometres north-north-west, were not significantly different from temperatures in surrounding terrain (Hogan and Ferrick 1998). Morning air temperatures near the frozen Connecticut River in the same area were measured and found to be systematically warmer than nearby air temperatures for several days, until a significant snowfall diminished the ice-growth rate (Hogan and Ferrick 1998). An examination of temperature profiles near the river showed that the increase in air temperature beneath the overnight inversion during this freezing period was proportional to the heat release resulting from river ice growth. In another study, the Mississippi River floodplain in the states of Arkansas, Tennessee, Mississippi, and Louisiana was found to appear as a heat island in weather satellite images (Raymond *et al.* 1994). In spring and early summer, the floodplain heat island was discernible as a daytime warm anomaly at infrared wavelengths, and as a bright reflective area at visible wavelengths; its nighttime remnants could sometimes be identified in the infrared satellite images.

The lake effect

Lakes may alter snowfall patterns owing to the 'lake effect'. In the Great Lakes region of North America, winter often sees westerly polar continental air (at a temperature of around −18°C) passing over the relatively warm Great Lakes (with a water temperature of about 1°C) and picking up moisture. On the lee of the lakes, where the air temperature has warmed to about −6°C, snow falls. Each Great Lake has its own lee snowbelt (e.g. Ellis and Leathers 1996). Some of the moisture remaining in the airstream is deposited as snow on the windward slopes of the Appalachian Mountains.

Small-scale mechanical airflow modification

Air flows as streamlines of parallel layers, as turbulent currents, and as vortices (helical or spiral motions). Topography may modify the aerodynamics of moving air. The effect of topographic obstacles on airflow depends crucially upon the vertical wind profile, the stability of the air, and the shape of the obstacle. Three cases will illustrate this point: the aerodynamic modification by ridges, by windbreaks, and by buildings.

Ridges

Airflow across relief barriers is betrayed by orographically formed clouds over windward slopes and lenticular, banner, and arched clouds over leeward slopes. The widespread use of hang-gliders, gliders, and powered aircraft from the middle of the twentieth century led to considerable research into relief-influenced airflow (e.g. Corby 1954; Scorer 1961). As demonstrated by Jiří Förchtgott (1949, 1969), the aerodynamic effects caused by a long ridge lying at right angles to airflow, and with stable air (potential temperature increases with height), depend upon wind speed (Figure 3.11). With light winds, laminar streaming occurs, a single shallow wave being uplifted symmetrically over the barrier. Slightly stronger winds lead to the downward displacement of the laminar-streaming crest and a lee-eddy forms. With stronger winds, wave streaming evolves, in which a series of lee-waves replaces the single lee-eddy. When winds are very strong, and particularly if the barrier is high compared with the airflow, the wave streaming

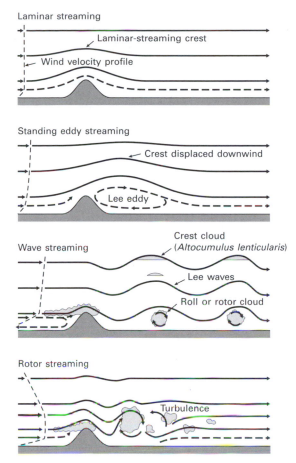

Laminar streaming

Laminar-streaming crest

Wind velocity profile

Standing eddy streaming

Crest displaced downwind

Lee eddy

Wave streaming

Crest cloud
(*Altocumulus lenticularis*)

Lee waves

Roll or rotor cloud

Rotor streaming

Turbulence

Figure 3.11 Airflow modification over a long ridge.
(Adapted from Förchtgott, 1949.)

breaks down and complicated turbulence or rotor-streaming develops. The four airflow-streaming patterns represent typical conditions, but the exact nature of airflow modification depends upon the vertical wind speed, the temperature profiles, and the width of the barrier. This was shown for a study of wave streaming over a long ridge. If stable air is forced over the ridge, it returns to its original level once it has crossed the ridge. This leeward descent of air, which produces a strong surface wind down the leeslope, often begins a series of downwind lee waves (or standing waves) and rotors (Wallington 1960). The wave form persists and holds its position relative to the ridge while air flows through it – the lee waves are 'trapped'. The wave pattern varies according to wind speed and the temperature profile of the air, with an upper stable layer as an important ingredient. Lee waves normally form only if a deep flow of air is directed

within about 30° of a perpendicular to the ridge line, and if there is little change of wind direction with height (Barry 1992, 125).

Later work has shown that the behaviour of trapped lee-waves is often non-stationary. Observations of strong lee-waves generated by a southerly wind crossing the Pyrenees revealed such behaviour (Caccia *et al.* 1997). The observations were made as part of the PYREX, a cooperative programme investigating airflow over the Pyrenees. They came from upstream radiosondes and measurements obtained by aircraft along the Pyrenean transect, and by constant-volume balloons launched near the mountain crest. Spatial lee-wave characteristics at different times and heights were derived from the observations. Horizontal lee-wave wavelength ranged from 7 to 14 km, and the vertical air velocity had a maximum amplitude of 3 to 5 m/s. The lee-waves, measured from the crest line, stretched from 30 to 55 km. Two very high-frequency, stratosphere–troposphere radars, one on the mountain mean axis and another downstream in the lee-wave field, recorded temporal variations of the vertical-velocity profiles (see also Caccia 1998). Lee-wave stationarity was studied using these data. The lee-wave was far from stationary during its lifetime, although it was quasi-stationary for periods lasting up to 1.5 hours. The airborne-instrument data also showed that wavelength, amplitude, and downstream wave extent went through temporal variations.

By subjecting radiosonde soundings to a technique called the horizontal projection method, the dominant horizontal wavelength over the entire depth of radiosonde soundings may be determined. The technique was applied to a well-defined family of lee-waves over southern Tasmania, Australia, on 18 June 1991 (Reeder *et al.* 1999). The lee-waves were plainly visible in the AVHRR imagery over southern Tasmania on that day. The analysis picked out a dominant horizontal wavelength of 8.249 km, which closely matched the wavelength estimated from an infrared satellite image of 8.9 ± 1.4 km. Applying the same technique to lee-waves over South Island, New Zealand, which were visible in a satellite image on 20 October 1996, yielded a horizontal wavelength of 11.3 km. This figure agreed quite well with the 13 ± 1 km horizontal wavelength determined from an AVHRR satellite. In the lower troposphere, the vertical wavelength was about 4 km and increased with height. Two types of waves are forced by airflow over a mountain range such as the Southern Alps of New Zealand. First, hydrostatic waves, which normally form the main contribution to the response, are found only above

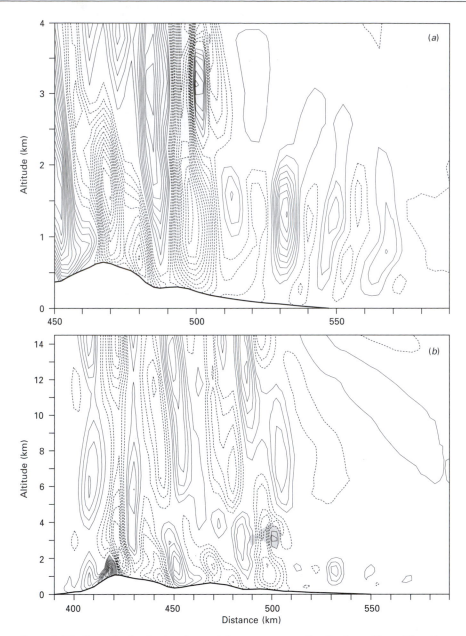

Figure 3.12 Numerically simulated mountain waves across the Southern Alps of New Zealand. (a) A coherent train of partially trapped lee-waves. The contour interval is 4 cm/s and the zero contour is omitted. (b) Hydrostatic mountain waves. The contour interval is 10 cm/s with the zero contour omitted. The shaded region lies within the 0.1 g/kg contour of liquid water (shaded). (After Lane *et al.*, 2000.)

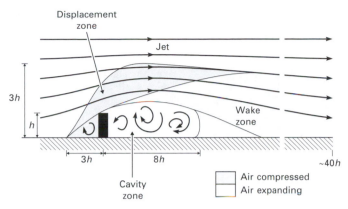

Figure 3.13 Airflow modification over a long, thin windbreak.

the mountain. Second, resonant lee-waves, the wavelengths of which are generally shorter than that of hydrostatic mountain waves, are able to propagate far downstream of the mountain. They may be described as the superposition of two non-hydrostatic gravity waves, one of them propagating upwards and the other downwards (Lane *et al.* 2000). A high-resolution numerical model was used to simulate the mountain waves observed in the New Zealand study. It broadly reproduced the partially trapped lee-waves observed on 20 October. The simulated, partially trapped lee-waves were aligned parallel to the mountain ridge, were confined to the lowest 2 km of air, and had wavelengths of about 15 km (Figure 3.12(a)). Well above the mountain tops, the hydrostatic wave field appeared to be determined mainly by the large valleys in the lee of the Southern Alps (Figure 3.12(b)).

Many other studies also report non-stationary behaviour of trapped lee-waves (e.g. Ralph *et al.* 1997; Worthington and Thomas 1998; Wurtele *et al.* 1999), a finding supported by modelling experiments. Two-dimensional mountain-wave simulations showed that trapped lee-waves displaying temporal variations in wavelength and amplitude may be generated by non-linear wave dynamics while the background flow remains perfectly steady (Nance and Durran 1998). For moderate amplitudes, a non-linear wave interaction, involving the stationary trapped wave and a pair of non-stationary waves, seemed to cause the development of non-stationary perturbations on the stationary trapped wave. The pair of non-stationary waves consisted of a trapped wave and a vertically propagating wave, both of which had horizontal wavelengths roughly twice that of the stationary trapped wave. As the flow became more non-linear, the non-stationary perturbations engaged a wider spectrum of horizontal wavelengths and could dominate the overall wave pattern at wave amplitudes significantly below the threshold required to produce wave breaking. The simulations suggested that strongly non-linear wave dynamics can generate a wider range of non-stationary trapped modes than that produced by temporal variations in the background flow.

Windbreaks

Small topographic features such as isolated hills, buildings, and shelterbelts affect the aerodynamics of airflow. Some of the aerodynamic modifications have important implications for the design and siting of buildings and of windbreaks. For this reason, as well as for pure scientific interest, they have been widely studied (e.g. Caborn 1955, 1957, 1965; Bird and Prinsley 1998; Nuberg and Prinsley 1998).

The modification to airflow and wind-speed patterns for a wind blowing at right-angles to a long, thin windbreak (height, h) is well established (Figure 3.13). Pressure builds ahead of the windbreak, in the upwind zone, and air is forced up and over it. The build-up begins at a distance some three times the height ($3h$) of the windbreak. As the air rushes over the windbreak, streamlines converge forming a zone of acceleration or a jet, but once over the windbreak the air expands again and decelerates. The pressure build-up, the convergence, and the divergence occur in the displacement zone. Once over the windbreak, the air cannot expand immediately to fill the available space. In consequence, the displaced air separates from the windbreak surface and turbulence occurs in the low pressure of the wake zone, which

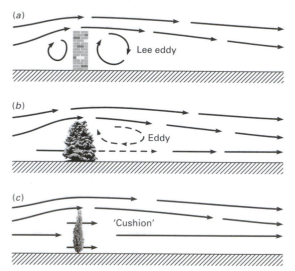

Figure 3.14 Airflow modification associated with (a) solid, (b) leaky, and (c) very leaky windbreaks.

Wind-speed reduction depends on the height of the barrier, h, its porosity, the distance downwind from the windbreak, the height above the ground surface, the roughness length in uninterrupted wind, and atmospheric stability (McNaughton 1988). A solid barrier forces all the air over it, but most windbreaks are porous (leaky) and allow flow to bleed though them (Figure 3.14). The effect of leakage is to reduce the air pressure difference between windward and leeward sides and to create a 'cushion' in the cavity zone, lessening turbulence and allowing the flow to assume a more aerodynamic shape. Windbreak height is the chief determinant of the downwind distance over which wind speed is reduced, but porosity affects turbulence as well as wind speed. High-porosity windbreaks, such as a row of planted white poplar (*Populus alba*) trees, bring about a minimal reduction of wind speed and a maximal reduction of turbulence. Low-porosity windbreaks, such as a wall or a dense evergreen hedge, are most effective in creating a short downwind zone of reduced wind speed, while generating high levels of turbulence. In many applications, a medium porosity windbreak is taken as the best compromise. Practical applications will be considered in Chapter 7.

Buildings

An isolated, flat-roofed building placed at right-angles to the wind modifies airflow in a similar way to a solid windbreak, but with flow also occurring around the sides (Figure 3.15). Interestingly, displacement, cavity, and wake zones are all present. The front of a building bears the brunt of the wind. Moving air is

extends from a downwind distance about eight times the height of the windbreak ($8h$) to as much as 40 times the height of the windbreak ($40h$). Just behind the windbreak is a pocket of low pressure, called the cavity (or quiet) zone, that tends to 'pull' the air into a more or less stationary lee-eddy or vortex. The cavity zone extends to a downwind distance some eight times the height of the windbreak ($8h$). The large lee-eddy breaks down into smaller eddies in the wake zone. Within the wake zone, the flow separation breaks the force of the wind on the ground.

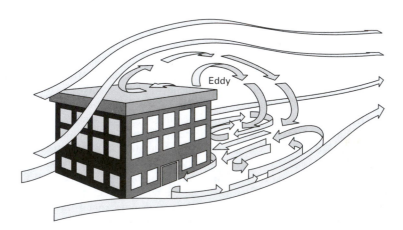

Figure 3.15 Airflow modification around an isolated, flat-roofed building lying at right-angles to the wind.

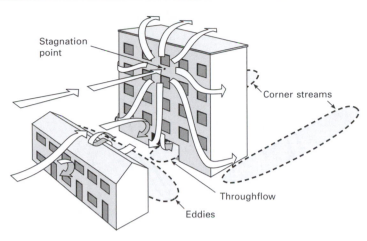

Stagnation point

Corner streams

Throughflow

Eddies

Figure 3.16 Airflow modification around a tall shopping-centre block rising above the roof level of a line of terraced houses.

forced over the top, down the front, and around the sides of the building. Maximum pressures occur near the upper middle part of the windward face, where the wind is brought to a standstill at the stagnation point. Near the outside edges of the windward face, pressure falls below that of the undisturbed atmosphere (i.e. there is suction). If the building has sharp edges (as opposed to a rounded building), flow separation takes place at the edges. This produces suction on the sides, the roof, and the leeward wall and leads to the formation of a cavity zone with a lee eddy circulation involving a double eddy at ground level. The suction of the wind on the building can remove roofing materials and window panes.

An interesting case is a tall shopping-centre block rising above the roof level of the surrounding houses (Figure 3.16). The wind hits the windward face of the building and produces a central stagnation point at about three-quarters of the building height. Airflow diverges away from the stagnation point, some passing over the top and adding to a lee-eddy in the cavity zone, much of the rest streaming down the windward face and either augmenting the lee-eddy of the upwind low building or flowing round the sides of the building as corner-streams, which wrap around the back to give a horseshoe shape. If the building is on pillars, or if there is a passageway under it, then air descending down the windward face contributes to a jetting throughflow. The building increases wind speeds at pedestrian height, giving strong and blustery conditions that batter shoppers and make doors difficult to open and close.

Small-scale thermal airflow modification

Local winds in mountains

Differential heating of topographic components creates characteristic systems of air motion, especially when the regional pressure gradients are weak, giving rise to thermo-topographic winds. Under certain conditions, winds are generated in mountains that fluctuate through a daily cycle. Three such winds are compensating winds, mountain and valley winds, and slope winds (Wagner 1938). Compensating winds occur on a regional scale and will be discussed in a later section.

Slope winds and mountain and valley winds are linked. Slope winds form owing to differential heating at the top and bottom of a slope. During the night, the ridges are cooler than the valley bottom and cold air drains downslope (there is a compensatory return of air overhead, thus forming a circulatory cell). Where slopes are steep, these downslope winds may be strong, 2 m/s or a shade more. In some cases, a cold-air pool forms at the top of the ridge, especially if there is a plateau at the slope top, which drains 'catastrophically' when it reaches a critical level. These are cold-air drops or 'air avalanches' and may be strong. Air avalanches on Mount Karisimbi, Rwanda (4507 m), north-east of Lake Kiwu, were almost strong enough to blow away a tent (Scaëtta 1935). During the day, upslope winds may develop. Owing to greater radiation exchange, these are commonly stronger than nighttime downslope winds, reaching up to 4 m/s.

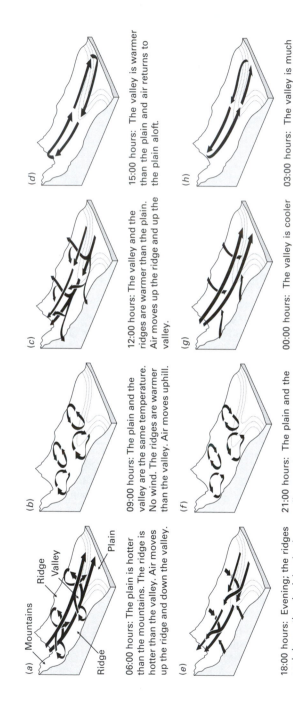

Figure 3.17 Defant's cycle of mountain winds, valley winds, and slope winds. (Adapted from Defant, 1951.)

(a) Mountains, Ridge, Valley, Plain, Ridge

06:00 hours: The plain is hotter than the mountains. The ridge is hotter than the valley. Air moves up the ridge and down the valley.

(b) 09:00 hours: The plain and the valley are the same temperature. No wind. The ridges are warmer than the valley. Air moves uphill.

(c) 12:00 hours: The valley and the ridges are warmer than the plain. Air moves up the ridge and up the valley.

(d) 15:00 hours: The valley is warmer than the plain and air returns to the plain aloft.

(e) 18:00 hours: Evening; the ridges cool down; the air sinks slowly towards the valley. The valley is still warmer than the plain, so the wind still blows up the valley.

(f) 21:00 hours: The plain and the valley are the same temperature again. The ridges are cooler than the valley. Air flows into the valley.

(g) 00:00 hours: The valley is cooler than the plain. Air moves down the valley and the ridges.

(h) 03:00 hours: The valley is much cooler than the plain. Air flows down the valley.

Mountain and valley winds flow between upper and lower parts of valleys:

> On the slopes of mountain ranges and the plains near the base, and especially in deep valleys extending from the slopes into the country below, there are often strong winds toward and up the mountain side during the day, and the reverse at night. This fact is well known to hunters and all who are accustomed to encamp in the mountains at night; and so the camp-fires are always placed below the tents on the slope, so that the smoke may be drawn away by the descending current, instead of being blown toward them. (Ferrel 1889, 223)

Mountain and valley winds are linked to slope winds, the two sets producing a system of periodic mountain winds. At nighttime, downslope winds are associated with downvalley winds (also called valley-exit and mountain winds). During the daytime, upslope winds are associated with upvalley winds. Defant (1951) argued that slope winds and mountain and valley winds pass through a succession of stages during a daily cycle (Figure 3.17). Defant's sequence is often cited in books, but some later work suggests that the component winds form almost at the same time. A study in an alpine valley near Davos, Switzerland, revealed that for 90 per cent of the time, the downslope breezes stop at sunrise, give or take 20 minutes, and the mountain wind stops about 25 minutes after sunrise. Upslope breezes start between sunrise and 40 minutes after, and the valley wind starts about an hour after sunrise (Urfer-Henneberger 1967; see also Urfer-Henneberger 1970).

Four local wind systems were detected within the TRACT (TRansport of Air pollutants over Complex Terrain) project (Löffler-Mang *et al.* 1997). The project considered an area of about 250×325 km covering south-west Germany, eastern France, and northern Switzerland. It involved 40 scientific teams from 10 countries working for three weeks. Slope winds, valley and mountain winds, and lake and sea breezes were all observed in the study. The nature of the TRACT project meant that the concurrent action of multiple local wind systems could be investigated. The wind systems in four valleys were looked at – the Rench Valley, the Kinzig Valley, the Dreisam Valley, and the Murg Valley (Table 3.4). The valley wind systems were divided into two groups. In the Dreisam and Murg Valleys, the daytime upvalley wind speeds were higher than the nocturnal downvalley winds. Conversely, in the Rench and Kinzig Valleys, the nighttime downvalley winds were the stronger, typically reaching 4 m/s in the Rench Valley and

Table 3.4 Typical valley-wind speeds in the TRACT project

Valley	Nighttime downvalley wind speed (m/s)	Daytime upvalley wind speed (m/s)
Rench	4	2
Kinzig	6	4.5
Dreisam	1	3–4
Murg	0.5	2

Source: Adapted from data in Löffler-Mang *et al.* (1997).

6 m/s or more in the Kinzig Valley compared with 2 m/s and 4.5 m/s, respectively, for the upvalley winds. The weak development of the valley-wind system in the Murg valley, especially at night, was attributed to the decelerating effect of dense settlements and industrial plants in the Gaggenau area.

Cold pockets and frost hollows

Cold air-drainage that produces cold pockets and frost hollows does not necessarily involve katabatic slope winds. Cold pockets are produced by the weak (less than 1 m/s) and sometimes shallow (up to 1 m) drainage of chilly air over small areas. Such small-scale air drainage does not set up the compensating flows that katabatic slope winds do. But it can form sizeable pools of cold air, as happens at Rickmansworth, in Hertfordshire, England:

> . . . on quiet clear autumn evenings when garden bonfires of damp leaves are burning, it is common to see during the hour after sunset, rivers of white smoke slowly winding their way down into the northern strip of the valley from points in all four quadrants of the compass . . . the estimated speed of the air flow seldom exceeds about 2 m.p.h. It does not take long for a lake of cold air 30–40 feet deep to accumulate. (Hawke 1944)

Rickmansworth, which lies at the southern end of the Chiltern dip slope, is a notorious frost hollow. The cold pools collect over a small area – about 1000 m long and 100 m wide – but uncommonly low temperatures are recorded. On clear nights, minima are commonly 10–12°F below those of nearby London suburbs. On 29 August 1936, when the weather was clear and calm, the minimum temperature was 34°F and the maximum temperature was 84.9°F, giving a range of 50.9°F, one of the greatest daily temperature ranges recorded in the British Isles (Hawke 1944).

Small closed basins may produce phenomenally strong temperature inversions. The Gstettneralm sinkhole near Lunz, Austria, has an elevation of 1270 m and a relief of 100–150 m. It shows temperature inversions of 20–30°C or more, and has a recorded minimum temperature for the period 1928–1942 of −52.6°C, a central European record (Geiger 1965, 398). Night frost can occur in the middle of summer.

Land and sea breezes

These are thermally induced winds (like mountain and valley winds). Sea breezes blow landwards, and land breezes blow seawards. Land and sea breezes are, in effect, miniature monsoons. They are generated by the thermal contrast between land and sea during the night and during the day when the weather is calm and clear, though the processes involved are complicated. They are developed most strongly in the tropics, where land–sea temperature differences are greatest. At Jakarta (Batavia), Indonesia, land and sea breezes blow on 7–8 days out of 10. They are common in temperate latitudes in summer, blowing on 2–3 days out of 10 in western Europe. They even occur in Greenland.

Sea breezes are about 1 km deep, thinning towards the advancing edge, and may push 50 km inland (Figure 3.18(a)). They blow at around 4–7 m/s and tend to be stronger in the tropics. In the Gulf of Guinea, West Africa, sea breezes can cause large swells on the sea and make sailing in estuaries and lagoons perilous. Land breezes are gentler, blowing at about 2 m/s, but are boosted by the presence of coastal hills (Figure 3.18(b)). In Great Britain, sea breezes are usually less well developed than those in the tropics, and occur intermittently, principally along southern and eastern coasts during spells of summer anticyclonic weather. They blew on several days in July 2000. Many coastal and near-coastal sites have a low frequency of sea breezes, and there are often few signs of an entire sea-breeze–land-breeze reversal over a 24-hour period (e.g. Elliott 1964; Stevenson 1961). However, there is evidence that at Manchester Airport (Ringway), which lies about 50 km from the sea, an extensive two-way diurnal exchange of sea and land air sometimes occurs aided by hill–valley circulations from the Pennines (Crowe 1962, 18).

Even small lakes are capable of inducing local circulations of air. Suwałki Landscape Park, in northeast Poland, contains several small lakes, mostly with areas of about 0.5 km². Lake Udziejek has a minimal effect on the local climate of areas adjacent to it, but

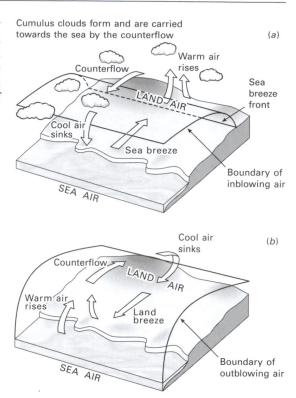

Figure 3.18 Schematic diagrams of land and sea breezes: (a) sea breeze; (b) land breeze.

a limited local transfer of air was observed and was associated with the formation, movement, and disappearance of mists (Błażejczyk and Grzybowski 1993). Under clear skies and moderately calm conditions in July 1988, mists formed a layer 1–2 m deep over the lake and wet meadows after sunset, as cool air drained down from the surrounding higher land. They disappeared from the meadows around midnight, but persisted over the lake until morning. After sunrise (around 3:30 a.m.), the lake started to warm but cold air still drained off the higher ground. The mists thickened, reaching a peak depth about 1.5 hours after sunrise. They then slowly moved towards the lakeside, and by about 7:00 a.m. they had disappeared.

Synoptic-scale climates

Synoptic-scale airflow modification

Mountains have two major effects on frontal systems (Barry 1992, 112): the structure of frontal cyclones is altered as they pass over them; and cyclogenesis is encouraged on their leeward sides.

'Cyclonic' winds

The winds associated with a cyclone (travelling depression) are modified by topography. In places, topographic constrictions channel cyclonic winds, causing them to blow more strongly and, in some cases, to cause either much colder or much hotter conditions than normal.

The mistral is a strong, dry, cold northerly to north-westerly wind that blows from the high central region of France towards a depression growing in the Gulf of Genoa, east of a high-pressure ridge from the Azores anticyclone. It also blows in Italy, where it is called the maestrale. It is often strongest in the Rhône Valley and the Golfe du Lion, due to topographic funnelling and katabatic drainage of cold air from the upper Rhone system. Some mistral events are similar to other fall winds (p. 74). The mistral may blow for several days.

Regional winds in the Mediterranean summer are characteristically associated with tropical continental airstreams that move northwards in advance of eastbound depressions. The sirocco (or scirocco) is a hot, southerly or south-easterly wind of Algeria and the Levant, southern Italy, Sicily, and other Mediterranean islands. It is thought to originate in the Sahara as a dry, dust-laden wind that becomes moist and oppressive as it crosses the Mediterranean Sea. It gives rise to 'blood' rains that sometimes reach northern Europe, including the United Kingdom. The leveche is similar to the sirocco. It is a hot, dry, and often dust-laden wind blowing from Morocco across the coast of southern Spain and heralds the coming of a depression. Under the right conditions, the dust from the leveche reaches northern Europe where, as with the dust carried on the sirocco, it falls as 'blood' rain. In Egypt, the equivalent wind is the khamsim. The levanter (levante) is a strong easterly or north-easterly wind that blows across south-east Spain, the Balearic Islands, and the Strait of Gibraltar, particularly in summer and autumn. It habitually blows strongest through the Strait of Gibraltar, where it may bring eastbound aircraft to a near-standstill. It brings mild, cloudy, foggy, and sometimes rainy weather to the southern Spanish coast.

Winds associated with cyclones also occur in the Americas. A norther is a sudden cold and dry north wind that causes large drops of temperature (up to 20°C in 24 hours) in Texas, the Gulf of Mexico coast, and other parts of southern North America. The winds normally come in behind a depression. The wind is often very strong (up to 100 km/hr) and may be accompanied by severe thunderstorms and hail. The pampero is a dry and bitterly cold south-westerly wind that blows from the Andes across the Argentinian and Uruguayan pampas, especially in summer. It is commonly associated with storms and severe drops of temperature. The equivalent wind in Australia is called a southerly buster, where it brings low temperatures to New South Wales when polar air is brought northwards behind a depression and suddenly crosses warmer regions during spring and summer months. To elucidate the circumstances under which southerly busters develop, Kathleen L. McInnes and her colleagues carried out numerical simulations using a weather prediction model (McInnes and McBride 1993; McBride and McInnes 1993; McInnes 1993). They came to two major conclusions. First, the chief reason for southerly winds moving northwards along the Australian east coast lies in synoptic-scale frontal dynamics. Second, the enhanced northwards movement associated with an S-shaped deformation of the frontal discontinuity is brought about exclusively by an eastbound front interacting with topography.

Mountains and cyclogenesis

Lee cyclogenesis, which may create a lee depression (cyclone), is very important downwind of barriers lying across the mid-latitude westerlies – the Western Cordillera of North and South America and the Tibetan Plateau – and in the lee of such smaller topographic features as the Alps.

The key to understanding lee cyclogenesis is cyclonic vorticity (spin), also called positive relative vorticity, ζ. Positive relative vorticity is part of the vertical vorticity (that is, the component of vorticity about a vertical axis). The other part is the vorticity about the local vertical resulting from the Earth's vorticity about its own spin axis, which is defined by the Coriolis parameter $f = 2\omega \sin\phi$ (where ω is the angular velocity of spin, which is 15°/hr, and ϕ is the latitude). For large-scale horizontal motion, the vertical component of absolute vorticity tends to be conserved, so that:

$$\frac{d(f + \varsigma)}{dt} = 0 \qquad (3.6)$$

Now, if the air moves northwards, the Coriolis parameter, f, increases while the cyclonic vorticity, ζ, decreases. The curvature becomes anticyclonic and the air turns back to lower latitudes. If the air moves equatorwards from its original latitude, then f decreases and ζ increases, the resulting cyclonic curvature deflecting the current polewards. The result is that large-scale horizontal motion tends to flow in a wave pattern (as seen in Rossby waves). Now, upper-level horizontal wave motion is connected to airflow near ground level. In particular, in the sector ahead of an upper trough, the decreasing cyclonic vorticity causes divergence (because the change in cyclonic vorticity outweighs that in the Coriolis parameter), which in turn favours surface convergence and low-level cyclonic vorticity – a depression is spawned. This process of cyclogenesis occurs without the assistance of topographic barriers. However, mountains can trigger cyclogenesis by encouraging the establishment of cyclonic vorticity. The basic mechanisms involved are fairly simple. An air column approaching a mountain from the west is forced into a reduced depth. To conserve vorticity, horizontal divergence with anticyclonic spin develops just ahead of, and above, the mountain. Once over the mountain, the air column is free to expand back to its previous depth. To conserve vorticity, horizontal convergence occurs with cyclonic spin to give a lee trough. It is in the lee trough that surface convergence favours the formation of lee depressions. In detail, the mechanisms involved are more complex, and can be studied using numerical atmospheric simulation models.

High-resolution numerical model simulations for the Alpine region suggested that lee cyclogenesis consists of two phases (Aebischer and Schar 1998). Phase 1 is the rapid formation of a low-level orographic vortex in the lee of the mountains. Phase 2 sees the orographic vortex interacting, baroclinically and diabatically, with an approaching upper-level trough. An intermediary phase, seen in the simulations, may involve the formation of potential vorticity streamers (or banners) caused by a splitting of the airflow around the entire Alpine Range (primary banners), or around individual massifs and peaks (secondary banners). A physical model, using a hydraulic turntable, indicated the importance of the Alps in blocking the impinging airflow, forcing it to split and delaying its arrival on the lee side (Longhetto et al. 1997). Other numerical models of lee cyclogenesis, if not producing a fully consistent picture (and given the varied nature of mountains, consistency is perhaps unlikely to be expected), at least largely agree as to the processes involved (e.g. Alpert et al. 1996a, 1996b; Mozer and Zehnder 1996).

Fall winds

When the synoptic situation allows, topography may alter airflow mechanically and thermodynamically to produce distinctive winds that blow, or fall, down the lee slopes of a mountain range. These fall winds include katabatic winds, the föhn (or the chinook), and the bora.

Katabatic winds

Strictly speaking, katabatic winds are local downslope gravity flows caused by nighttime radiative, near-surface cooling under clear skies (Barry 1992: 158). However, nocturnal gravity drainage over an area broader than an individual slope is also called a katabatic (or drainage) wind.

Katabatic winds may be studied using remotely sensed data. Synthetic aperture radar (SAR) images of Mediterranean coastal waters (from the ERS-1 satellite) showed evidence of katabatic wind fields in sea-surface roughness patterns (Alpers et al. 1998). The katabatic winds that caused the sea surface to roughen blew from 1800-m-high mountains through a broad valley at the western coast of Calabria (southern Italy) into the Gulf of Gioia. The images indicated that the size and shape of the katabatic winds over the sea varied strongly, depending on the meteorological conditions. The roughness pattern varied, its shape taking various forms: a mushroom, an elongated tongue, a broad blob, and a narrow truncated band.

Katabatic winds are common on Antarctica and mainly result from strong radiative cooling of air over icy slopes. They may blow with considerable ferocity. When Sir Douglas Mawson and his party explored King George V Land and Adélie Land during their 1911–1913 expedition, they were astounded by the strength of the katabatic wind that is prevalent in the area. On reaching the ocean, the winds are strong enough to move the sea-ice away from the shore, so forming coastal polynyas, in summer and even in midwinter (Wendler et al. 1997). Antarctic katabatic winds are in places strengthened by topography. Cape Adare is notorious for gusting, high-velocity winds. In 1911, the Northern Party of Robert Falcon Scott's British Antarctic expedition (1910–1913) built a wooden hut there, which, during the expedition, suffered superficial damage from 18 storms blowing at force 11–12 on the Beaufort scale. The topography of the Adare Peninsula forces the strong south-east wind to change to an east-south-east wind on meeting high ground. By the time the east-south-east wind reaches

Ridley Beach, it assumes the character of a high-velocity katabatic wind (Harrowfield 1996). The topographic reinforcement of the Cape Adare winds might have been largely responsible for the destruction of the historic hut.

Atmospheric models have aided the investigation of Antarctic katabatic winds. By way of example, a hydrostatic mesoscale atmospheric model, coupled to a snow model, was used to simulate katabatic wind events observed in the coastal zone of Adélie Land on 27 November and 3 December 1985 (Gallée and Pettré 1998). The diurnal insolation cycle was strong on both dates, but the large-scale (regional) wind was different, being weak on 27 November and moderate on 3 December. For both events, temperature and wind displayed striking diurnal cycles, with katabatic winds blowing during the night and upslope winds during the day. In addition, during both events, the katabatic airstream slowed progressively over the ocean. This slowing of the winds had an important result – continental air accumulated to form a pool of cold air that generated a pressure gradient force opposing the katabatic wind. The pooling of cold air thus augmented the retardation of the katabatic winds. In the morning, when insolation increased, the surface inversion weakened but the influence of the cold-air pool became even stronger. The katabatic flow started to decay over the coastal zone and then retreated progressively towards the ice-sheet interior. With a weak large-scale wind, the surface warming was sufficient to produce an additional upslope buoyancy force that led to anabatic flow over the ice sheet in the afternoon. With a moderate and downslope large-scale wind, the pooling of cold air was important and had a dramatic impact on the flow. A sharp spatial transition was generated between downslope and upslope winds over the ocean. This discontinuity moved towards the ice-sheet interior in the morning, and was responsible for the sudden ending of the katabatic flow seen by static observers.

There is concern about how global warming might affect Antarctic katabatic winds. One study investigated this problem using a high-resolution general circulation model (ECHAM-3 T106, which has a spatial resolution of $1.1° \times 1.1°$) (Van den Broeke *et al.* 1997). In one simulation, the model was forced to a new steady state corresponding to IPCC Scenario A – a doubling of atmospheric carbon dioxide concentration. Under this new regime, summertime (June, July, August) katabatic windspeed decreased by up to 15 per cent in the lower parts of the ice sheet near coastal areas, due to the destruction of the surface

inversion by increased absorption of solar radiation (temperature–albedo feedback). However, wintertime (December, January, February) near-surface wind speed increased by up to 10 per cent, due to a deepening of the circumpolar low-pressure belt, which is locally enhanced by the removal of sea ice. In consequence, the model predicted that the annual mean wind speed will remain within 10 per cent of its present value in a doubled carbon dioxide climate, but that the amplitude of the annual cycle will increase.

Föhn and chinook

The föhn is a warm, southerly wind that frequently crosses the Alps

> . . . melting the snow even in mid-winter, and causing great discomfort to man and beast. The frozen torrents break out afresh, and, swollen from the sudden access of water from the dissolving snow, often overflow, and tear raging into the valleys below. Houses are dismantled, and animals are enervated. The chamois retreats to the deepest gorges and the coolest exposures, and the birds desert the mountain woods. Fires are extinguished, lest sparks should ignite the wooden châlets, and generally its appearance is unwelcome. Yet in the spring it is regarded as the pleasant harbinger of summer, and for weeks at a time it brings joy and plenty to many a chilly glen. It sweeps away the accumulated snow and ice . . . while the Alpine pastures are, under its warm breath, sooner ready for cattle than they would be if left to ordinary atmospheric conditions. (Brown 1889, 327)

The chinook is a föhn wind that occurs in western North America. It is named after the Columbia River tribe of Chinook Indians. Its warming effect in early spring is remarkable and is sometimes referred to as a 'snow-eater'. In the Canadian Rockies, a chinook is

> . . . a strong westerly wind, becoming at times almost a gale, which blows from the direction of the mountains out across the adjacent plains. It is extremely dry, and, as compared with the general winter temperature, warm. Such winds occur at regular intervals during the winter, and are also not infrequent in the summer, but, being cool as compared with the average summer temperature, are in consequence then not commonly recognized by the same name. When the ground is covered with snow, the effect of the winds in its removal is marvellous, as, owing to the extremely desiccated condition of the air, the snow may be said to vanish rather than to melt, the moisture being licked up as fast as it is produced. (Dawson 1886, 33)

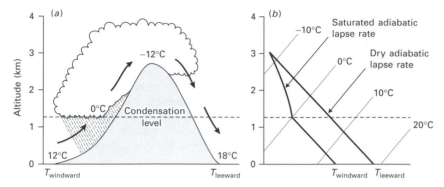

Figure 3.19 Classic explanation of föhn wind. (a) Schematic representation of forced ascent and temperature changes in an air parcel. (b) A tephigram tracing temperature changes in the air parcel while crossing a mountain range. Notice that, on crossing a mountain range, an air parcel ascends the windward side, cooling at the dry adiabatic lapse rate until it reaches the condensation level, after which it cools at the saturated adiabatic lapse rate. On descending the leeward side of the range, the air parcel warms at the dry adiabatic lapse rate and ends up warmer at the leeward foot than at the windward foot. (Adapted from Barry and Chorley, 1968, 108.)

The classic mechanism invoked to explain the föhn and chinook winds is elegant (Figure 3.19). When moist air meets a mountain range, it is forced to rise, leading to cloud build-up and precipitation on the windward slope. Above the cloud base, the rising air cools at the saturated adiabatic lapse rate (about 5–6°C/km) because latent heat of condensation is released. Over the leeward slope, the air descends and warms at the dry adiabatic lapse rate (9.8°C/km) because it is cloud-free. In consequence, potential temperatures are higher and relative humidities are lower on the leeward side than on the windward side of the mountain range. Other mechanisms produce föhn-type temperature changes but are probably special cases of the basic downslope föhn wind (see Beran 1967; Barry 1992, 145–6).

The Santa Ana of southern California is a föhn-type wind. It involves the outward flow of air from an anticyclone over the interior desert regions of California that moves over the San Bernardino Mountains to the coastal regions of the Pacific coast. It is typically hot and dry, leading to relative humidities falling below 10 per cent and raising the risk of forest fires, with gusts of 10–50 m/s.

Bora

The bora is a cold, dry, and gusty wind that blows in winter over the Dalmatian Mountains towards the Adriatic Sea. Similar downslope winds occur on the Black Sea coast of the Crimea and elsewhere in parts of Russia, the northern Norwegian fjords, and along the east slope of the Colorado Front Range (see Barry 1992, 149). An analogous wind also occurs at Crossfell in the northern Pennines, England, where it is called the north-easterly helm wind (Manley 1945); and in the Kanto Plain, inland of Tokyo, Japan, where it is called the oroshi (Yoshino 1975, 368–72). In its type area, the bora occurs on the eastern shore of the Adriatic from Trieste, Italy, and southward for about 500 km. Strong north-easterly gusts, which may exceed 40 m/s, can be felt up to 60 km offshore. The strongest gusts occur at night, with a peak between about 5:00 and 8:00 a.m. Each bora event lasts 12–20 hours, on average, but spells lasting around a week or more occur at least once each winter. Temperatures fall to freezing on the coast. The sudden arrival of a severe bora can cause problems for people: 'Near Präwald, in 1805, a division of the Austrian army, then retreating from Italy, was taken unawares by a severe Bora, so that it was impossible for the soldiers to continue the march, and many of them were frozen to death' (Brown 1889, 323).

Compensating winds

Winds resulting from cold-air drainage may be part of a larger regional circulation set up by mountainous areas. Winds blowing from mountains to plains near the ground may be compensated by flows from plains to mountains aloft – mountain winds are counterbalanced by anti-mountain winds. The basic mechanism that creates a compensating wind system is straightforward. During the day, greater heating on the plain leads to a horizontal pressure gradient, and so to a large-scale flow of air, between plain and mountain. At night, the temperature gradient is reversed and a return flow occurs from mountain to plain. Snow in the mountains exacerbates the temperature gradient and boosts compensating winds. In narrow passes between mountains and plains, the winds may reach gale force. Compensating winds affect cloud formation, precipitation, and vegetation. They are almost invariably associated with other local wind systems (valley winds, land and sea breezes, and monsoons).

Regional winds in the Rocky Mountains of Colorado

Diurnal wind changes on a regional scale were observed in summit wind data, collected over four summers, for the Colorado Rocky Mountains, USA (Bossert *et al.* 1989). Two diurnal wind regimes were noted according to the synoptic situation. The first regime occurs under weak synoptic conditions (clear summer days with strong radiative heating). Under these conditions, thermal differences drive simple wind reversals with daytime inflow and nighttime outflow, with attenuated transition periods lasting 6–7 hours in the evening and 4–5 hours in the morning. The nighttime outflow takes place in a shallow layer at mountain-top level above a deep layer of stable air in the valleys, which arise from radiative cooling. The second regime occurs with convective thunderstorms and latent-heat forcing. Under these conditions, the smooth morning transition from outflow to inflow is disrupted in mid-afternoon by a switch from inflow to outflow, and by another reversal in late afternoon or early evening. These inflow and outflow patterns are regional phenomena. Over lower ranges or individual mountains, the return flows occur at a much lower level and are local anti-valley and anti-mountain winds.

When examined in more detail using observed data and a numerical simulation with the Regional Atmospheric Modeling System (RAMS), the basic flow systems in the Rocky Mountains – daytime inflow towards the highest terrain, and nocturnal outflow away from it – displayed more unusual behaviour, especially west of the barrier crest (Bossert and Cotton 1994a, 1994b). The main thermally forced flow within the simulation was the mountains–plains

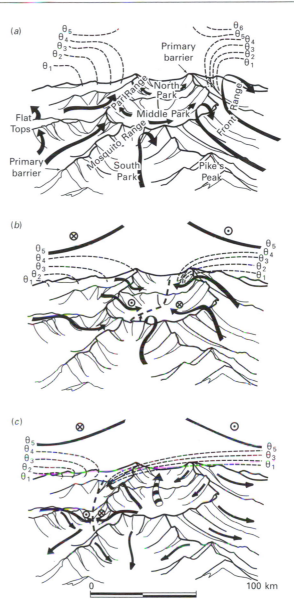

Figure 3.20 Conceptual models of the regional-scale circulation of air over the north–central Colorado Rocky Mountains: (a) daytime inflow circulation system; (b) transition phase; (c) nocturnal outflow circulation. (Adapted from Bossert and Cotton, 1994a.)

circulation along the east slope of the Front Range (Figure 3.20(a)). This circulation was a pressure–density solenoid with a low-level upslope branch, vertical branch, and return branch at 5–6 km above ground level. The vertical branch marks a boundary with the heated air mass over the intermontane region west of the Front Range. The boundary is a baroclinic zone that intensifies throughout the late morning and early afternoon. In the late afternoon, convergent flow in the low-level branch of the circulation system weakened due to reduced surface forcing (Figure 3.20(b)). The low-level winds in the Front Range mountains–plains circulation by this time blew from the south-east, owing to the prolonged influence of Coriolis force. The regional-scale temperature contrast across the Front Range barrier smothered the heat-induced convergence over each slope, forcing the potentially cooler air within the Front Range mountains–plains circulation to ride over the Front Range crest and move westwards down the western slope. Thus, in late afternoon, the Front Range mountains–plains circulation system changed into a shallower, westwards-propagating density current. During the evening, the density-current moved farther westwards across the mountains, leaving in its wake strong south-easterly flow at the mountaintop level. In the final late-night adjustment phase, the density-current dissipated near the western edge of the Colorado mountains. At the same time, a steady southerly flow evolved over the high mountain terrain, which was a steady response to the differential heating that develops between the low-lying plains and the intermontane region, with topographic channelling and the continuing action of the Coriolis force exerting their influence. Late at night, nocturnal flow over north–central Colorado consisted of shallow downslope flows along the major slopes of the complex terrain, and a regional-scale southerly jet through the intermontane region (Figure 3.20(c)).

A series of two- and three-dimensional numerical experiments examined further the effects of different physical processes on the evolution of thermally driven, regional-scale circulations over north–central Colorado (Bossert and Cotton 1994b). In the experiments, a westwards-propagating disturbance developed as a robust summertime feature over a range of low-level ambient wind speeds, directions, and shears. It was initiated by differential heating across the Front Range between the plains and the intermontane region. It was suppressed with low-level westerly flow, which also weakened the development of its progenitor, the Front Range mountains–plains solenoid. The unique topography of the region – with a low plain to the east, a high dividing range (the Front Range) in the middle, and a high plateau on the west – boosted the development of westwards-propagating current. Low-level stratification affected the depth and strength

of the Front Range mountains–plains solenoid, which was most energetic in summertime conditions of near-neutral stability below 50 kPa. High stability in the lower troposphere stifled the vertical development of the solenoid. However, it increased the baroclinicity across the Front Range generated by surface heating, so producing a significant density-current disturbance. Indeed, under conditions of intense summer heating and weak upper-level airflow, almost any pattern of stratification in the lower troposphere will lead to the growth of the westwards-propagating density current. Soil moisture had a potent impact on the development of the Front Range mountains–plains circulation. Dry soil on the eastern slope of the Front Range and on the eastern plains, combined with wet soil over the high terrain and the western slope of Colorado, was not conducive to the growth of the westwards-propagating density-current and arrested the formation and strength of the Front Range solenoid. Conversely, wet soil along the eastern slope of the Front Range and eastern plains, coupled with drier conditions over the high mountain terrain and western Colorado, greatly promoted the baroclinicity within the solenoid and the subsequent formation of a density-current. Interestingly, latitude dramatically affected the propagating density-current feature, which was powerful and long-lived in tropical latitudes but almost non-existent at high latitudes. The density-current failed to appear at high latitudes, owing to the overwhelming influence of the Earth's rotation there bringing the thermally driven circulation into geostrophic balance much faster than at lower latitudes, so minimising the formation of an ageostrophic current.

Thermo-topographic winds in southern Africa

Compensating wind systems were observed on the south-east side of the Drakensberg Escarpment, South Africa (Tyson 1968; Tyson and Preston-Whyte 1972). The escarpment edge sits at 3000 m, from where the land falls sharply to a sloping plateau at 950 m that eventually hits the Indian Ocean, some 180 km away. The sloping plateau is cut into by deep valleys (250–500 m). The local mountain and valley winds are overlain by regional air movements involving compensating flows. At night, the regional wind blows from the Lesotho Massif and the Drakensberg towards the ocean (the mountain–plain wind). During the day, the regional wind blows from the coast towards the escarpment (the plain–mountain wind).

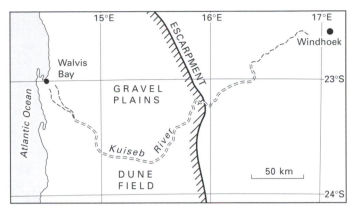

Figure 3.21 Topography of the central Namib region, showing the escarpment, the Kuiseb River, the gravel plains, and the dune field. The study transect runs from a point on the escarpment some 50 km south of the Kuiseb River to a little south of Walvis Bay in the west. (Adapted from Lindesay and Tyson, 1990.)

In a later study, the same kind of regional wind system was observed over the central Namib Desert (Lindesay and Tyson 1990). The topography of the central Namib is seen in Figure 3.21. The desert is bounded to the west by the cold Atlantic Ocean (bathed by the cold Benguela Current), and to the east by a dissected plateau, which averages 1500 m, lying beneath the eastern Escarpment, some 170 km from the coast. The Kuiseb River is deeply incised and separates fairly flat gravelly terrain to the north from sand dunes to the south, which form ridges running parallel to the coast. In summer, daytime south-westerly sea-breezes and north-westerly valley and plain–mountain winds are dominant when calms are less frequent and regional pressure gradients boost the strong thermal contrast between land and sea (Figure 3.22). In winter, nighttime north-easterly land-breezes and south-easterly mountain and mountain–plain winds tend to occur in association with cold air drainage. The plain–mountain winds seem to be purely antitriptic winds produced by the thermal gradient between the gravel plains to the north and the Escarpment to the east. They are strongest (10–15 m/s), deepest (1000–1600 m), and most persistent in summer, when surface heating differences are greatest. In winter, when nighttime cooling is strongest, they occur only during the day, are less strong and deep than their summertime counterpart, and are more disturbed by local winds. The mountain–plain wind is the reverse of the plain–mountain wind. It dominates the boundary layer structure during summer when nighttime south-easterly winds regularly blow from the Escarpment zone towards the coast, except when disturbed by synoptic-scale weather systems. Mountain–plain winds are about 1000 m deep and blow at 5–10 m/s. They occur in summer but for only a few hours at a time in a fairly shallow layer (up to 500 m).

In summer, a typical daily sequence of airflow runs as shown in Figure 3.22(a). In the late afternoon, a strong plain–mountain wind blows with a sea breeze over it. After sunset, the plain–mountain wind and sea-breeze decay rapidly. During the night, airflow is reversed as cold air draining from the Escarpment pushes the plain–mountain wind and sea breeze back towards the coast and replaces them with a mountain–plain wind and localised land breeze. After sunrise, the mountain–plain wind decays and the plain–mountain wind is re-established during the next morning. In winter, the same local winds develop as in summer, but in a different way (Figure 3.22(b)). In late afternoon, a strong sea breeze (with a clear return current aloft) displaces a decaying plain–mountain wind. The sea breeze weakens during the evening. By 10:00 p.m., airflow has reversed as cold air drains seawards during the night as a mountain–plain wind. Between 09:00 and 11:00 the next day, the mountain–plain flow reverses swiftly to a plain–mountain wind, which advances inland as it strengthens during the day and is joined by a developing sea breeze.

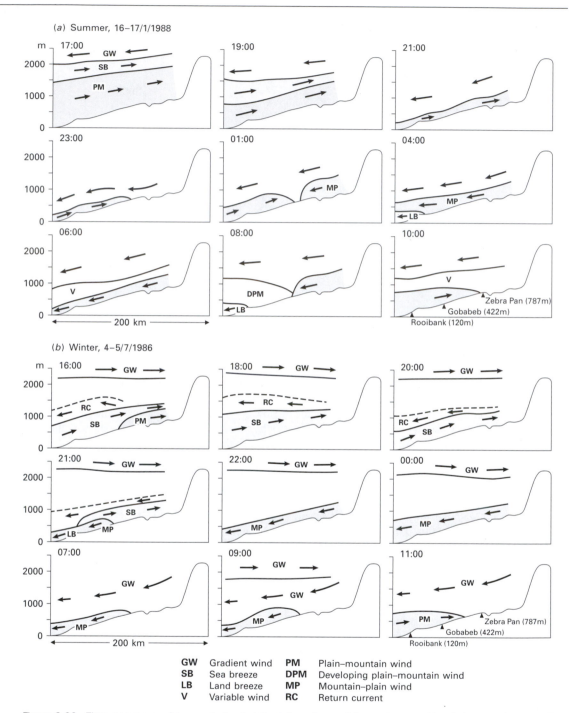

GW Gradient wind   PM   Plain–mountain wind
SB Sea breeze   DPM Developing plain–mountain wind
LB Land breeze   MP   Mountain–plain wind
V   Variable wind   RC   Return current

Figure 3.22 Time sequence of transect showing the interactions of thermo-topographic winds over the central Namib Desert between the coast and the Escarpment: (a) summertime circulations; (b) wintertime circulations. (Adapted from Lindesay and Tyson, 1990.)

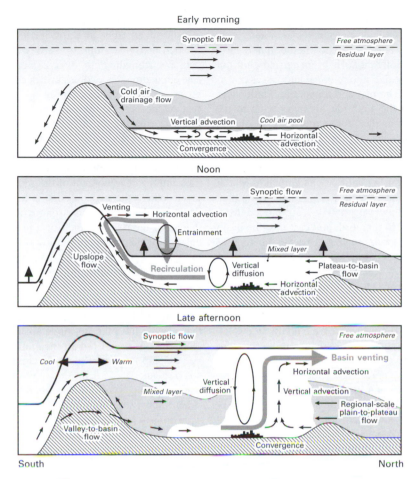

Figure 3.23 Important meteorological processes contributing to pollutant transport in the Mexico Basin. Details are given in the text. (After Edgerton *et al.*, 1999.)

Regional winds and pollution in Mexico City

Mexico City, Mexico, is one of the world's biggest metropolitan areas. Some 20 million people live in the Valle de Mexico (Mexico City Basin), which occupies about 1300 km^2 at an elevation of 2240 m. To the east and west of the Valle de Mexico lie mountains rising 1000 m above the valley floor; to the north and south-east are low points (Figure 3.23). In common with many large cities sited in valleys with limited ventilation, Mexico City suffers from polluted air, and especially from ozone and suspended particle pollution. Tight controls since 1990 have attenuated emissions associated with the city's growth, which runs at 3 per cent per year, but suspended-particle problems persist. A persistent haze blankets the city,

especially in winter, and residents and visitors are worried about the effects of the particulates on their health.

In 1996, the US Department of Energy and Mexico's Petróleos Mexicanos started to sponsor a major collaborative effort to characterise the nature and causes of particulate air pollution in Mexico City, which followed an earlier collaborative effort to study gas-phase pollutants and photochemical oxidants in the Valle de Mexico (e.g. Fast and Zhong 1998; Edgerton *et al.* 1999). A major field campaign was carried out from 23 February to 22 March 1997. Properties were measured that helped to understand the concentration and chemical composition of the city's particulate matter (see Whiteman *et al.* 2000 for details). Meteorological measurements were also taken and these revealed thermally and topographically in-duced airflow patterns that produce a complete vent-ing of the city air during the late afternoon and early evening, so that polluted air is often carried aloft out of the Valle de Mexico into regional air masses down-wind of the city. Thermal and topographic influences are prominent in the Valle de Mexico owing to the topographic setting, the moderately high insolation levels associated with the tropical latitude and high elevation, and weak prevailing synoptic winds.

Three daytime airflows were observed in the Valle de Mexico during February and much of March in the study period (Figure 3.23). First, a regional plain-to-plateau flow was noted – air moved from lower-lying areas to the north and east into the basin in the late afternoon, impelled by the heating of elevated terrain in central Mexico. Second, a local valley-to-basin flow was observed as southerly winds developing and propagating through mountains to the south-east and over the ridge that forms the southern boundary of the Valle de Mexico. Third, local upslope flows, driven by the heating of mountain sidewalls, were detected around noon and into the afternoon. On about half the days studied, late afternoon drain-age flows from the north-east, which started blowing around 18:00 hours and lasted for several hours, ven-tilated the valley. Figure 3.23 illustrates the chief day-time circulations and their effects on atmospheric pollution. However, data analysis and modelling sug-gest that the airflow patterns are very complex and that upper-level synoptic systems, even though they are moderately weak, affect the thermally powered flows in the region. In addition, the evolution of the boundary layer and the circulation patterns varied from day to day, despite the existence of recurrent flow patterns.

The Valle de Mexico, according to the study, does not behave as a typical valley or basin as far as mete-orological characteristics are concerned. As a basin, it fails to develop strong temperature inversions at night, and as a valley it does not generate diurnally reversing valley-wind systems. Further work probed this aberrant behaviour using data analysis, two- and three-dimensional numerical simulations with the Regional Atmospheric Modeling System (RAMS), and a Lagrangian particle dispersal model (Whiteman *et al.* 2000). This work examined the effects of topography and regional diurnal circulations on the boundary layer over the Valle de Mexico and surrounding areas. Results suggested that, during fair weather in winter, changes in the Valle de Mexico's boundary layer are driven first and foremost by regional diurnal circula-tions. These regional-scale circulations are forced by a deep mixed layer of warm air above the Valle de Mexico that grows during the day and the generally cooler surrounding air over adjacent coastal areas. The circulation that evolves brings cool air from ocean areas on to the plateau during the day. But the key feature seems to be the timing of the diurnal circula-tions. During the morning, the air over the Valle de Mexico heats up quickly before noon. In the after-noon, diurnal winds bring cool air into the Valle de Mexico through topographic gaps around its rim, sup-pressing the development of the convective boundary layer. Where the plateau edge is high, the inflow of cool air is held back and strong rising motions occur that form the rising branch of a plain-to-plateau cir-culation, which has a return flow aloft that sinks again over the oceans. During the day, this circulation pro-duces an intense baroclinic zone around the edge of the plateau that separates warm connective air in the boundary layer from the cooler air surrounding the plateau. Just before sunset, the strong rising motions that anchor the plain-to-mountain circulation to the plateau's rim stop. Cool air then rather suddenly con-verges onto the warm plateau. This rapid horizontal convergence of cool air into the Valle de Mexico leads to rising motions that quickly cool the atmo-sphere above the Valle de Mexico and the plateau. This process tends to even out the temperature differ-ence of air over the plateau and the cooler air around it. The equilibration of temperature is assisted by the dissipation of gravity waves that propagate away from the domed entrainment zone sitting on top of the day-time convective boundary layer over the plateau. It produces horizontal isentropes (lines of equal entropy) by sunrise, when the temperature and stability of the air over the plateau are little different from those of air at the same elevation surrounding the plateau.

Topography and climate: global connections

Mountain-building and global climate

Planetary airflow modification

On a planetary scale, mountains act as barriers to airflow by transferring angular momentum to the surface (through friction and from drag), by blocking and deflecting airflow, and by modifying energy fluxes (notably through changes in cloud cover and precipitation resulting from the modified airflow). Planetary waves are germane to the formation and movement of pressure systems. Mid-latitude depressions tend to form beneath the eastern limb of an upper-wave trough, as over the eastern seaboard of North America and off the east coast of Asia. The effect of mountains on atmospheric circulation systems and global climates depends partly on latitudinal extent of the barrier. Mountain ranges that run north to south are far more effective as modifiers of global airflow than mountain ranges that run east to west. Thus the Rocky Mountains, the Andes, and the Himalayas (because of their large latitudinal extent) are effective modifiers, while the Southern Alps of New Zealand and the European Alps are not.

Computer experiments using the Community Climate Model revealed the large effect that mountain ranges – particularly those with a north–south trend – exert upon on atmospheric circulation patterns and world climates (Kutzbach *et al.* 1989). The experiments explored the climatic effects of the Alpine mountain-building episode. The Alpine orogeny has seen the Tibetan Plateau rise by up to 4 km over the last 40 million years, and at least 2 km in the last 10 million years. It has seen two-thirds of the uplift in the Sierra Nevada, North America, occur during the past 10 million years. It has witnessed comparable changes in other mountainous areas of the North American west, in the Bolivian Andes, and in the New Zealand Alps (Ruddiman *et al.* 1989). The experiments were conducted against this backdrop of substantial orogenic change.

Three different 'orographies' were investigated:

1. A no-mountain case, in which present-day mountains were sliced off at 400 m.
2. A half-mountain case, in which the added uplift was half that known to have occurred.
3. A mountain case, representing present conditions.

The results were given for January and July in the Northern Hemisphere. With no mountains, average wind velocity was uniform and symmetrical and planetary waves tended to be anchored at the boundaries of continents and oceans. On the contrary, with mountains, the average wind velocity was non-uniform and asymmetrical, and planetary waves anchored over high plateaux. On the basis of the simulations, several features of the present climate appeared attributable to the effect of mountains (Ruddiman and Kutzbach 1989). For example, the Mediterranean climate, with its dry and hot summers on the western seaboard of the USA, is caused by the conversion of the westerly

winds that would flow over the region if the Rockies were not there, into northerly winds blowing southwards from British Columbia to the Mexican border and associated with a deepened low-pressure cell over the Colorado Plateau. On the other side of the Rockies, both seasons are presently wetter than earlier in the Cenozoic era. Winter is wetter because the jet stream is forced south and winds that were formerly westerly are north-westerly. Summer is wetter because monsoon flows are created by the Colorado low-pressure centre. Some of the simulated climatic changes suggested that orogeny may affect global climates. European winters are now colder than once they were because the Icelandic low has been displaced westwards. Mediterranean summers are now drier than once they were owing to the development of cyclonic flow around the Tibetan Plateau and the development of a high-pressure cell over the subtropical Atlantic.

Mountains and the global impact of ENSO events

On a much larger scale, the effects of El Niño–Southern Oscillation (ENSO) events occurring in the south Pacific Ocean may be transmitted globally through the agency of mountain chains. If they are so transmitted, then here is a case of topography affecting global climates. ENSO events are driven by short-term instabilities in the atmosphere–ocean system. Every few years, exceptionally warm waters appear off Ecuador and Peru and extend far westwards in equatorial regions. This is an El Niño event, and is associated with an air-pressure drop over much of the south-eastern Pacific Ocean, and an air-pressure rise over Indonesia and northern Australia. When cold water returns to the western seaboard of South America – a La Niña event – the pressure gradient reverses over the Pacific and Indian Oceans. This flip-flopping of pressure is called the Southern Oscillation. The combined oceanic and atmospheric changes are ENSO events. ENSO events affect climates worldwide in part by causing wavelike disturbances of airflow in the upper troposphere that extend into mid-latitudes. Mid-latitude mountains, in particular the Himalayan–Tibetan region, tend to amplify these wave disturbances and transmit a tropical region disturbance to the rest of the world (DeWeaver and Nigam 1995), causing major disruption to global weather patterns. The 1982–1983 El Niño event caused heavy rains and disastrous flooding in Ecuador, Peru, central Pacific islands, and the westernmost United States; and droughts in interior Australia,

Indonesia, the Philippines, southern Africa, Central America, and northern India.

Climate and the arrangement of land and sea

The distribution of continents and oceans exerts a powerful influence on climates. Land and sea have different thermal properties and this leads to differences in the continentality and oceanicity of places. It also drives regional circulations of air, as in the monsoons.

Monsoons, which are large-scale reversals of wind regimes, are partly created by the differential heating of land and sea. The Asiatic monsoon, which affects a sizeable portion of Asia, is perhaps the best known. The classic explanation of monsoons envisaged summer winds, in the manner of a gigantic sea breeze, being 'pulled' into a low pressure cell sitting over a continent and formed by intense heating. This explanation, though attractively simple, has been superseded by more complicated explanations that recognise the involvement of interacting global and regional factors acting through the full depth of the troposphere in the formation of most monsoon systems.

The arrangement of continents and oceans affects global climates in an even profounder way. Charles Lyell, in the first volume of his *Principles of Geology* (1830, 104–24), was the first to fully explore the notion that changes in palaeogeography would have a radical impact on climates. The theme was taken up by Henry Hennessy (1859, 1860) who fancied that changes in the distribution of land masses had exercised a basic control over global climate through geological time. In essence, he showed that 'if no great continents existed, but a great number of islands without any remarkable preponderance of land towards the tropical or the polar regions, the mean temperature of the earth would be increased, and the distribution of heat over its surface rendered far more uniform' (Hennessy 1860, 385–6). The role of palaeogeography in climatic change was occasionally raised during the first half of the twentieth century, but it was not until the 1970s, following the advent of the plate-tectonic theory and the first modern attempts to reconstruct the geography of the geological past, that a string of papers was published in which climatic changes, and especially the contrasts between 'greenhouse' and 'icehouse' states, were ascribed to changes in the palaeogeography (e.g. Crowell and Frakes 1970; Frakes and Kemp 1972; Tarling 1978). Each of these papers explored one of a few highly

plausible ways in which geography might affect climates (Barron 1989a). One argument is that, as land and sea have different reflective and thermal properties, their placement and the relative areas of continents and oceans will modulate climates. Moreover, each arrangement of land masses would have produced atmospheric and oceanic circulations that were different from present circulations, and would have an altered poleward transport of heat. Another possibility is that land provides a surface on which snow, with a high albedo, may accumulate and so the more land there is around the poles, the greater the chances of a glaciation occurring. The often subtle effects of geography on climates are only just beginning to be appreciated (Barron 1989b; Useinova 1989; Wright 1990; Moore *et al.* 1994; Fawcett and Barron 1998).

References

Aebischer, U. and Schar, C. (1998) Low-level potential vorticity and cyclogenesis to the lee of the Alps. *Journal of the Atmospheric Sciences* **55**, 186–207.

Alpers, W., Pahl, U., and Gross, G. (1998) Katabatic wind fields in coastal areas studied by ERS-1 synthetic aperture radar imagery and numerical modeling. *Journal of Geophysical Research C: Oceans* **103**, 7875–86.

Alpert, P., Tsidulko, M., Krichak, S., and Stein, U. (1996a) A multi-stage evolution of an ALPEX cyclone. *Tellus* **48A**, 209–20.

Alpert, P., Krichak, S. O., Krishnamurti, T. N., Stein, U., and Tsidulko, M. (1996b) The relative roles of lateral boundaries, initial conditions, and topography in mesoscale simulations of lee cyclogenesis. *Journal of Applied Meteorology* **35**, 1091–9.

Aniello, C., Morgan, K., Busbey, A., and Newland, L. (1995) Mapping micro-urban heat islands using Landsat TM and a GIS. *Computers and Geosciences* **21**, 965–9.

Barron, E. J. (1989a) Studies of Cretaceous climate. In A. Berger, R. E. Dickinson, and J. W. Kidson (eds) *Understanding Climatic Change*, pp. 149–57. American Geophysical Union, Washington DC, Geophysical Monograph 52 (International Union of Geodesy and Geophysics, Vol. 7). Washington DC: American Geophysical Union.

Barron, E. J. (1989b) Climate variations and the Appalachians from the Late Paleozoic to the present: results from model simulations. *Geomorphology* **2**, 99–118.

Barros, A. P. and Kuligowski, R. J. (1998) Orographic effects during a severe wintertime rainstorm in the Appalachian Mountains. *Monthly Weather Review* **126**, 2648–72.

Barros, A. P. and Lettenmaier, D. P. (1993) Dynamic modeling of the spatial distribution of precipitation in remote mountainous areas. *Monthly Weather Review* **121**, 1195–214.

Barros, A. P. and Lettenmaier, D. P. (1994) Dynamic modeling of orographically induced precipitation. *Reviews of Geophysics* **32**, 265–84.

Barry, R. G. (1981) *Mountain Weather and Climate*, 1st edn. New York: Methuen.

Barry, R. G. (1992) *Mountain Weather and Climate*, 2nd edn. London and New York: Routledge.

Barry, R. G. and Chorley, R. J. (1968) *Atmosphere, Weather and Climate*, 1st edn. London: Methuen.

Beran, D. W. (1967) Large amplitude lee waves and chinook winds. *Journal of Applied Meteorology* **6**, 865–77.

Bird, P. R. and Prinsley, R. T. (1998) Tree windbreaks and shelter benefits to pasture in temperate grazing systems. (Special issue. Windbreaks in Support of Agricultural Production in Australia: Review Papers from the Australian National Windbreaks Program.) *Agroforestry Systems* **41**, 35–54.

Black, R. F. (1954) Precipitation at Barrow, Alaska, greater than recorded. *Transactions of the American Geophysical Union* **35**, 203–6.

Bland, W. L. and Clayton, M. K. (1994) Spatial structure of solar radiation in Wisconsin. *Agricultural and Forest Meteorology* **69**, 75–84.

Błażejczyk, K. and Grzybowski, J. (1993) Climatic significance of small aquatic surfaces and characteristics of the local climate of Suwałki Landscape Park (north-east Poland). *Ekologia Polska* **41**, 105–21.

Bleasdale, A. and Chan, Y. K. (1972) Orographic influences on the distribution of precipitation. In *Distribution of Precipitation in Mountainous Areas*, Vol. 2, pp. 322–33. World Meteorological Organization Publication No. 326. Geneva: World Meteorological Organization.

Bogdan, O. (1988) Un modèle concepuel du topoclimat. *Revue Roumaine de Géologie, Géophysique et Géographie: Géographie* **32**, 13–19.

Bonacina, L. C. W. (1945) Orographic rainfall and its place in the hydrology of the globe. *Quarterly Journal of the Royal Meteorological Society* **71**, 41–55.

Bossert, J. E. and Cotton, W. R. (1994a) Regional-scale flows in mountainous terrain. Part I: a numerical and observational comparison. *Monthly Weather Review* **122**, 1449–71.

Bossert, J. E. and Cotton, W. R. (1994b) Regional-scale flows in mountainous terrain. Part II: simplified numerical experiments. *Monthly Weather Review* **122**, 1472–89.

Bossert, J. E., Sheaffer, J. D., and Reiter, E. R. (1989) Aspects of regional-scale flows in mountainous terrain. *Journal of Applied Meteorology* **28**, 590–601.

Brown, R. (1889) *Our Earth and Its Story: A Popular Treatise on Physical Geography*, Vol. III. London, Paris, New York and Melbourne: Cassell.

Browning, K. A. and Hill, F. F. (1981) Orographic rain. *Weather* **36**, 326–9.

Caborn, J. M. (1955) The influence of shelter-belts on microclimate. *Quarterly Journal of the Royal Meteorological Society* **81**, 112–15.

Caborn, J. M. (1957) *Shelterbelts and Microclimate*. Forestry Commission Bulletin 29. London: Her Majesty's Stationery Office.

Caborn, J. M. (1965) *Shelterbelts and Windbreaks*. London: Faber & Faber.

Caccia, J. L. (1998) Lee wave vertical structure monitoring using height-time analysis of VHF ST radar vertical velocity data. *Journal of Applied Meteorology* **37**, 530–43.

Caccia J. L., Benech, B., and Klaus, V. (1997) Space–time description of nonstationary trapped lee waves using ST radars, aircraft, and constant volume balloons during the PYREX experiment. *Journal of the Atmospheric Sciences* **54**, 1821–33.

Chua, S. H. and Bras, R. L. (1982) Optimal estimators of mean areal precipitation in regions of orographic influence. *Journal of Hydrology* **57**, 23–48.

Chuan, G. K. and Lockwood, J. G. (1974) An assessment of topographical controls on the distribution of rainfall in the Central Pennines. *Meteorological Magazine* **103**, 275–87.

Cooter, E. J. and Dhakhwa, G. B. (1995) A solar radiation model for use in biological applications in the south and southeastern USA. *Agricultural and Forest Meteorology* **78**, 31–51.

Corby, G. A. (1954) The airflow over mountain: a review of the state of current knowledge. *Quarterly Journal of the Royal Meteorological Society* **80**, 491–521.

Corradini, C. and Melone, F. (1989) Spatial structure of rainfall in mid-latitude cold frontal systems. *Journal of Hydrology* **105**, 297–316.

Court, A. and Bare, M. T. (1984) Basin precipitation estimates by Bethlahmy's two axis method. *Journal of Hydrology* **68**, 149–58.

Creutin, J. D. and Obled, C. (1982) Objective analyses and mapping techniques for rainfall fields: an objective comparison. *Water Resources Research* **18**, 413–31.

Crowe, P. R. (1962) Climate. In C. F. Carter (ed.) *Manchester and Its Region*, pp. 17–46. (British Association Scientific Survey.) Manchester: Manchester University Press.

Crowe, P. R. (1971) *Concepts in Climatology*. New York: St Martin's Press.

Crowell, J. C. and Frakes, L. A. (1970) Phanerozoic glaciation and the causes of ice ages. *American Journal of Science* **268**, 193–224.

Daly, C., Neilson, R. P., and Phillips, D. L. (1994) A statistical topographic model for mapping climatological precipitation over mountainous terrain. *Journal of Applied Meteorology* **33**, 140–58.

Dawson, G. M. (1886) Chinook winds. *Science: An Illustrated Journal* **7**, 33–4.

Defant, F. (1951) Local winds. In T. F. Malone (ed.) *Compendium of Meteorology*, pp. 655–72. Boston, MA: American Meteorological Society.

DeWeaver, E. and Nigam, S. (1995) Influence of mountain ranges on the mid-latitude atmospheric response to El Niño events. *Nature* **378**, 706–8.

Dingman, S. L. (1994) *Fluvial Hydrology*. New York: W. H. Freeman.

Dingman, S. L. and Johnson, A. R. (1971) Pollution potential of New Hampshire Lakes. *Water Resources Research* **7**, 1208–15.

Dingman, S. L., Seely-Reynolds, D. M., and Reynolds, R. C. (1988) Application of kriging to estimating mean annual precipitation in a region of orographic influence. *Water Resources Bulletin* 24, 329–39.

Dubayah, R. (1992) Estimating net solar radiation using Landsat Thematic Mapper and digital elevation data. *Water Resources Research* 28, 2469–84.

Dubayah, R. and Rich, P. M. (1995) Topographic solar radiation models for GIS. *International Journal of Geographical Information Systems* 9, 405–19.

Dubayah, R. and Van Katwijk, V. (1992) The topographic distribution of annual incoming solar radiation in the Rio Grande River basin. *Geophysical Research Letters* 19, 2231–4.

Dubayah, R., Dozier, J., and Davis, F. W. (1989) The distribution of clear-sky radiation estimates with a topographic solar radiation model during FIFE. *Proceedings of the ASPRS–ACSM Annual Convention*, Vol. 3, pp. 44–53. Bethesda, MD: American Society for Photogrammetry and Remote Sensing.

Duguay, C. R. (1993) Radiation modelling in mountainous terrain: review and status. *Mountain Research and Development* 13, 339–57.

Edgerton, S. A., *et al.* (1999) Particulate air pollution in Mexico City: a collaborative research project. *Journal of Air and Waste Management Association* 49, 1221–9.

Elliott, A. (1964) Sea breezes at Porton Down. *Weather* 19, 147–50.

Ellis, A. W. and Leathers, D. J. (1996) A synoptic climatological approach to the analysis of lake-effect snowfall: potential forecasting applications. *Weather and Forecasting* 11, 216–29.

Fast, J. D. and Zhong, S. (1998) Meteorological factors associated with inhomogeneous ozone concentrations within the Mexico City basin. *Journal of Geophysical Research* 103, 18 927–46.

Fawcett, P. J. and Barron, E. J. (1998) The role of geography and atmospheric CO_2 in long term climate change: results from model simulations for the Late Permian to the Present. In T. J. Crowley and K. C. Burke (eds) *Tectonic Boundary Conditions for Climate Reconstructions*, pp. 21–36. New York: Oxford University Press.

Ferrel, W. (1889) A *Popular Treatise on The Winds: Comprising the General Motions of the Atmosphere, Monsoons, Cyclones, Tornadoes, Waterspouts, Hail-storms, etc. etc.* London: Macmillan.

Folland, C. K. (1988) Numerical models of the raingauge exposure problem, field experiments and an improved collector design. *Quarterly Journal of the Royal Meteorological Society* 114, 1485–516.

Förchtgott, J. (1949) Vlnové proudění v závětří horských hřebenů. *Meteorologické Zprávy* 3, 49–51. (Translated as Wave streaming in the lee of mountain ridges, *Meteorological Magazine* 94, 1965, p. 11.)

Förchtgott, J. (1969) Evidence for mountain-sized lee eddies. *Weather* 24, 255–60.

Fourcade, H. G. (1942) Some notes on the effects of the incidence of rain on the distribution of rainfall over the surface of unlevel ground. *Transactions of the Royal Society of South Africa* 29, 235–54.

Frakes, L. A. and Kemp, E. (1972) Influence of continental positions on Early Tertiary climate. *Nature* 240, 97–100.

Frei, C. and Schär, C. (1998) A precipitation climatology of the Alps from high-resolution rain-gauge observations. *International Journal of Climatology* 18, 873–900.

Gallée, H. and Pettré, P. (1998) Dynamical constraints on katabatic wind cessation in Adélie Land, Antarctica. *Journal of the Atmospheric Sciences* 55, 1755–70.

Garen, D. C., Johnson, G. L., and Hanson, C. L. (1994) Mean areal precipitation for daily hydrologic modeling in mountainous regions. *Water Resources Bulletin* 30, 481–91.

Garnett, A. (1937) *Insolation and Relief: Their Bearing on the Human Geography of Alpine Regions*. The Institute of British Geographers Publication No. 5. London: George Philip & Son.

Garnett, A. (1939) Diffused light and sunlight in relation to relief and settlement in high latitudes. *Scottish Geographical Magazine* 55, 271–84.

Geiger, R. (1965) *The Climate Near the Ground*. Translated by Scripta Technica, Inc., from the fourth German edition of *Das Klima der bodennahen Luftschicht* (1961), Brunswick: Friedrich Vieweg & Sohn. Cambridge, MA: Harvard University Press.

George, J. J. (1960) *Weather Forecasting for Aeronautics*. New York: Academic Press.

Goldreich, Y. (1995) Urban climate studies in Israel – a review. *Atmospheric Environment* 29, 467–78.

Graham, E. (1993) The urban heat island of Dublin City during the summer months. *Irish Geography* 26, 45–57.

Gregory, S. (1982) Review of White and Smith 1982. *Weather* 38, 284.

Hahn, J. von (1906) *Lehrbuch der Meteorologie*. Leipzig: C. H. Tauchnitz.

Hanson, C. L. (1982) Distribution and stochastic generation of annual and monthly precipitation on a mountainous watershed in southwest Idaho. *Water Resources Bulletin* 18, 875–83.

Harrison, S. J. (1993) Differences in the duration of snow cover on Scottish ski-slopes between mild and cold winters. *Scottish Geographical Magazine* 109, 37–44.

Harrowfield, D. L. (1996) The role of the wind in the destruction of an historic hut at Cape Adare in Antarctica. *Polar Record* 32, 3–18.

Hawke, E. L. (1944) Thermal characteristics of a Hertfordshire frost hollow. *Quarterly Journal of the Royal Meteorological Society* 70, 23–48.

Helvey, J. D. and Patric, J. H. (1983) Sampling accuracy of pit vs. standard rain gauges on the Fernow Experimental Forest. *Water Resources Bulletin* 19, 87–9.

Hennessy, H. (1859) Terrestrial climate as influenced by the distribution of land and water at different geological epochs. *American Journal of Science*, Second Series, 27, 316–28.

Hennessy, H. (1860) Change of climate. *The Athenaeum* 1717, 384–6.

Hetrick, W. A., Rich, P. M., Barnes, F. J., and Weiss, S. B. (1993a) GIS-based solar radiation flux models. *Proceedings of the ASPRS–ACSM Annual Convention*, Vol. 3, pp. 132–43. Bethesda, MD: American Society for Photogrammetry and Remote Sensing.

Hetrick, W. A., Rich, P. M ., and Weiss, S. B. (1993b) Modelling insolation on complex surfaces. *Proceedings of the Thirteenth Annual User Conference*, Vol. 2, pp. 447–58. Redlands, CA: ESRI Press.

Hevesi, J. A., Istok, J. D., and Flint, A. L. (1992a) Precipitation estimation in mountainous terrain using multivariate geostatistics. Part I: Structural analysis. *Journal of Applied Meteorology* **31**, 661–76.

Hevesi, J. A., Flint, A. L., and Istok, J. D. (1992b) Precipitation estimation in mountainous terrain using multivariate geostatistics. Part II: Isohyetal maps. *Journal of Applied Meteorology* **31**, 677–88.

Hogan, A. W. and Ferrick, M. G. (1998) Observations in nonurban heat islands. *Journal of Applied Meteorology* **37**, 232–9.

Hudson, G. and Wackernagel, H. (1994) Mapping temperature using kriging with external drift: theory and an example from Scotland. *International Journal of Climatology* **14**, 77–91.

Hutchinson, M. F. (1995) Interpolating mean rainfall using thin plate smoothing splines. *International Journal of Geographical Information Systems* **9**, 385–403.

Hutchinson, M. F. and Gessler, P. E. (1994) Splines – more than just a smooth interpolator. *Geoderma* **62**, 45–67.

Hutchinson, M. F., Booth, T. H., McMahon, J. P., and Nix, H. A. (1982) Estimating monthly mean values of daily total solar radiation for Australia. *Solar Energy* **32**, 277–90.

Hutchinson, M. F., Nix, H. A., McMahon, J. P., and Ord, K. D. (1996) The development of a topographic and climate database for Africa. In *Third International Conference/Workshop on Integrating GIS and Environmental Modeling*, unpaginated. University of California, Santa Barbara: National Center for Geographical Information Analysis.

Ishida, T. and Kawashima, S. (1993) Use of cokriging to estimate surface air temperature from elevation. *Theoretical and Applied Climatology* **47**, 147–57.

Journel, A. G. and Huijbregts, C. J. (1978) *Mining Geostatistics*. New York: Academic Press.

Kim, H. H. (1992) Urban heat island. *International Journal of Remote Sensing* **13**, 2319–36.

Kira, T. (1976) *Terrestrial Ecosystems*. Tokyo: Kyoritsu.

Kitayama, K. (1992) An altitudinal transect study of the vegetation on Mount Kinabalu, Borneo. *Vegetatio* **102**, 149–71.

Koch, H. G. (1934) *Temperaturverhältnisse und Windsystem eines geschlossen Waldgebietes*. Veröffentlichungen des Geophysikalischen Instituts der Universität Leipzig, 3, No. 3.

Kodama, K. and Barnes, G. M. (1997) Heavy rain events over the south-facing slopes of Hawaii: attendant conditions. *Weather and Forecasting* **12**, 347–67.

Kondratyev, K. Ya. (1965) *Radiative Heat Exchange in the Atmosphere*. New York: Pergamon Press.

Kondratyev, K. Ya. (1969) *Radiation in the Atmosphere*. New York: Academic Press.

Köppen, W. (1936) *Das geographische System der Klimate*. Handbuch der Klimatologie, Band I, Teil C. Berlin: Gebrüder Borntraeger.

Krige, D. G. (1966) Two-dimensional weighted moving average trend surfaces·for ore-evaluation. *Journal of the South African Institute of Mining and Metallurgy* **66**, 13–38.

Kuligowski, R. J. and Barros, A. P. (1999) High-resolution short-term quantitative precipitation forecasting in mountainous regions using a nested model. *Journal of Geophysical Research* **104**, 31 533–64.

Kumar, L., Skidmore, A. K., and Knowles, E. (1997) Modelling topographic variation in solar radiation in a GIS environment. *International Journal of Geographical Information Science* **11**, 475–97.

Kutzbach, J. E., Guetter, P. J., Ruddiman, W. F., and Prell, W. L. (1989) Sensitivity of climate to Late Cenozoic uplift in southern Asia and the American west: numerical experiments. *Journal of Geophysical Research* **94D**, 18 393–407.

LaBaugh, J. W. (1985) Uncertainty in phosphorus retention, Williams Fork reservoir, Colorado. *Water Resources Research* **21**, 1684–92.

Lane, T. P., Reeder, M. J., Morton, B. R., and Clark, T. L. (2000) Observations and numerical modelling of mountain waves over the Southern Alps of New Zealand. *Quarterly Journal of the Royal Meteorological Society* **126**, 2765–88.

Larson, L. W. and Peck, E. L. (1974) Accuracy of precipitation measurements for hydrologic modeling. *Water Resources Research* **10**, 857–63.

Lebel, T., Bastin, G., Obled, C., and Creutin, J. D. (1987) On the accuracy of areal rainfall estimation: a case study. *Water Resources Research* **23**, 2123–34.

Lennon, J. J. and Turner, J. R. G. (1995) Predicting the spatial distribution of climate: temperature in Great Britain. *Journal of Animal Ecology* **64**, 370–92.

Leuschner, C. (1998) Vegetation an der Waldgrenze auf tropischen und subtropischen Inseln. *Geographische Rundschau* **50**, 690–7.

Lindesay, J. A. and Tyson, P. D. (1990) Thermo-topographically induced boundary layer oscillation over the central Namib, southern Africa. *International Journal of Climatology* **10**, 63–77.

Löffler-Mang, M., Kossmann, M., Vögtlin, R., and Fiedler, F. (1997) Valley wind systems and their influence on nocturnal ozone concentrations. *Contributions to Atmospheric Physics (Beiträge zur Physik der Atmosphäre)* **70**, 1–14.

Longhetto, A., Briatore, L., Chabert-d'Hieres, G., Didelle, H., Ferrero, E., and Giraud, C. (1997) Physical modelling of baroclinic development in the lee of the Alps. *Annali di Geofisica* **40**, 1293–302.

Lourens, U. W., Van Sandwyck, C. M., De Jager, J. M., and Van Den Berg, J. (1995) Accuracy of an empirical model for estimating daily irradiance in South Africa from METEOSTAT imagery. *Agricultural and Forest Meteorology* **74**, 75–86.

Lyell, C. (1830) *Principles of Geology, Being an Attempt to Explain the Former Changes of the Earth's Surface, by Reference to Causes Now in Operation*, Volume 1, 1st edn. London: John Murray.

Mandeville, A. N. and Rodda, J. C. (1970) A contribution to the objective assessment of areal rainfall amounts. *Journal of Hydrology (New Zealand)* **9**, 281–91.

Manley, G. (1945) The helm wind of Crossfell, 1937–1939. *Quarterly Journal of the Royal Meteorological Society* **71**, 197–219.

Manley, G. (1952) *Climate and the British Scene* (The New Naturalist, 22). London: Collins.

Matheron, G. (1963) Principles of geostatistics. *Economic Geology* **58**, 1246–66.

McBride, J. L. and McInnes, K. L. (1993) Australian southerly busters. Part II: The dynamical structure of the orographically modified front. *Monthly Weather Review* **121**, 1921–35.

McInnes, K. L. (1993) Australian southerly busters. Part III: The physical mechanism and synoptic conditions contributing to development. *Monthly Weather Review* **121**, 3261–81.

McInnes, K. L. and McBride, J. L. (1993) Australian southerly busters. Part I: Analysis of a numerically simulated case study. *Monthly Weather Review* **121**, 1904–20.

McKendry, I. G. (1993) Mesoclimatology: present themes and future prospects. *New Zealand Geographer* **49**, 56–63.

McKenney, D. W., Mackey, B. G., and Zavitz, B. L. (1999) Calibration and sensitivity analysis of a spatially-distributed solar radiation model. *International Journal of Geographical Information Science* **13**, 49–65.

McNaughton, K. G. (1988) Effects of windbreaks on turbulent transport and microclimate. *Agriculture, Ecosystems and Environment* **22/23**, 17–39.

Molnau, M., Rawls, W. J., Curtis, D. L., and Warnick, C. C. (1980) Gage density and location for estimating mean annual precipitation in mountainous areas. *Water Resources Bulletin* **16**, 428–32.

Moore, G. T., Barron, E. J., and Hayashida, D. N. (1994). Kimmeridgian (Late Jurassic) general lithostratigraphy and source rock quality for the western Tethys Sea inferred from paleoclimate results using a general circulation model. In A. Y. Huc (ed.) *Paleogeography, Paleoclimate and Source Rocks*, pp. 157–72. American Association of Petroleum Geologists, Studies in Geology 40. Tulsa, OK: The American Association of Petroleum Geologists.

Moore, I. D. (1992) *SRAD: Direct, Diffuse, Reflected Shortwave Radiation, and the Effects of Topographic Shading* (Terrain Analysis Programs for the Environmental Sciences (TAPES) Radiation Program Documentation). Canberra: Centre for Resource and Environmental Studies, Australian National University.

Moreno-Garcia, M. C. (1994) Intensity and form of the urban heat island in Barcelona. *International Journal of Climatology* **14**, 705–10.

Mozer, J. B. and Zehnder, J. A. (1996) Lee vorticity production by large-scale tropical mountain ranges. Part I: Eastern North Pacific tropical cyclogenesis. *Journal of the Atmospheric Sciences* **53**, 521–38.

Nance, L. B. and Durran, D. R. (1998) A modeling study of nonstationary trapped mountain lee waves. Part II: Nonlinearity. *Journal of the Atmospheric Sciences* **55**, 1429–45.

Neff, E. (1977) How much rain does a rain gauge gauge? *Journal of Hydrology* **35**, 213–20.

Nuberg, I. K. and Prinsley, R. T. (1998) Effect of shelter on temperate crops: a review to define research for Australian conditions. (Special issue. Windbreaks in Support of Agricultural Production in Australia: Review Papers from the Australian National Windbreaks Program.) *Agroforestry Systems* **41**, 3–34.

Obrębska-Starklowa, B. (1995) *Differentiation of topoclimatic conditions in a Carpathian Foreland valley based on multiannual observations* (Zeszyty Naukowe Uniwersytetu Jagiellońskiego, Prace Geograficzne, Zeszyt 101). Kraków, Poland: Uniwersytet Jagielloński.

Oke, T. R. (1978) *Boundary Layer Climates*, 1st edn. London: Methuen.

Oke, T. R. (1982a) *Boundary Layer Climates*, 2nd edn. London and New York: Methuen.

Oke, T. R. (1982b) The energetic basis of the urban heat island. *Quarterly Journal of the Royal Meteorological Society* **108**, 1–24.

Oke, T. R., Johnson, G. T., Steyn, D. G., and Watson, I. D. (1991) Simulation of surface urban heat islands under 'ideal' conditions at night. Part 2: Diagnosis of causation. *Boundary Layer Meteorology* **56**, 339–58.

Ono, H. S. P., Yasunari, T., Oki, R., and Oda, T. (1994) Detection of the urban climatic component based on the seasonal variations of surface air temperature anomaly. *Geographical Review of Japan, Series A*, **67**, 561–74.

Price, J. C. (1979) Assessment of the urban heat island effect through the use of satellite data. *Monthly Weather Review* **107**, 1554–7.

Ralph, F. M., Neiman, P. J., Keller, T. L., Levinson, D., and Fedor, L. (1997) Observations, simulations, and analysis of nonstationary trapped lee waves. *Journal of the Atmospheric Sciences* **54**, 1308–33.

Raymond, W. H., Rabin, R. M., and Wade, G. S. (1994) Evidence of an agricultural heat island in the Lower Mississippi River floodplain. *Bulletin of the American Meteorological Society* **75**, 1019–25.

Reeder, M. J., Adams, N., and Lane, T. P. (1999) Radiosonde observations of partially trapped lee waves over Tasmania, Australia. *Journal of Geophysical Research D: Atmospheres* **104**, 16 719–27.

Rich, P. M., Hughes, G. S., and Barnes, F. J. (1993) Using GIS to reconstruct canopy architecture and model ecological processes in pinyon–juniper woodlands. *Proceedings of the Thirteenth Annual ESRI User Conference*, Vol. 2, pp. 435–45. Redlands, CA: ESRI Press.

Rodda, J. C. (1962) An objective method for the assessment of areal rainfall amounts. *Weather* **17**, 54–9.

Rorison, I. H., Sutton, F., and Hunt, R. (1986) Local climate, topography and plant growth in Lathkill Dale NNR. I. A twelve-year summary of solar radiation and temperature. *Plant, Cell, and Environment* **9**, 49–56.

Ruddiman, W. F. and Kutzbach, J. E. (1989) Forcing of late Cenozoic Northern Hemisphere climate by plateau uplift in southern Asia and the American west. *Journal of Geophysical Research* **94D**, 18 409–27.

Ruddiman, W. F., Prell, W. L., and Raymo, M. E. (1989) Late Cenozoic uplift in southern Asia and the American west: rationale for general circulation modeling experiments. *Journal of Geophysical Research* **94D**, 18 409–27.

Running, S. W., Nemani, R. R., and Hungerford, R. D. (1987) Extrapolation of synoptic meteorlogical data in mountainous terrain and its use for simulating forest evapotranspiration and photosynthesis. *Canadian Journal of Forest Research* **17**, 472–83.

Salter, M. de Carle S. (1921) *The Rainfall of the British Isles.* London: University of London Press.

Saving, S. C., Rich, P. M., Smiley, J. T., and Weiss, S. B. (1993) GIS-based microclimate models for assessment of habitat quality in natural reserves. *Proceedings of the ASPRS–ACSM Annual Convention*, Vol. 3, pp. 319–30. Bethesda, MD: American Society for Photogrammetry and Remote Sensing.

Scaëtta, H. (1935) Les avalanches d'air dans les Alpes et dans les hautes montagnes de l'Afrique centrale. *Ciel et Terre* **51**, 79–80.

Scorer, R. S. (1961) Lee waves in the atmosphere. *Scientific American* **204**, 124–34.

Sevruk, B. (1982) *Methods of Correcting for Systematic Error in Point Precipitation Measurement for Operational Use* (Operational Hydrology Report No. 21, WMO No. 589). Geneva: World Meteorological Organization.

Sevruk, B. (ed.) (1986) *Correction of Precipitation Measurements* (Proceedings of the ETH/IAHS/WMO Workshop on the Correction of Precipitation Measurements, Zürich, 1–3 April 1985; WMO/TD No. 104). Zürich: ETH Geographisches Institut.

Sharon, D. (1980) The distribution of hydrologically effective rainfall incident on sloping ground. *Journal of Hydrology* **46**, 165–88.

Sharon, D., Morin, J., and Moshe, Y. (1988) Micro-topographical variations of rainfall incident on ridges of a cultivated field. *Transactions of the American Society of Agricultural Engineers* **31**, 1715–22.

Shaw, E. M. and Lynn, P. P. (1972) Areal rainfall evaluation using two surface fitting techniques. *Bulletin of the International Association of Hydrological Sciences* **17**, 419–33.

Shearman, R. J. and Salter, P. M. (1975) An objective rainfall interpolation and mapping technique. *Hydrological Sciences Bulletin* **3**, 353–63.

Simpson, R. H. (1952) Evolution of the Kona storm, a subtropical cyclone. *Journal of Meteorology* **9**, 24–35.

Smith, R. B. (1979) The influence of mountains on the atmosphere. *Advances in Geophysics* **21**, 87–233.

Spreen, W. C. (1947) A determination of the effect of topography upon precipitation. *Transactions of the American Geophysical Union* **28**, 285–90.

Stevenson, R. E. (1961) Sea-breezes along the Yorkshire coast in the summer of 1959. *Meteorological Magazine* **90**, 153–62.

Stoutesdijk, P. (1977) High surface temperatures in the winter and their biological significance. *International Journal of Biometeorology* **21**, 325–31.

Stoutesdijk, P. and Barkman, J. J. (1992) *Microclimate, Vegetation and Fauna.* Knivsta, Sweden: Opulus Press.

Sumner, G. (1988) *Precipitation.* Chichester: John Wiley & Sons.

Swift, L. W. (1976) Algorithm for solar radiation on mountain slopes. *Water Resources Research* **12**, 108–12.

Tabios, G. Q. and Salas, J. D. (1985) A comparative analysis of techniques for spatial interpolation of precipitation. *Water Resources Bulletin* **21**, 365–81.

Tabony, R. C. (1985) The variation of surface temperature with altitude. *Meteorological Magazine* **114**, 37–48.

Tarling, D. H. (1978) The geological–geophysical framework of ice ages. In J. Gribbin (ed.) *Climatic Change*, pp. 3–24. Cambridge: Cambridge University Press.

Tyson, P. D. (1968) Nocturnal local winds in a Drakensberg valley. *South African Geographical Journal* **50**, 15–32.

Tyson, P. D. and Preston-Whyte, R. A. (1972) Observations of regional topographically-induced wind systems in Natal. *Journal of Applied Meteorology* **11**, 643–50.

Urfer-Henneberger, C. (1967) Zeitliche Gesetzmässigkeiten des Berg und Talwindes. *Veröffentlichungen Schweiz. Met. Zentralanstalt* **4**, 245–52.

Urfer-Henneberger, C. (1970) Neuere Beobachtungen über die Entwicklung des Schönwetterwindsystems in einem V-förmigen Alpental (Dischma bei Davos). *Archiv für Meteorologie, Geophysik und Bioklimatologie, B: Allgemeine und biologische Klimatologie* **18**, 21–42.

Useinova, I. (1989) The astounding continental factor. *Geographical Magazine* **61**, 24–6.

Van den Broeke, M. R., Van de Wal, R. S. W., and Wild, M. (1997) Representation of Antarctic katabatic winds in a high-resolution GCM and a note on their climate sensitivity. *Journal of Climate* **10**, 3111–30.

Vysoudil, M. (1996a) Bioclimate and air quality assessment in the culture landscape by use of topoclimatic maps. (Proceedings of the 14th International Congress of Biometeorology, 1–8 September 1996, Ljubljana, Slovenia.) *Biometeorology* **14**, Part 2, Vol. 3, 311–16.

Vysoudil, M. (1996b) Adaptation of appropriate GIS techniques for application in climatology by the use of GeoPackage©. *Acta Universitatis Palackianae Olomucensis, Facultas Rerum Naturalium (1996), Geographica* **34**, 41–9.

Vysoudil, M. (1998) Enhanced topoclimatic mapping using satellite images: possibilities and suitability. *Acta Universitatis Palackianae Olomucensis, Facultas Rerum Naturalium (1998), Geographica* **35**, 51–5.

Wagner, A. (1938) Theorie und Beobachtungen der periodischen Gebirgswinde. *Gerlands Beiträge zur Geophysik* **52**, 408–49.

Wallington, C. E. (1960) An introduction to lee waves in the atmosphere. *Weather* **15**, 269–76.

Washington, R. (1996) Mountains and climate. *Geography Review* **9**, 2–7.

Wendler, G., Gilmore, D., and Curtis, J. (1997) On the formation of coastal polynyas in the area of Commonwealth Bay, eastern Antarctica. *Atmospheric Research* **45**, 55–75.

White, E. J. (1979) The prediction and selection of climatological data for ecological purposes in Great Britain. *Journal of Applied Ecology* **16**, 141–60.

White, E. J. and Smith, R. I. (1982) *Climatological Maps of Great Britain.* Midlothian: Institute of Terrestrial Ecology.

Whiteman, C. D., Zhong, S., Bian, X., Fast, J. D., and Doran, J. C. (2000) Boundary layer evolution and regional-scale diurnal circulations over the Mexico Basin and Mexican Plateau. *Journal of Geophysical Research* **105**, 10 081–102.

Wigley, T. M. L., Lough, J. M., and Jones, P. D. (1984) Spatial patterns of precipitation in England and Wales and a revised, homogeneous England and Wales precipitation series. *Journal of Climatology* **4**, 1–25.

Williams, L. D., Barry, R. G., and Andrews, J. T. (1972) Application of computed global radiation for areas of high relief. *Journal of Applied Meteorology* **11**, 526–33.

Wilson, W. T. (1954) Discussion of 'Precipitation at Barrow, Alaska, greater than observed' (by R. F. Black). *Transactions of the American Geophysical Union* **35**, 203–7.

Worthington, R. M. and Thomas, L. (1998) The frequency spectrum of mountain waves. *Quarterly Journal of the Royal Meteorological Society* **124**, 687–703.

Wright, V. P. (1990) Equatorial aridity and climatic oscillations during the Early Carboniferous, southern Britain. *Journal of the Geological Society of London* **147**, 359–63.

Wurtele, M. G., Datta, A., and Sharman, R. D. (1999) Unsteadiness and periodicity in gravity waves and lee waves forced by a fixed rigid boundary. *Journal of the Atmospheric Sciences* **56**, 2269–76.

Yoshino, M. M. (1975) *Climate in a Small Area*. Tokyo: University of Tokyo Press.

CHAPTER 4

Water

Running water is the dominant force in producing changes in the physical landscape with widespread effects. The chemical and physical disintegration of rock, together with mass wasting, supply material that is subsequently removed by rivers to the sea. The work of rivers may be more significant than that of other agents, such as glaciers, wind, and waves, because rivers may operate over very long periods of time. Topography plays a crucial role in the distribution and flux of water within the natural landscape.

The hydrological cycle

The natural circulation of water near the surface of the Earth is known as the hydrological or water cycle (Figure 4.1). The cycle is driven by radiant energy received from the Sun. Put simply, ocean water is heated by solar radiation, causing it to evaporate, that is to change from its liquid to its gaseous state to form part of the atmosphere. Once in the atmosphere, water vapour is transported by winds to the land where, under particular conditions, it changes back to the liquid state through the process of condensation to form clouds and then falls as rain. Of course, some water vapour condenses over the oceans and falls straight back. So, the precipitation will either fall directly back into the oceans or begin a more convoluted route to the ocean via the land surface through the hillslope hydrological system. The hydrological cycle is a closed system as there are no effective gains or losses in this cycle – the atmosphere and lithosphere together have a fixed amount of water.

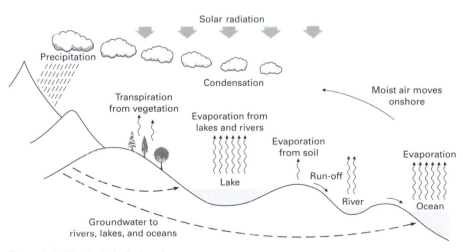

Figure 4.1 The hydrological cycle.

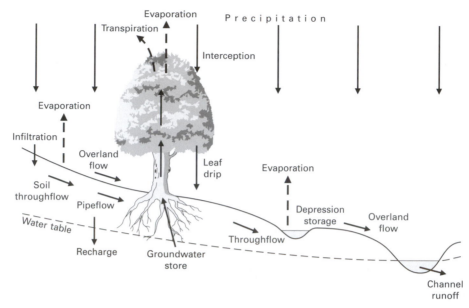

Figure 4.2   A hillslope as a hydrological system.

Hillslope hydrology

The hillslope is part of the hydrological cycle and is an open system. Figure 4.2 shows the main components of the hillslope hydrological system and these can also be represented schematically as a system of inputs, outputs, storages, and transfers (Figure 4.3). It is the storages and transfer elements of this system over which topography exerts a significant sway, although it also influences the inputs and outputs. Precipitation is an input and may fall in the form of snow, which in polar regions or on high mountains may accumulate and consolidate to form ice that may be stored naturally for very long periods. However, in more temperate regions the precipitated water will enter the basin or hillslope hydrological cycle. A detailed discussion of the influence of topography upon precipitation was provided in Chapter 3. The precipitated water infiltrates the soil and is distributed through the system via a number of processes to become output. Evaporation, transpiration, and channel runoff are the main output processes.

Evapotranspiration

Some precipitation that falls may be intercepted by vegetation, the rainwater then runs down the stems and branches (stemflow) and some drips from the leaves (leaf drip) to the ground. Some intercepted water may be evaporated. Evaporation also occurs from the surfaces of soil, rock, lakes, and rivers. In addition, transpiration, which is the process of water uptake by vegetation from the soil and subsequent evaporation through the leaf stomata, takes place. Evaporation (E_o) and transpiration are usually lumped together as evapotranspiration (E_t). Evapotranspiration is then the total loss of water by both evaporation and transpiration from the land surface and its vegetation. The rate of evapotranspiration varies according to the vegetation type, its ability to transpire, and the availability of water in the soil (Shaw 1994). Evapotranspiration is much more difficult to quantify than evaporation because transpiration rates can vary considerably over an area and the source of water from the ground for the plants needs careful definition.

Both evaporation and evapotranspiration are influenced by the general climatic conditions and are therefore indirectly controlled by topography (see Chapter 3). The actual physical process of the change of state from liquid to vapour occurs in both evaporation and evapotranspiration, which means that the general physical conditions influencing evaporative rates are the same for both processes. Shaw (1994) identified these as solar radiation, temperature (of both the air and the evaporating surface), the saturation deficit of the air, wind speed, atmospheric pressure, the nature of the evaporating surface, and the availability of

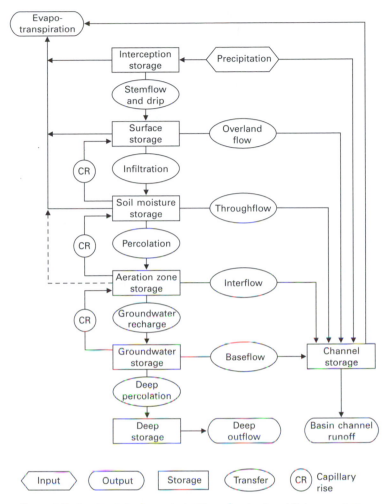

Figure 4.3 Inputs, outputs, and storages of water in a hillslope system.

moisture. While topography is not explicitly stated, it does influence many of the above factors and is therefore an indirect influence on evaporation and evapotranspiration and sometimes appears as a variable in evaporation and evapotranspiration models.

In the UK a significant amount of research has been focused on upland catchments where there has been major land-use change from moorland to coniferous forest. Many upland catchments are an important part of regional water supplies, and an understanding of the implications of such land-use change upon catchment balances is key for planning water resources. To address this issue (as well as a number of other research questions) several research catchments were established across the UK uplands,

including Balquhidder (Scotland), Plynlimon (Wales), Coalburn (northern England), and Stocks (Lancashire).

Work at Balquhidder found that evaporation from upland grassland (measured using weighing lysimeters) at high elevations (595 m) was significantly below that of the Penman evaporation, E_t (Wright and Harding 1993; Hall and Harding 1993). The Penman formula (Penman 1948) allows the calculation of open-water evaporation, E_o, based upon fundamental physical principles, with some empirical concepts incorporated, using standard meteorological observations. Owing to the relative ease of obtaining the input parameters, Penman's formula is probably the most popular approach for calculating evaporation worldwide. Furthermore, many adjustments have been made to the formula to allow the calculation of evapotranspiration of a number of different vegetation types, as in the UK Meteorological Office Rainfall and Evaporation Calculation System (MORECS) (Meteorological Office 1981). The Penman formula is often used as a check for lysimeter measurements or for E_t model estimates.

The discrepancies between the Penman estimates and the lysimeter rates at Plynlimon were considered to be due to the fact that the Penman equation is calibrated against lysimeter observations at Rothamsted, Hertfordshire, England. The vegetation at Rothamsted is lowland grass, which generally has an annual evaporation rate below the Penman E_t value as a result of seasonal soil-water stress. It has been found that the use of E_t with a suitable soil-moisture model can provide good agreement with soil moisture observations (Calder *et al.* 1983). Additionally, Ward (1981) has shown that catchment losses from UK lowlands are equal to, or below, the Penman E_t value. In the UK uplands, however, there are high rainfall totals and a high number of rain days (over 200 per year) that together do not allow a significant soil-water stress to develop (Wright and Harding 1993). A variety of studies in upland catchments has shown evaporative losses from grassland areas that are significantly below E_t calculated from local measurements. For instance, water use of the grassland Wye catchment (Plynlimon) is 77 per cent of E_t (1976–1983), the Coalburn catchment has a water use of 85 per cent of E_t (Robinson 1986), and the losses of the mixed grassland (61 per cent) and forest (39 per cent) Kirkton catchment (Balquhidder) are 73 per cent of E_t (Wright and Harding 1993). Moreover, drainage lysimeter experiments at Plynlimon and Stocks have shown losses of 76 per cent and 81 per cent of E_t, respectively (Calder 1990). Wright and Harding (1993) believe that the reduction from potential in these cases was less likely to be due to soil-water stress than to the reduction of transpiration from upland grassland caused by annual dieback and low temperatures. At upland sites, grass is 'dead' during the winter and early spring, and even when the grass begins to become active, transpiration will be suppressed by low temperatures. When the grass is 'dead' or dormant the calculated E_t value can be substantial, but it may well increase with altitude as a result of higher wind speeds and prolonged sunshine hours at higher elevations (Johnson 1985; Blackie 1987). It can therefore be concluded that the actual evaporation will be below the E_t value in the early part of the year, when the grass is dormant (Wright and Harding 1993).

Both Wright and Harding (1993) and Hall and Harding (1993) developed models which attempted to realistically describe the actual evaporation in terms of commonly measured meteorological variables. In particular, one of Hall and Harding's models had an elevational element. This Altitudinal Zone (AZ) model allowed for the effects of altitude on evaporation and rainfall. In the model, each catchment was divided into three altitudinal zones with classification of the vegetation within each zone. Evaporation from each zone was calculated as the sum of the evaporation from each class of vegetation in that zone, and the catchment evaporation was the sum of the evaporation from the three zones. The input data for the E_t calculations for each altitudinal zone was taken from three weather stations, one located in each altitude zone. The results of the model were compared to the results of another model that separated the catchment only into vegetation zones, not into altitudinal zones. It was found that splitting the catchment into the three altitudinal ranges resulted in an improved annual water use estimate of approximately 7 per cent (44 mm).

Wright and Harding (1993) showed that, over a year, the evaporation from high-altitude grass at Balquhidder is about $0.75E_t$. Hall and Harding (1993) explained that the shortfall arises from low temperatures limiting growth, and therefore transpiration rates, and also probably reducing transpiration rates directly. Because the Penman equation is calibrated for unstressed grass, its use for predicting evaporation from upland grass is reasonable as long as it is adjusted by a time-dependent or temperature-dependent variable (Wright and Harding 1993).

In another study, topography and land-use conditions were used to produce an improved Penman model to estimate actual evapotranspiration (Kotoda 1989). The distribution of topography (altitude, slope, and aspect) and land use were considered. Direct solar radiation, sky-diffuse radiation, and ground-reflected diffuse radiation were combined to calculate the total shortwave radiation on a sloping surface. Kotoda (1989) found that the model provided a fairly reliable estimate of monthly and annual river basin evapotranspiration.

Channel runoff

Where slopes meet at the foot of a valley, water is concentrated and may form a river channel. Some of these channels are permanent due to a reasonably constant supply of water from the slopes above, and some, storm channels, are temporary, fed by water during and immediately after rainfall events. Water-gathering slopes concentrate water very efficiently and permanent streams often have their source at the base of such slopes. Water landing on a slope may take a number of different routes before reaching the stream channel.

When the precipitation lands on the ground it will either run off the surface, lie in surface depressions, or infiltrate through the slope surface. Infiltration capacity is the rate at which a surface can absorb water. Apart from on impermeable rock slopes, some proportion of the precipitation landing on a slope will invariably infiltrate the ground to reach the water table and will contribute to groundwater storage. Some infiltrated water may move laterally down through the slope, more or less parallel to the surface, where the ground becomes saturated. Some infiltrated water may be taken up by vegetation, from which it may be transpired back into the atmosphere. The lateral movement of water occurs due to compaction (that is, the filling of voids by fine particles washed down from above, and, sometimes, the precipitation of iron oxide, silica, calcium carbonate, and other compounds, which reduces the ability of the soil to transmit water at depth). Laterally diverted water is known as throughflow, and sometimes the term interflow is used for water moving laterally below the actual soil but above the water table.

The total amount of precipitation that flows downslope, as opposed to infiltrating, depends on both the intensity and duration of the precipitation as well as the properties of the slope surface. The most important slope property is the infiltration capacity; this is influenced by a number of factors, the chief among which include particle size, the abundance of organic matter, and the amount of faunal activity. These factors all promote the development of an open soil structure capable of absorbing water efficiently from a slope surface. Arid and semi-arid environments are the most likely to have low infiltration capacities, because arid and semi-arid soils have a low organic content. Low infiltration capacities also occur in environments where human activities lead to the artificial removal of vegetation cover, the compaction of the surface, and the construction of concrete and tarmac surfaces. When the intensity of precipitation exceeds the infiltration and storage capacity of the surface, some of the precipitation will flow over the surface as infiltration-excess overland flow, which is sometimes called Hortonian overland flow after the hydrological researcher Robert E. Horton. This process is the basis of the partial-area model of streamflow

generation. Very large volumes of water can be transported to river channels by overland flow – up to 500 m^3/hr.

In catchments that are well vegetated, the soils usually have infiltration capacities far in excess of the rainfall intensity; however, peak discharges can still be observed during storms. In such catchments, peak flows are generated from rain falling directly into stream channels and from saturated overland flow contributed from zones adjacent to channels where the soil has become completely saturated. Areas that contribute saturated overland flow change during a storm as the saturated zones initially expand with the onset of rain, and then contract when the rain stops. This is referred to as the variable-source area model of streamflow generation. The distinction between infiltration-excess and saturated overland flow is that while the former is influenced by soil infiltration capacity, the latter is related to antecedent soil-moisture conditions and location within a catchment. This type of overland flow predominates in humid environments.

In humid environments, the generally thick soils provide another route for quickflow (see below) via lateral movement of water through the soil itself. This lateral movement is most pronounced immediately above the soil horizon with a lower hydraulic conductivity (that is, a lower permeability or capacity to transmit water). In such conditions, a higher proportion of the water percolating through the soil will be moved laterally downslope. Such subsurface flow, or throughflow, is usually much slower than overland flow, and generally reaches a maximum velocity of only 0.4 m/hr. In some soils, much more rapid subsurface flow, at velocities of up to 200 m/hr, may occur through natural pipes created by the rotting of roots or the burrowing of animals. These pipes develop to create well-integrated conduits allowing rapid downslope transmission of water. If there is downslope thinning of permeable soil horizons, subsurface flow may emerge at the surface as return flow towards the base of the slope. While the majority of subsurface flow is too slow to contribute to the main peak flows accompanying storm events, it is important since it primes areas adjacent to the stream channel and enhances the area of saturation overland flow in later storms.

Surface runoff and groundwater flow combine in surface streams and rivers to form channel or streamflow. Streamflow may be temporarily held up in lakes but the portion that escapes evaporation will finally flow into the ocean. Water contributing to streamflow can be separated into two types in terms of the speed at which it enters the stream channel after a rainfall event. In the majority of streams there is a relatively low level of baseflow that is maintained between significant rainfall events; this is supplied by delayed flow from the groundwater or from slow percolation of water through the soil. At the other extreme is water that enters stream channels soon after a storm; this is called quickflow and is generated in a number of ways depending upon the infiltration capacity which, as discussed above, is controlled by the nature of the topography, vegetation, and soils in the basin.

The drainage basin

From the point at which precipitation first hits the ground to the point at which it re-enters the sea is the effective part of the hydrological cycle in producing changes in the land surface. Water always takes the steepest downslope route available. Water falling on either side of a watershed divide will flow in opposite directions. The watershed line extends along the crests of hills to separate the area draining to a river, which flows through the outlet, from areas which contribute their water to other rivers and other outlets. The area enclosed by the watershed and draining through the outlet is known as the catchment or drainage basin. Catchment area is a measure of surface or shallow subsurface runoff at a given point in a landscape (Moore *et al.* 1993a). It integrates the effects of upslope contributing area and catchment convergence and divergence on runoff. Sizes of drainage basins can vary greatly. The Coalburn Catchment in the UK is 1.5 km^2, while one of the world's biggest rivers, the Amazon, drains an area of 5 776 000 km^2.

The catchment is used as the fundamental unit for both hydrology and fluvial geomorphology, since within the drainage divide, boundaries, surface or near-surface flows of water, and associated movements of sediment and solutes are contained. Catchments are used as the natural unit for water resources studies when calculating runoff and budgeting regional water-supply allocations. Many distributed-parameter hydrological models are based upon the subdivision of catchments into subcatchments, each with a single soil and land-use type. These subdivisions are sometimes referred to as hydrological response units. Examples include the Stanford Watershed Model (SWM) (Crawford and Linsley 1966), the Système Hydrologique Européen (SHE) (Bathurst and O'Connell 1992), TOPMODEL (Beven and Wood 1983; Quinn *et al.* 1995), ANSWERS (Beasley *et al.* 1980), the Limburg Soil Erosion Model (LISEM) (De Roo *et al.* 1996), and the Water Erosion Prediction Model (WEPP) (Savabi *et al.* 1996).

The catchment or drainage basin is also used as a basic unit when carrying out analysis of fluvially eroded landscapes. The transfer of materials causes changes in elevation and land-surface form over time, and the outlet of a drainage basin provides a convenient point at which to monitor the movement of the water, sediments, and solutes. Additionally, there are many quantitative geomorphological properties of basins that can be used for comparative analysis of two or more basins. Such quantitative description is known as drainage basin morphometry and can be applied to the areal and relief properties of a basin and to the characteristics of river channel networks.

Catchment runoff

Catchment runoff is the quantity of water that, through the processes discussed earlier, enters stream channels in a drainage basin over a specified period of time. It may be computed using the water-balance equation:

$$\text{runoff} = \text{precipitation} - (\text{evaporation} + \text{transpiration}) \qquad (4.1)$$

This equation is adequate for long-term analysis, but for shorter-term analysis changes in the volume of water stored as groundwater, soil water, and channel water may occur and so the equation should be:

$$\text{runoff} = \text{precipitation} - (\text{evaporation} + \text{transpiration}) \pm \text{changes in storage} \qquad (4.2)$$

Since it is very difficult to measure evapotranspiration and water storage, water balance experiments are usually started and finished at the time of minimum storage. In the UK, this occurs at the end of the summer, when evapotranspiration has been high and rainfall low. However, in limestone terrains, the movement of groundwater may bear little relation to surface patterns of water flow and a complex hydrological system may occur.

Drainage basin morphometry

A long-running theme in geomorphology has been the provision of a description of landform geometry, particularly for landforms shaped by fluvial erosion. One of the most dominant aspects of this landform description has been the search for a basic unit within which these land-form data could be collected, organised, and analysed. There are three chief conceptions of, or approaches to, these basic units. Regional delimitation was the earliest approach to be developed (Fenneman 1914). With this, physiographic regions were identified largely upon structural geology, although other morphometric attributes – relief and degree of dissection – were used. The US Corps of Engineers employed a similar approach using four terrain factors – slope, relief, occurrence of steep slopes greater than 26.5°, and plan profile (which involved the 'peakedness', areal extent, elongation, and orientation of topographic highs) – to divide the gross landscape of a region into component landscapes in a simple taxonomic manner. The second approach attempted to identify the 'physiographic atoms out of which the matter of regions is built' (Wooldridge 1932, 33). The 'atoms' were defined as the facets of 'flats' and 'slopes' forming the intersecting surfaces characteristic of polycyclic landscapes. Later, the definition was relaxed to include segments of smoothly curved surface (Savigear 1965) and to enable the grouping of facets into landscape patterns, for instance a 'mature river valley' (Beckett and Webster 1962). However, the subjective nature of this morphometric division limited its usefulness (Gregory and Brown 1966). The third approach to morphometric division came from the unitary features of both geometry and process exhibited by the erosional drainage basin (Chorley *et al.* 1964). This topographic, hydraulic, and hydrological unity of a basin served as the basis for the morphometric system developed by Horton (1945), which was in turn elaborated by Strahler (1964).

Three basic landscape properties can be recognised: linear properties – stream channel networks (one-dimensional); areal properties – catchment characteristics (two-dimensional); and relief properties (three-dimensional).

Channel network characteristics

A river system can be regarded as a network consisting of a set of links that connect nodes. Stream networks possess two major sets of properties: topological properties, which refer to the interconnections of the system; and geometrical properties, which include measures of length, area, shape, relief, and orientation.

The fundamental unit of a stream network is the stream segment or link. This is a section of stream between two stream junctions, or between a junction and the upstream termination of a channel. Stream order is used to denote the hierarchical relationship between stream segments and also allows the classification of drainage basins according to size. Stream order is a basic property of stream networks because it is related to the relative discharge of a channel segment.

Several stream-ordering systems exist, the most commonly used being those devised by Arthur N. Strahler and by Ronald L. Shreve (Figure 2.10) (p. 29). In Strahler's system, a stream segment with no tributaries and flows from source is denoted as a first-order segment. A second-order segment is created by joining two first-order segments, a third-order segment by joining two second-order segments, and so on. There is no increase in order when a segment of one order is joined by another of a lower order. Strahler's system takes no account of distance and all fourth-order basins are considered as similar. Shreve's ordering system, on the other hand, defines the magnitude of a channel segment as the total number of tributaries that feed it. Stream magnitude is closely related to the proportion of the total basin area contributing runoff, and so provides a good estimate of relative stream discharge for small river systems.

Strahler's stream order has been applied to many river systems and has been statistically proven to be related to a number of drainage-basin morphometry elements. For instance, the mean stream gradients of each order approximate an inverse geometric series, in which the first term is the mean gradient of first-order streams. A commonly used topological property is the bifurcation ratio, that is the ratio between the number of stream segments of one order and the number of the next highest order (cf. p. 30). A mean bifurcation ratio is usually used because the ratio values for different successive basins will vary slightly. With relatively homogeneous lithology, the bifurcation ratio is not normally more than 5 nor less than 3. However, a value of 10 or more is possible in very elongated basins where there are narrow, alternating outcrops of soft and resistant rocks.

The main geometrical properties of stream networks are listed in Table 4.1. The most important of these is probably drainage density, which is the average length of channel per unit area of drainage basin. Drainage density is a measure of how frequently streams occur on the land surface. It reflects a balance between erosive forces and the resistance of the ground surface, and is therefore related closely to climate, lithology, and vegetation. Drainage densities can range from less than 5 km/km^2 when slopes are gentle, rainfall low, and bedrock permeable (e.g. sandstones), to much

Table 4.1   Selected morphometric properties of drainage basins

Property	Symbol	Definition
Network properties		
Drainage density	D	Mean length of stream channels per unit area
Stream frequency	F	Number of stream segments per unit area
Length of overland flow	L_g	Mean upslope distance from channels to watershed
Areal properties		
Texture ratio	T	Number of crenulations in the basin contour having the maximum number of crenulations divided by the basin perimeter length. Usually bears a strong relationship to drainage density
Circulatory ratio	C	Basin area divided by the area of a circle with the same basin perimeter
Elongation ratio	E	Diameter of a circle with the same area as the drainage basin divided by the maximum length of the drainage basin
Lemniscate ratio	k	The square of basin length divided by four times the basin area
Relief properties		
Basin relief	H	Elevational difference between the highest and lowest points in the basin
Relative relief	R_{hp}	Basin relief divided by the basin perimeter
Relief ratio	R_h	Basin relief divided by the maximum basin length
Ruggedness number	N	The product of basin relief and drainage density

larger values of more than 500 km/km^2 in upland areas where rocks are impermeable, slopes are steep, and rainfall totals are high (e.g. on unvegetated clay 'badlands'). Climate is important in basins of very high drainage densities in some semi-arid environments that seem to result from the prevalence of surface runoff and the relative ease with which new channels are created. Vegetation density is influential in determining drainage density, since it binds the surface layer, preventing overland flow from concentrating along definite lines and from eroding small rills, which may develop into stream channels. Vegetation slows the rate of overland flow and effectively stores some of the water for short time periods. Drainage density is also related to the length of overland flow, which is approximately equal to the reciprocal of twice the drainage density.

Early studies of stream networks indicated that fluvial systems with topological properties similar to natural systems could be generated by purely random processes (Shreve 1975). Such definitions of network randomness concern topology alone and are descriptive rather than causative in that they assume nothing regarding network growth processes. Explanation of the random topology model's approximate prediction properties is still unestablished and over the years has generated much debate (Abrahams 1984; Watson 1969; Smart 1978; Kirchner 1993; Costa-Cabral and Burges 1997). Shreve (1975) has proposed that the explanation is that random factors dominate network

development. However, another explanation is that these geomorphic laws are insensitive to topological variability. Kirchner (1993) tested the sensitivity of the Horton ratios to particular types of topological distribution. Costa-Cabral and Burges (1997) took this work further to test the sensitivity of the topological analogue of each law using non-topologically random network test samples. One type of the samples was created with a stochastic variable parameter 'Q model' of network growth in which the probability of tributary development on interior and exterior links is allowed to vary. They demonstrated that the coefficients of all channel network plan-form laws vary with model parameter Q, and suggested that it is possible that these laws are sensitive also to other models of network growth. They concluded that the success of the random model in approximately predicting geomorphic laws may not be due to the insensitivity of these laws (which has been previously suggested) but is a result that remains unexplained. Such random-model thinking has been extremely influential in channel network studies. However, later research has identified numerous regularities in stream network topology (Summerfield 1991). These systematic variations appear to be a result of various factors, including the need for lower-order basins to fit together, the sinuosity of valleys and the migration of valley bends downstream, and the length and steepness of valley sides. These elements are more pronounced in large basins, but they are present in small catchments. The debate over the nature of the network-forming processes and resulting network form looks set to continue.

Areal characteristics

Areal properties express the overall plan-form and dimensions of a drainage basin, whereas relief properties express elevation differences (Table 4.1). The shape of a basin is very important since this will influence the rainfall runoff rate. For instance, very long basins will take longer to produce output of water from a rainstorm. The most efficient basin would be circular with all the water draining out of a hole in the middle. This shape minimises the channel length within the basin. Basin circularity and elongation are two simple measures that have been devised to determine shape.

Basin circularity (circularity ratio), C, compares the area of a basin to the area of a circle with the same circumference. With P as the length of the watershed (catchment perimeter), D as the diameter of a circle equivalent in circumference to P, A_0 as the area of a circle of diameter D, and A as the total area of the basin, the basin circularity is derived as

$$P = \pi D \qquad (4.3a)$$

Rearranging yields

$$D = P/\pi \qquad (4.3b)$$

which leads to

$$A_0 = \pi \left(\frac{D}{2}\right)^2 = \pi \left(\frac{P}{2\pi}\right)^2 = \frac{P^2}{4\pi} \qquad (4.4)$$

and

$$C = \frac{\Sigma A}{A_0} = \frac{\Sigma A (4\pi)}{P^2} \qquad (4.5)$$

Basin elongation (elongation ratio), E, compares the longest dimension of the basin to the diameter of a circle of the same area as the basin. With D_L as the longest dimension of the basin, D_C as the diameter of the circle of the same area as the basin, and $R = D_C/2$, that is, the radius of the circle of the same area as the basin, the elongation ratio is derived as

$$\Sigma A = \pi r^2 \qquad (4.6)$$

therefore

$$r = \sqrt{\frac{\Sigma A}{\pi}} \qquad (4.7)$$

and

$$D_C = 2\sqrt{\frac{\Sigma A}{\pi}} \qquad (4.8)$$

so

$$E = \frac{D_L}{D_C} = \frac{D_L}{2\sqrt{\dfrac{\Sigma A}{\pi}}} \qquad (4.9)$$

Relief characteristics

Relief, or height, differences can be expressed using maximum and minimum elevation values. For instance, the difference between maximum and minimum elevations within a given area is known as the basin or local relief, H. Relief is related to the slope and stream gradients in a basin, and therefore has an influence on the rates of slope processes and sediment transport by streams and rivers. However, if an index such as the relief ratio, R_h, is applied to a large basin, it is likely to conceal significant variations of relief within the basin. A more useful measure of relief can be obtained by averaging the local relief over cells of a given size across an entire basin, and such average relief estimates have been found to be closely correlated with fluvial denudation rates.

Long profiles of streams can be created by measuring the vertical drop from the head of each stream segment to the junction where it joins the higher-order stream and dividing the total by the number of streams of that order. This process yields the average vertical fall. If this value is plotted against the average stream length of the order, the average gradient is obtained. When the average gradients of streams of each order are linked, they create a long profile of the basin (Figure 4.4). In general, these long profiles

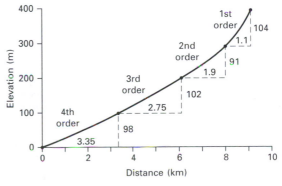

Figure 4.4 Long profile of a river basin.

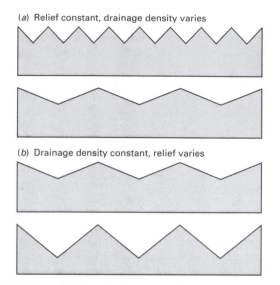

(a) Relief constant, drainage density varies

(b) Drainage density constant, relief varies

Figure 4.5 Relationships between drainage density, mean slope-angle, and relief. (Adapted from Summerfield, 1991, 210.)

illustrate that lower-order tributaries are steeper than those of higher-order tributaries.

A close relationship exists between drainage density, mean slope-angle, and relief (Figure 4.5). When drainage density is constant and stream channels maintain a constant spacing through time, an increase in relief due to stream incision must cause an increase in mean slope-angles in the basin. This effect is limited, however, as a progressive increase in slope angles cannot continue indefinitely. At a certain point, erosion rates will be so high that interfluves will be lowered as rapidly as stream channels and local relief will attain a constant value.

Hypsometric analysis is another approach to the description of drainage basin relief in which elevation is related to basin area (Figure 4.6). It allows the calculation of a hypsometric integral that summarises the drainage basin form in a single value. For most basins, the hypsometric integrals lie between 25 and 75 per cent, high values indicating a relatively large proportion of land lying at high elevation within the

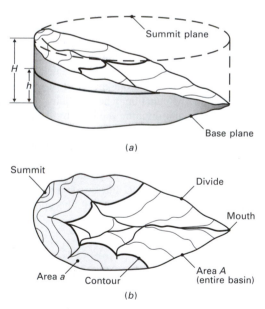

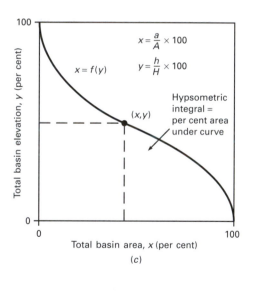

Figure 4.6 The elements of hypsometric analysis. (a) The drainage basin to be analysed. H is basin relief and h is local relief. (b) The calculation of area–elevation data. (c) The hypsometric curve. The procedure is to compute the proportion of total basin elevation for contours at appropriate intervals by dividing the contour elevation by the maximum basin elevation (with respect to the basin outlet). The proportion of the total basin area is produced by dividing the area of the basin above each contour interval by the total basin area. These two ratios, which are usually expressed as percentages, are plotted on the rectangular co-ordinates to define the hypsometric curve for the basin. The hypsometric integral is the area under the hypsometric curve as a percentage of the total area. It is defined by the two graph axes and may be estimated by overlaying graph paper and counting the squares. (Adapted from Strahler, 1957.)

basin and low values denoting a small proportion. So basins with stream channels that have deeply incised the valley, leaving large areas of relatively high elevation, have high hypsometric integrals. Hypsometric integrals have been used as an index of the 'stage of development' of a landscape.

Channel networks

Channel initiation

Channels can be created on a newly exposed surface or can develop by the expansion of an existing channel network. To understand how they are created, the conditions under which water flowing on a slope becomes sufficiently concentrated for channel incision to occur must be considered. It is also important to consider how, once a channel is created, channels are maintained and grow to form permanent features. The initiation and extent of channel networks are controlled by processes occurring at the channel head. These processes are incision by overland flow, seepage and tunnel erosion by subsurface flow, and mass failure or landsliding (Prosser and Abernethy 1996; Dietrich and Dunne 1993). Topography exerts a strong influence on these processes through gradient and by concentrating surface and subsurface flow. A number of studies have demonstrated the role of topography in controlling the generation control of throughflow, overland flow, saturation zones, and possible channel development (Speight 1980; Anderson and Burt 1978;

Kirkby and Chorley 1967; O'Loughlin 1981, 1986). Zones of high water potential, likely to lead to saturation, are associated with concave contours and concave slope profiles, as well as locally thin soils and proximity to a perennial stream. Topography is clearly a strong influence upon the extent and distribution of channel networks (Prosser and Abernethy 1996), but the physical processes driving these are not well understood (Abrahams 1984).

The actual shape of the land surface is important with both surface and subsurface flows converging in areas of contour concavity (Figure 4.7), this convergence being an important factor in channel development. Furthermore, infiltration-excess overland flow can lead to the creation of rills, although exactly how this occurs is not clear. In humid environments, the presence of vegetation means that rills usually develop only on artificially disturbed surfaces, whereas in arid and semi-arid environments rill development occurs on natural surfaces. Microtopography on slopes tends to disrupt sheetflow and promote the concentration of water movement and rill formation. However, this can be counteracted by the lateral shifting of flowlines, or by rainsplash erosion, which has the effect of evening out the surface. Rills located in favourable positions may eventually grow to form gullies and become a permanent part of a channel network. Contour curvature has a significant impact on slope wash – where contours are convex in plan, sheetflow will be dispersed downslope, thereby minimising erosion and the initiation of rills. On the other hand, at valley heads and other locations where contours are concave in plan, the flow will be

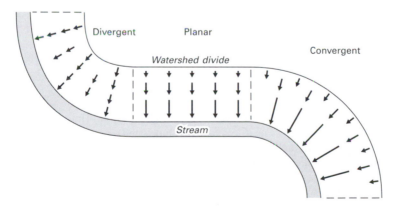

Figure 4.7 The influence of topography on the spatial variability of surface and subsurface erosion. Longer arrows indicate more flow per unit catchment area. Converging flow can be seen in concave topography. (Adapted from Woods *et al.*, 1997.)

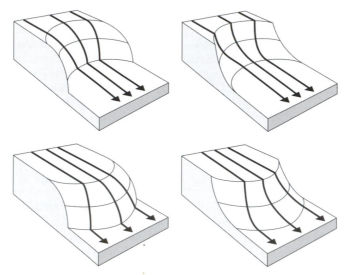

Figure 4.8 Convergence and divergence of flowpaths caused by contour curvature.

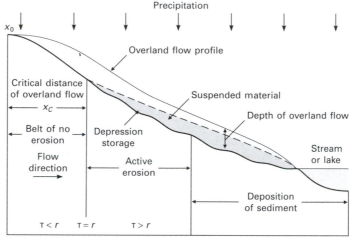

τ Eroding stress
r Shearing resistance of soil surface

Figure 4.9 Horton's model of overland flow production. (Adapted from Horton, 1945.)

concentrated downslope and rill erosion will therefore be more significant (Figure 4.8).

Horton (1945) was the first to formalise the importance of topography to hillslope hydrology by proposing that a critical hillslope length was required to generate a channel. The critical length was identified as that required to generate a sufficient overland flow depth to overcome the surface resistance and result in erosion (Figure 4.9). In his model, Horton proposed that a 'belt of no erosion' is present on the upper part of slopes because here the flow depth is not sufficient

to cause erosion. However, subsequent work has demonstrated that some surface wash is possible even on slope crests, although here it does not lead to rill development because the rate of incision is slow and incipient rills are filled by rainsplash.

The infiltration-excess overland-flow model provides a reasonable framework for explaining channel initiation for semi-arid and arid environments, but not for humid regions. In the latter, channel initiation is more related to the location of surface and subsurface flow convergence, usually in slope concavities and adjacent to existing drainage lines, than to a critical distance of overland flow. Rills can develop as a result of a sudden outburst of subsurface flow at the surface close to the base of a slope. So, channel development in humid regions is very likely to occur where subsurface pipes are present. Pipe networks can help initiate channel development, either through roof collapse or by the concentration of runoff and erosion downslope of pipe outlets. Piping can also be important in semi-arid regions.

Open channel flow

Energy variations in rivers

The potential energy of a still body of water at any point above sea level is proportional to the mass of the water body and to the height through which it has to fall to return to the sea. Originally, the energy is from the Sun, which evaporates water from the sea, enabling it to be deposited at a higher level as precipitation over land. Two opposing forces work on water flowing in an open channel. Gravity is the driving force that acts in a downslope direction and is governed by gravitational acceleration and channel gradient. Friction is the resisting force from within the water body and between the flowing water and the channel surface. The ability of the water to carry out geomorphic work – that is, to entrain and transport materials – is essentially controlled by the relationship between these two forces. The amount of work that the river does depends on its velocity and on the mass of water, the velocity in turn depending on the steepness of the channel. The steeper the channel the higher the river velocity. The volume of water passing a point in a given time, usually a second, is known as the discharge. The higher the discharge, the larger the total energy of the river. So, larger channels have a greater total energy than small ones. The kinetic energy of a river, represented by its velocity, is determined by the internal friction of the water, the bed friction, the slope of the channel, the discharge, and the size of the channel.

Laminar and turbulent flow

The velocity of flow fluctuates in all directions within the fluid, that is, vertically and laterally, as well as in its overall downstream direction. The velocity of an individual particle is much greater than the mean velocity in a downstream direction. Water is constantly interchanged in eddies between adjacent zones of flow, and local changes in velocity occur that work against the mean velocity gradient and result in a loss of energy. The resulting additional resistance to shear is called eddy viscosity. Flow of this type is known as turbulent flow, and it predominates in most natural channels and accounts for their efficiency in eroding and transporting sediments. Laminar flow, a second type, can be distinguished by the nearly linear trajectories of water particles. It can be found at lower flow velocities and is of very limited occurrence in natural river channels.

The shapes of the velocity profiles of turbulent and laminar flow are very different (Figure 4.10). The profile of laminar flow is a smooth parabola curving up from the stream bed, whereas the turbulent flow profile is sharply angular in the zone immediately above the stream bed. This sharp change in the profile reflects the transition from laminar flow at the bottom of the profile, where water slides or shears over itself in layers, to turbulent flow, where velocity distribution is more random and the velocity profile more uniform. The narrow zone of laminar flow in the profile is thought to occur around most objects on the river bed. In these areas, velocity is low because of the friction between the static bed and the moving water. Water overcomes this frictional resistance by shearing over itself, producing a laminar flow pattern and exerting a shear force on the obstacle or bed.

In plan, the velocity variation is affected by the same factors as the vertical velocity profile. Friction is greatest at the channel edges, which are therefore the zones of slowest flow. The fastest water is usually in the centre of the channel. This simple pattern of flow is disrupted if the channel shape is more complex. For instance, a large boulder in the bed may divide the channel into two, producing two velocity peaks in the plan.

Whether flow is laminar or turbulent depends on the mean flow velocity, the molecular viscosity and density of the fluid, and the dimensions of the flow section. With stream channels, the dimensions of the flow section are taken as either the depth of flow or the hydraulic radius. The hydraulic radius of a stream channel, R, is defined as the cross-sectional area of flow in a channel divided by the wetted perimeter, P, so it is the length of the boundary along which water is in contact with the channel (Figure 4.11). In broad, shallow channels the hydraulic radius can be approximated by the flow depth.

The Reynolds number, Re, defines the conditions under which laminar or turbulent flow occurs. A Reynolds number is essentially a dimensionless measure of flow rate and is calculated by multiplying the mean flow velocity, v, and the hydraulic radius, R,

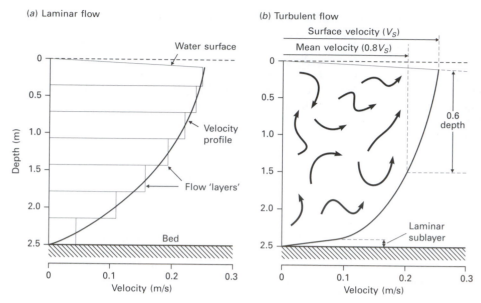

Figure 4.10   Velocity profiles of (a) laminar and (b) turbulent flow in a river.

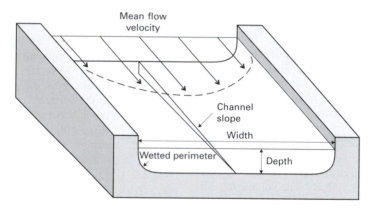

Figure 4.11   Defining the hydraulic radius of a stream channel.

and dividing by the kinematic viscosity, ν, which represents the ratio between molecular viscosity, μ, and fluid density, ρ ($\nu = \mu/\rho$):

$$Re = \frac{\rho \nu R}{\mu} \qquad (4.10)$$

In stream channels, the maximum Reynolds number at which laminar flow is sustained is about 500. Above values of about 2000, flow is turbulent, and between 500 and 2000 laminar and turbulent flow are both present.

Flow regimes

In natural channels, local variations in the depth of flow are caused by irregularities on the channel bed;

these create waves that exert a weight or gravity force. The Froude number, F, of the flow is defined by the ratio of the mean flow velocity to the velocity of the gravity waves, and can be used to distinguish different flow states:

$$F = \frac{\upsilon}{\sqrt{gd}} \qquad (4.11)$$

where υ is the flow velocity, g the acceleration due to gravity, and d the depth of flow. When $F < 1$ the wave velocity is greater than the mean flow velocity and the flow is known as subcritical or tranquil. Under these conditions, ripples propagated by a pebble dropped into a stream can move upstream. When

$F = 1$ flow is critical, and when $F > 1$ it is supercritical or rapid. These different types of flow occur because changes in discharge can be accompanied by changes in depth and velocity of flow. In other words, a given discharge can be transmitted along a stream channel either as a deep, slow-moving, subcritical flow or as a shallow, rapid, supercritical flow. Changes between subcritical and supercritical flow are determined by the velocity of flow. A sudden change from super-critical to subcritical flow is known as a hydraulic jump and produces a stationary wave and an increase in water depth. When flow changes from subcritical to supercritical, water depth decreases to produce a hydraulic drop. Such sudden changes in flow regimes may happen where there is an abrupt change in chan-nel bed-form. This is common in mountain streams where there may be large obstructions in the channel such as boulders.

In natural channels, mean Froude numbers are not usually higher than 0.5 and supercritical flows are only temporary, since the large energy losses that occur with this type of flow promote bulk erosion and channel enlargement. This erosion results in a lower-ing of flow velocity and a consequential reduction in the Froude number of the flow through negative feed-back. Froude and Reynolds numbers can be combined to produce four types of flow regimes: subcritical turbulent flow, subcritical laminar flow, supercritical turbulent flow, and supercritical laminar flow.

Velocity of flow

Velocity of streamflow is influenced by the gradient roughness and cross-sectional form of the channel. Natural channels almost always have a rough surface that creates significant frictional energy losses and causes a reduction in flow velocity, especially close to the channel boundary. Direct measurement of stream flow velocity is very time-consuming, so empirical equations have been developed to estimate mean flow velocities. The Chézy equation estimates velocity in terms of the hydraulic radius and gradient of the stream channel, and a coefficient expressing the gravitational and frictional forces acting on the water. It defines mean flow velocity, $\bar{v}$, as

$$\bar{v} = C\sqrt{Rs} \qquad (4.12)$$

where R is the hydraulic radius, s is the channel gradient, and C is the Chézy coefficient representing gravitational and frictional forces.

The Manning equation is a more commonly used estimator that incorporates an index of channel-bed roughness. It defines the mean flow velocity as

Table 4.2 Manning's coefficient of bed roughness

Surface	Typical value for Manning's n
Very smooth (e.g. glass)	0.010
Artificial concrete channels	0.013
Winding natural channels	0.017
Mountain streams with rock beds and rivers with variable sections	0.04–0.05
Alluvial channels with large channel-bed dunes	0.02–0.35

Source: Adapted from Clowes and Comfort (1990, 105).

$$\bar{v} = \frac{R^{2/3} s^{1/2}}{n} \qquad (4.13)$$

where R is the hydraulic radius, s the channel gradi-ent, and n the Manning roughness coefficient, which is usually estimated from tables (e.g. Table 4.2) or by comparison with photographs of channels of known roughness.

Operation of the Manning equation is complicated by the fact that resistance to flow varies with discharge and flow depth. The Manning roughness coefficient decreases as flow depth increases up to bankfull dis-charge. Once channel capacity is exceeded, the flow spreads over a large area and the Manning roughness coefficient increases. The equation is further complic-ated by adjustments in bed form, and hence channel roughness, that happen when changes in flow regime occur in alluvial channels. Manning's formula can be useful in estimating the discharge in flood conditions. The height of the water can be determined from debris stranded in trees and high on the bank. Only the cross-section and the slope have to be measured.

Automatic derivation of catchment characteristics

In the past, many of the above catchment characteristics were obtained from topographic maps, aerial photo-graphs, or field surveys. Unfortunately, these methods are subjective, prone to errors, time-consuming, and costly. Over the last two decades, such characteristics as catchment boundaries, channel length and slope, configuration of the channel network and stream ordering, valley heads, and subcatchment geometric properties, and such aerial properties as statistical measures of the distribution of elevation within regions or catchments (e.g. skew, kurtosis, and hypsometric analysis) have been derived directly from DEMs using automated procedures (Jenson and Domingue 1988; Mark 1984; Moore *et al.* 1991; Martz and Garbrecht 1992; Gallant and Wilson 2000; Tribe 1992; Band 1986; O'Callaghan and Mark 1984). As well as being more reliable, reproducible, and less costly and time-consuming, automated procedures allow the digital results to be linked to other meaningful datasets of the study area, for instance land-use or soils using GIS. These automated procedures have provided new tools for hydrologists to estimate the flow of water and sediment over landscapes and to link models of hydrological processes to GIS (Beven and Moore 1994; Maidment 1993, 1996; Moore *et al.* 1991, 1993b; Moore 1996). Automated methods for deriving catchment boundaries and stream networks have in particular received considerable attention.

Catchment boundary and stream network delineation

Early algorithms (e.g. Peuker and Douglas 1975) were relatively simple and based on identifying concave upward portions of a DEM, which are places where surface runoff will tend to be concentrated. Thus a map or image of the concave-upward portions of a DEM could be considered to be an approximation of the drainage network. The algorithm simply flagged the highest point among each square of four mutually adjacent points in the grid, and all of those points which were not flagged were considered to be part of the drainage network. The map output looked reasonable, but detection was strictly local and the features had no continuity (O'Callaghan and Mark 1984). Furthermore, many pits appeared as isolated dots, and channels could be broken by possible errors.

More recently, a number of different and somewhat more sophisticated algorithms, which attempt to yield a set of connected drainage lines for computing catchment boundaries and channel networks, have been developed (e.g. Band 1986, 1989; Hutchinson 1989; Jenson and Domingue 1988; Marks *et al.* 1984; McCormack *et al.* 1993; Morris and Heerdegen 1988; Olivera and Maidment 2000). All these algorithms usually follow a standard set of procedures (Figure 4.12). To carry out the subdivision of a landscape into catchments and subcatchments, the identification of the stream networks is required, followed by the areas defined on the basis of the drainage network above the catchment outlet and above the interior nodes or junctions where the upstream channels join (Moore *et al.* 1993a). To achieve this using a DEM,

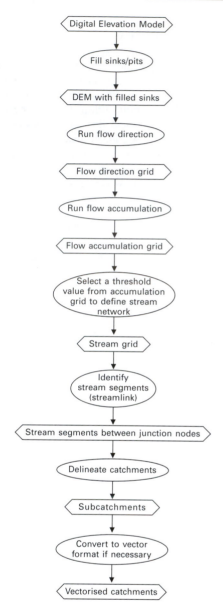

Figure 4.12 Standard procedures for computing catchment delineations summarised as a flowchart.

the first step is to deal with depressions and flat areas within the DEM, the importance and process of doing this being covered in Chapter 2 (p. 19). Then the flow direction or routing of the DEM must be calculated. This is the procedure that creates a grid of flow direction from each cell in the elevation grid to its steepest downslope neighbouring cell; in other words, it defines paths of steepest downhill descent. There

32	64	128
16	✳	1
8	4	2

The centre cell is labelled
with the flow direction value
according to the direction
of flow out of the cell

Digital Elevation Model

75	79	80	80	80	85	90
68	73	70	78	77	83	88
55	54	72	75	74	79	84
41	43	45	67	70	76	78
31	32	44	55	60	65	69
20	22	35	48	54	60	62
18	25	39	50	57	62	63

Flow direction grid

4	8	4	8	4	8	8
2	4	8	16	4	8	8
4	2	8	8	8	8	8
4	8	8	8	8	8	8
4	4	8	8	8	8	8
4	8	16	16	16	16	16
8	16	32	32	32	32	32

Flow accumulation grid

0	0	0	0	0	0	0
2	0	3	0	2	1	0
0	8	0	0	5	1	0
1	0	10	6	2	1	0
2	11	7	3	2	1	0
3	40	18	12	7	3	0
46	0	0	0	0	0	0

Stream network

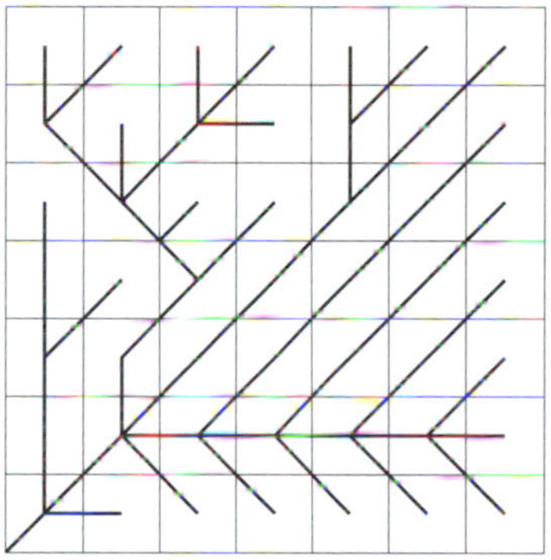

Figure 4.13 Steps in the D8 deterministic algorithm for estimating flow direction across a DEM image.

are a number of different flow direction methods and each will produce different results from the same input data.

The single-flow-direction D8, eight pour-point algorithm, slope-weighted algorithms, and stream-tube (DEMON) algorithms are all examples of different flow direction algorithms. The D8 deterministic algorithm developed by O'Callaghan and Mark (1984) is the most simple and frequently used, estimating flow direction by the direction of steepest downhill slope within a 3 × 3 window of cells that traces across the DEM image (Figure 4.13). The algorithm calculates a new attribute of flow direction which can take eight different directional values; these are usually expressed as numeric codes and sometimes as degrees (Figure 4.13). Each pixel in the output grid contains a value

indicating direction. The D8 algorithm allows flow to accumulate into a cell from several upslope cells but allows flow out of only one cell. Irrespective of which flow direction algorithm is used, the output is often referred to as a set of local drainage directions (ldd).

After flow direction, flow accumulation is calculated to produce a grid of accumulated flow to each cell from all other cells in the flow direction grid (Figure 4.13). Definition of the stream network can be achieved by selecting a subset of cells with a very high threshold flow accumulation value. This process will identify all the cells in the flow accumulation grid that are greater than the threshold value and produce a new grid representing the stream network, for instance: streams = if(flow accumulation > or = 100 then 1 else 0). This Boolean operation will result in a binary image in which all cells with 100 or more upstream elements are extracted as part of the set 'streams'. The threshold value holds no specific geomorphic meaning – it simply provides a means of catchment partitioning. A higher threshold value will result in a less dense network with fewer internal subcatchments, while a lower threshold will produce a denser network with more subcatchments.

The output from stream-network extraction is usually a string of raster cells. Such images are not useful for hydrological purposes unless the individual channel links and adjacent contributing areas are explicitly identified and associated with topological information for upstream and downstream connections. This is not a straightforward process with raster data. However, the spatial organisation and connectivity of the channel network are essential for flow routing and automated linkage of raster GIS information for network and traditional surface-runoff modelling. Algorithms have been developed that can interpret a raster image of a network, index the channel links and network nodes, and organise the channels into a sequence for cascade flow routing (e.g. Garbrecht and Martz 1997; Garbrecht 1988). These methods use a cell-by-cell trace along the channels of the network to identify for each channel link the Strahler order, the index, the sequence for cascade flow routing, the upstream inflow tributaries, the downstream connection, and the channel longitudinal slope and length. This information is provided in tabular format and can subsequently be used as input data for hydrological modelling.

The next stage is to identify the surface water intake locations and convert them into catchment outlet (pour-point) grid cells. The pour-points must be located directly over a grid cell from the drainage network. Pour-points may be located from maps or collected using a GPS (e.g. Saunders 2000). Catchment delineation algorithms then use the pour-points, the flow direction grid, and the stream networks to determine contributing areas. Upslope contributing area, A, is the total area above a certain contour length that contributes flow to the contour. Specific catchment area, A_s, is the ratio of the contributing area to the contour length, A/l. When using a DEM to calculate these areas, the contour length is approximately the size of a grid cell, and the most basic contributing area is produced by the number of cells contributing flow to a cell (Gallant and Wilson 2000). Calculation of flow direction is necessary for the determination of both the upslope contributing area and the specific catchment area.

An enormous number of studies have utilised these automated techniques for hydrological purposes (O'Callaghan and Mark 1984; Band 1986; Perez 2000; Saunders 2000; Doan 2000; Olivera and Maidment 2000). Perez (2000), Saunders (2000), Doan (2000), and Olivera and Maidment (2000) demonstrate the implementation of the catchment delineation techniques in the USA. Olivera and Maidment (2000) subsequently use the delineated catchments as input into the HEC–HMS (Hydrologic Modeling System developed by the Hydrologic Engineering Center of the US Army Corps of Engineers). This software package is used to model rainfall–runoff processes in a catchment or region. Doan (2000) used the catchments in the development of the Water Control Data System (WCDS) for Buffalo Bayou watershed in Houston, Texas, an area prone to flooding. The WCDS allows real-time monitoring, active warning, and forecasting of floods in the region. For the forecasting of floods, the Hydrologic Modeling System (HMS) was developed. This employs a quasi-distributed rainfall-to-runoff transformation procedure called Modified Clark (Mod-Clark) and grid-based NEXRAD radar rainfall in the Standard Hydrologic Grid (SHG) format to model the watershed response to a storm. The delineation of the subcatchments and major stream confluences are two of the many inputs into the model.

Band (1989) modelled runoff production using the stream network and catchment boundary produced from digital topography analysis along with TOPMODEL (Beven and Wood 1983; Quinn et al. 1995). TOPMODEL computes the topographic parameter $\ln(a/\tan b)$, where a is the contributing area and b is the surface gradient, directly from the information produced by the digital topographic analysis. Distance from the nearest stream channels and drainage distance to basin outlet were produced by a modification of a recursive branching program. Soils and vegetation cover were assumed constant so that the only effects were that of the basin's topography and the chosen velocities.

Other studies concerned with channel or gully networks have demonstrated that it is possible to predict the extent of channel networks to a reasonable accuracy using high resolution DEMs and in some instances have also predicted steady-state saturation overland flow, channel incision by overland flow, and channel incision by shallow landsliding (Prosser and Abernethy 1996; Dietrich et al. 1993; Montgomery and Dietrich 1988, 1989, 1992, 1994a, 1994b; Moore et al. 1988). Moore et al. (1988) employed a DEM to demonstrate that ephemeral gully erosion in an agricultural catchment was restricted to locations with relatively large upslope contributing area and steep gradient. It was considered that this topographic control reflected the conditions necessary for soil saturation and was also related to the unit power of overland flow.

Topographic controls on channel networks were examined in detail using field measurements and DEMs by Dietrich *et al.* (1992, 1993) and Montgomery and Dietrich (1988, 1989, 1992, 1994a, 1994b). These researchers found a strong inverse relationship between contributing area and slope at channel heads from field measurements over a number of catchments and suggested that there is a threshold contributing area required to support a channel (Montgomery and Dietrich 1988). In areas of steep terrain, this topographic threshold was defined in terms of a process threshold for the onset of landsliding using field and DEM data for three catchments (Montgomery and Dietrich 1989, 1994a; Dietrich *et al.* 1992, 1993). In lower-gradient hollows, the threshold contributing area required to support a channel was inversely proportional to the square of the slope. This is consistent with control of the channel network of a critical shear-stress for incision by saturation or Hortonian overland flow (Montgomery and Dietrich 1988, 1989, 1992, 1994b). The research team successfully simulated the channel network of a small catchment in Marin County, California, USA, by combining a DEM with the process threshold for incision by saturation overland flow (Dietrich *et al.* 1992, 1993). Ground-truth field experiments demonstrated that the modelled critical shear-stress was realistic if convergence of flow along the axis of hollows was accounted for (Prosser *et al.* 1995; Prosser and Dietrich 1995). Other variables, such as soil and vegetation, which also influence erosion processes, were considered spatially uniform and this highlights the strong control of topography on the channel network.

Prosser and Abernethy (1996) used the above model to test whether there was a topographic threshold to the extent of channels in a typical gullied catchment of south-eastern Australia. They also used the model to interpret the extent of the channel network in terms of thresholds for incision by Hortonian and saturation overland flow and to compare their predictions to field observations. They found that the limits to gully erosion in the catchment were strongly controlled by a topographic threshold that had an inverse relationship between upslope catchment area and local gradient. In addition, they found that, despite its simplicity, the model for incision by overland flow appeared

capable of distinguishing the hydrological processes responsible for channel incision when these were reflected in the relationship between channel network and landscape morphology. As the model has relatively modest input requirements, Prosser and Abernethy (1996) suggested that it is useful for mapping gully erosion in actively eroding catchments.

It is possible to extract drainage networks from a DEM with an arbitrary drainage density or resolution (Tarboton *et al.* 1991). The definition of channel sources influences the character of the extracted network. After the channel sources are identified, the essential topology and morphometric characteristics of the drainage network are implicitly defined due to their close dependence on channel source definition. Therefore, the correct identification of channel sources is essential for extracting a representative drainage network from DEMs. Montgomery and Foufoula-Georgiou (1993) and Tribe (1992) presented, respectively, the constant threshold-area and the slope-dependent critical support area methods to do this. The constant threshold-area approach assumes that channel sources represent the transition between the concave profiles of the channel slope (channel discharge dominated) and the convex profile of the hillslope (sheetflow dominated). It has had many applications (Band 1986; Morris and Heerdegen 1988; Tarboton *et al.* 1991; Gardner *et al.* 1991). With the slope-dependent critical support area method, an assumption that the channel source represents an erosion threshold is made. This implies that the channel source is the result of a change in sediment transport processes from sheetflow to concentrated flow, rather than a spatial transition in the longitudinal slope profiles. The main difference between the two methods and the defined networks are the spatial variability of slope. With the slope-dependent threshold method, drainage density is greater in the steeper areas of the catchment, as with natural catchments (Montgomery and Foufoula-Georgiou 1993). However, it is the constant-threshold method that is more practical to implement and has had more widespread use (Garbrecht and Martz 2000) because local slope values are difficult to obtain from DEMs.

Derivation of the other hydrological parameters

While the delineation of catchment boundaries and the definition of stream networks have been extremely useful to hydrologists, derivatives of the procedures such as the ldd grid have further uses. Once the ldd network has been defined, it is very useful for computing other properties of a DEM because it explicitly contains information about the connectivity of different cells. For instance, accumulating fluxes of materials over a network can be obtained. In a topologically correct network, each cell is connected to a downstream neighbour. This allows the calculation of attributes such as the cumulative amount of material that passes through each cell. The accumulation algorithm calculates the new value of a cell as the sum of the original value of the cell $S(c_i)$ plus the sum of the upstream elements draining to other cells (c_u) (Burrough and McDonnell 1998):

$$S(c_i) = S(c_i) + \sum_u^n(c_u) \qquad (4.14)$$

The calculated output grid could simply have cell values that represent the total number of cells flowing into each cell. Alternatively, a material value can be supplied from another grid (e.g. precipitation or sediment loss), in which case the accumulation algorithm will calculate the cumulative flow or sediment flux from a catchment. In such a situation the accumulative operator will compare the cumulative flow over an ideal surface. For instance, in the calculation of mass balance for each cell in terms of

$$S = P - I - F - E \qquad (4.15)$$

where S is surplus water per cell, P is input precipitation, I is interception, F is infiltration, and E is evaporation, cumulative flow over the net is calculated by accumulating S over the linked cells (Burrough and McDonnell 1998).

The ldd network can also be used for calculating a number of other dynamic processes. The upstream element map can be used to calculate indices of terrain, e.g. a wetness index map:

$$\text{wetness index map} = \ln(A_s/\tan b) \qquad (4.16)$$

where A_s is the contributing catchment area in m^2 (number of upstream elements × area of each cell) and b is the slope measured in degrees (Beven and Kirkby 1979). Other indices include the stream power index (Moore et al. 1993c) and the sediment transport index, which resembles the length–slope factor of the Universal Soil Loss Equation (Wischmeier and Smith 1978).

Slope length and maximum flow length can also be determined using the ldd. Slope length algorithms calculate a new value for a cell as the sum of the original cell value and the upstream cells multiplied by the distance travelled over the network, d_u:

$$S(c_i) = S(c_i) + \sum_u^n(C_u \times d_u) \qquad (4.17)$$

The distance travelled can be the Euclidean distance depending on the size of the cells (1 × the unit cell-size for N–S and E–W; 1.414 × the unit cell-size for diagonals), or it can be a frictional term dealing with resistances within the cells of the network (Burrough and McDonnell 1998).

Maximum flow length is the maximum length of all flow paths from the catchment boundary to a given point in a DEM. It can be determined by flow direction algorithms D8 or Rho8, but rather than accumulating areas the algorithm accumulates flow distances across cells, and only the largest flow-path length of all upslope cells is passed on to the downslope cell, instead of the sum (Gallant and Wilson 2000). The cardinal flow direction ($i = 2, 8, 32,$ or 128) has a cell distance of h, and the diagonal flow direction ($i = 1, 4, 16,$ or 64) has a flow distance of $\sqrt{2h}$. The problem with flowpaths calculated by these methods is that they have a zigzag appearance due to the restriction of cardinal and diagonal flow directions. The DEMON algorithm does, however, avoid this drawback.

Accuracy of derived catchment attributes

While automated techniques are extremely advantageous for hydrological and geomorphic studies, the DEM data source, grid resolution, and flow routing methods must be chosen carefully since they can have a large impact upon the magnitude and spatial pattern of computed topographic attributes (Wilson et al. 2000). The implication of data source upon DEM quality for water resource applications depends largely on the production techniques (Garbrecht and Martz 2000). For instance, USGS 7.5 minute DEMs were produced before 1988 mainly using manual profiling of photogrammetric stereomodels (United States Geological Survey 1990). This can often result in the DEM displaying systematic striping patterns in areas of low relief, making them unsuitable for parametrising drainage features. This was the situation found by Garbrecht and Starks (1995), who had intended to use a DEM for investigation of depressional wetlands in south-central Nebraska.

Garbrecht and Martz (2000) explain that the implications of striping upon drainage studies are threefold. First, the outlines of drainage features, e.g. drainage paths or depressions, are poorly defined. Boundaries of shallow features, which are oriented perpendicular to the striping, are often represented in the DEM as ragged lines with indentations orthogonal to the striping orientation. Second, drainage pathways are systematically biased in a direction orthogonal to the striping because of flow draining into and following the artificial elevation stripes. Finally, the striping may cause drainage blockages in the same orientation as the striping. These drainage blockages may produce artificial depressions of various sizes. Striping occurs due to a combination of human and algorithmic errors associated with manual profiling. These striping errors are well recognised (Garbrecht and Starks 1995), but in USGS-derived DEMs they lie within the established accuracy levels (see p. 18).

Other studies have considered the sensitivity of computed terrain attributes to DEM grid resolution, and some have attempted to find the optimum resolution required to represent accurately the major hydrological and geomorphic processes (Quinn et al. 1991, 1995; Wolcock and Price 1994; Moore 1996; Wilson et al. 2000). Zhang and Montgomery (1994) looked at the implications of using DEMs with different resolutions (2, 4, 10, 30, and 90 m) to determine flow across a landscape using the D8 single-flow-direction algorithm. DEM grid resolution significantly influenced the frequency distributions for slope, catchment area, and topographic wetness index attributes and the hydrographs produced with TOPMODEL. For instance, Saulnier et al. (1997) applied the TOPMODEL to the Maurets catchment of Réal Collobrier located in southern France, using DEMs with resolutions ranging from 20 to 100 m. They found that DEM data with increasing grid size affects the shape of the topographic index distribution because the topographic index values of pixels that were crossed by a channel were being overestimated by allowing area to accumulate down the valley axis. It was considered that, in many cases, river width would be small relative to DEM grid size, and in such situations it would be more appropriate to treat pixels crossed by a channel as hillslope pixels rather than channel pixels. Although there may be some contribution of subsurface from up valley, the inclusion of all the up-valley accumulated area would generate a bias in the calculated index.

Elsheikh and Guercio (1997) investigated the implications of DEM resolution on two other hydrological models, the distributed unit hydrograph model and a lumped geomorphic instantaneous unit hydrograph (IUH) applied to a sub-basin of the Tiber River basin in Italy. For the distributed unit hydrograph model, they examined the sensitivity of its morphologic parameters to DEM resolution in terms of their statistical distribution. For the lumped IUH, the sensitivity of its peak discharges and the time to peak parameters of the two scale features were investigated. The

analysis showed that the parameters of both models are significantly affected by DEM resolution. For the distributed unit hydrograph, the scale dependence on DEM grid size was particularly significant and arose mainly due to the scale influence on the cells of the upper basin area, whose frequency is related directly to the basin response. For the study area, it was recommended that a DEM resolution of 50 m or less be used, because beyond this limit strong scale-dependence occurs. The lumped IUH model was found to be influenced by DEM resolution but to a lesser extent than the distributed model.

In general, a 10-m DEM is recommended for geomorphic and hydrological applications because it performs significantly better than a 30-m or 90-m DEM and only slightly better than a 2-m or 4-m DEM (Wilson et al. 2000). Grid resolutions of 50 m or greater have been found to ignore the occurrence of lower-order streams and to artificially smooth landforms in complex landscapes so that topographic features that do affect hydrological processes are lost (Dikau 1989; Quinn et al. 1991, 1995). Wilson et al. (2000) found that flow-path lengths varied with grid resolution and DEM source. Larger grids produced fewer short flowpaths and overall flowpath lengths increased with an increase in grid resolution. Likewise, specific catchment area was found to increase with the increase in grid resolution (holding DEM source constant).

Miller and Morrice (1996) considered the implications of DEM error on catchment boundary delineation. They found that a consequence of not identifying erroneous peaks in a DEM was that streams that would normally be derived automatically using hydrologically based algorithms remained unidentified, which in turn caused errors in catchment boundary delineation. Wise (1998, 2000) considered the implication of using DEMs created from the same source data but created using different algorithms upon hydrological derivatives including flow direction, drainage networks, catchments, and TOPMODEL (a topographic index developed by Beven and Kirkby in 1979 that essentially measures the tendency of water to accumulate at any point on a slope). Wise found that for the drainage networks and catchment areas there was little difference between results produced by different DEMs at a broad scale, but that there were important differences in detail. For instance, some DEMs predicted drainage lines that at times crossed the original contours. For the calculation of TOPMODEL, far greater variation was found because the index is calculated for each pixel in the area, rather than being an aggregate result derived from numerous pixels. Wise's main conclusion was that care should be taken to assess a DEM's quality before using it, and that results should always be checked so that they appear reasonable. In another study, Lee *et al.* (1992) carried out controlled simulations to document the impact of errors in a DEM on the accuracy of extracting floodplain cells from a DEM. They found that the magnitudes and the spatial patterns of error in a DEM significantly affect the results.

Many studies have investigated the effects of different flow-routing methods upon computed topographic attributes (Quinn *et al.* 1991, 1995; Costa-Cabral and Burges 1994; Wolcock and McCabe 1995; Desmet and Govers 1996; Moore 1996). In particular, the D8 method has received much criticism. The method is able to model flow convergence in valleys but not flow divergence in ridge areas, and this is considered by some as a drawback (Gallant and Wilson 2000; Freeman 1991; Quinn *et al.* 1991; Costa-Cabral and Burges 1994). Another problem with this method is that it can generate a bias in flow path orientation in that it automatically discretises flow direction into units of 45°, which means that it tends to produce flow in parallel lines along preferred directions that agree with the aspect only when in a multiple of 45° (Fairfield and Leymarie 1991). Despite these limitations, the D8 algorithm has been incorporated into many commercial GIS packages. While it is incapable of computing flow divergence, it is suitable for catchment boundary delineation. Gallant and Wilson (2000) recommend that for calculating the distribution of contributing area and specific catchment area that cross hillslopes, sophisticated methods such as the randomised single-flow direction (Rho8) method (Fairfield and Leymarie 1991), FD8 (Moore *et al.* 1993b), or DEMON (Lea 1992; Costa-Cabral and Burges 1994) are utilised. A multiple-flow-direction algorithm provides better results in the headwater region of a source channel, while a single-flow-direction algorithm is superior in zones of convergent flow and along well-defined valleys (Freeman 1991; Quinn *et al.* 1991; Desmet and Govers 1996). So, on hillslopes, the multiple-flow method is better for overland flow analysis. However, if the delineation of the drainage network for large drainage areas with well-developed channels is the main objective, then the single-flow-direction algorithm is considered most appropriate (Martz and Garbrecht 1992).

Garbrecht and Martz (2000) have identified another problem with drainage networks extracted from DEMs, namely the precise location of channels in the digital landscape. Discrepancies have been found when comparisons were made with maps and aerial photographs, particularly in areas of low relief. The main reason for this is that digital representations of the landscape cannot capture critical topographic information below the DEM resolution. While the channel position in the DEM is consistent with the digital topography it may not reflect the actual drainage path in the field. This can be overcome by 'burning in' the route of the channels along pre-digitised pathways. This is carried out by artificially lowering the DEM cells along digitised lines or by raising the whole DEM except along stream lines (Cluis *et al.* 1996; Maidment *et al.* 1996). However, this procedure may produce flow-paths that are not consistent with the digital topography.

Flat surfaces, which may result from sink filling (see p. 19), from too low a vertical and/or horizontal DEM resolution to simulate adequately the landscape, or from a truly flat landscape, can cause problems for surface drainage identification. A method has been presented by Martz and Garbrecht (1995) that recognises that surface drainage in natural landscapes is towards lower terrain and away from higher terrain. To reproduce this trend on a flat surface, two shallow gradients can be imposed on flat surfaces to force flow away from higher terrain surrounding the flat surface and to attract flow to the lower terrain on the edge of the flat surface. This method will produce a convergent flow-direction pattern over the flat surface that is also consistent with the topography surrounding the flat surface (Garbrecht *et al.* 1996). However, any method of drainage identification in flat areas is an approximation and is unlikely to reflect accurately the actual drainage pattern that may follow channel incisions that are too small to be resolved at the resolution of the digital landscape.

The global cycle

The global circulation of water can be thought of as a series of storages and fluxes (Figure 4.14). The movement between the stores and the residence time of any molecule of water in each store varies throughout the cycle. The atmosphere and then rivers have the most rapid turnover of water. This rapidity both maintains the natural distribution system of water and also causes the extremes experienced by the terrestrial environment – floods and droughts. Residence times are also important in regard to pollution: damage

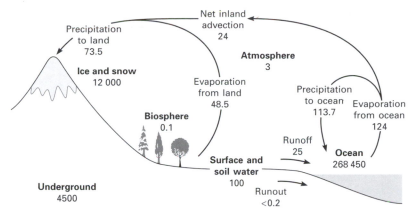

Figure 4.14 The global hydrological cycle, showing storages (in bold) and fluxes in thousands of cubic kilometres. (Adapted from Ward, 1975.)

to lake and groundwater stores lasts much longer than damage to rivers. A complete natural turnover in water volume is necessary for self-purification.

The actual quantification of a global water balance has yet to be achieved, although it is vital in order to fully understand climatic patterns, for instance cloud cover patterns that may have a crucial effect on the progress of the enhanced 'greenhouse' effect (Newson 1996). The period from 1965 to 1974 was known as the Hydrological Decade, during which a co-ordinated international effort at gauging precipitation and run-off enabled refinement to the estimates of global averages for both elements of the cycle. Newson (1996) reported estimates of global coverage of 50 000–60 000 rain-gauges and 20 000 streamflow gauges. Distributions are, however, extremely uneven and the outcome of global quantifications will depend largely on the techniques chosen for interpolating and extrapolating the basic data. A number of studies have been carried out looking at the interpolation of precipitation gauges, usually at a local or catchment scale (e.g. Court and Bare 1984; Creutin and Obled 1982; Daly *et al*. 1994; Dingman *et al*. 1988; Garen *et al*. 1994; Tabios and Salas 1985), although sometimes at a national scale (e.g. Hutchinson 1989). The results of such studies emphasise the variability of the results that can be achieved through using different interpolation methods.

Keller (1984) provided one of the most recent concise estimates of magnitude for the major elements of the global hydrological cycle (Table 4.3). Most of the fresh water (approximately 70 per cent) is locked up as polar ice or in glaciers. This has great significance climatically. Both groundwater and natural lakes are important storages but both are vulnerable to over-exploitation by over-abstraction or diversion of flows and pollution (e.g. the Aral Sea, the Great Lakes).

While humans are unable to alter the total volume of water in the global system, they can disturb the flux between storages in both time and space. For instance, afforestation and deforestation both alter the evapotranspiration rates of a region (e.g. Lean and Warrilow 1989; Lean and Rowntree 1997); and wetland drainage or river-flood protection will change the seasonal and short-term nature of storage.

Global hydrological change

To keep the global hydrological cycle going and to spread the input geographically (on average, 1030 mm of rain falls across the world each year), a total turnover takes place every nine days (Keller 1984). This process requires energy in the form of solar radiation to drive it. The way in which the global atmospheric system routes energy is extremely sensitive. The implications of 'greenhouse' gases on the climate have been observed and predicted. However, hydrologists are actually poorly equipped to predict the hydrological consequences resulting from future climate change. Global atmospheric models have been constructed to improve our predictions, but they are very poorly calibrated for the main components of the hydrological cycle (Newson 1996). McBean (1989) has pointed out that so far it has not been possible to measure precipitation, evaporation or soil moisture on continental scales. To progress further on global weather and climate predictions these climatological databases must be acquired.

Table 4.3   The world's water resources

Resource zones	Catchment area (km²)	Water volume (km³)	Depth of layer (m)	Percentage of world's water resources	
				Total water resources	Fresh water resources
Oceans	361 300 000	1 338 000 000	3700	96.5	—
Land	148 800 000	47 971 710	322	3.5	
Groundwater[a]	134 800 000	23 400 000[b]	174	1.7	
Fresh water	134 800 000	10 530 000	78	0.76	30.1
Soil moisture	82 000 000	16 500	0.2	0.0001	0.05
Polar, ice, glaciers, snow	16 232 500	26 600 000	1483	1.74	68.7
Antarctic	13 980 000	21 600 000	1545	1.56	61.7
Greenland	1 802 400	2 340 000	1298	0.17	6.68
Arctic	226 100	83 500	369	0.006	0.24
Mountains	224 000	40 600	181	0.003	0.12
Permafrost	21 000 000	300 000	14	0.022	0.86
Freshwater lakes	1 236 400	91 000	73.6	0.007	0.26
Saltwater lakes	822 300	85 400	103.8	0.006	—
Marshland	2 682 600	11 470	4.28	0.0008	0.3
Watercourses	148 800 000	2 120	0.014	0.00002	0.006
Biological water	510 000 000	1120	0.002	0.00001	0.003
Atmospheric water	510 100 000	12 900	0.025	0.001	0.04

[a] Gravitational and capillary water.
[b] This does not include groundwater resources in the Antarctic, which are estimated at 2 000 000 km³ of fresh water.
Source: Adapted from Keller (1984, 8).

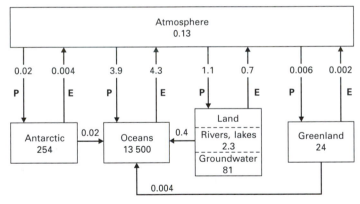

Figure 4.15   The role of the Antarctic and Greenland ice sheets in controlling the global water cycle during rapid climatic change. Units are 10^{14} m³; fluxes are annual. (Adapted from McBean, 1989.)

McBean (1989) also illustrated the global water balance (Figure 4.15), emphasising the controlling role of the oceans and ice sheets on the gross circulation of water. So, while humans have altered the quantity and quality of only a relatively very small part of the world's water (the freshwater component), its even-tual distribution under climate change scenarios is modified by vapour and energy exchanges occurring in other regions and at other times. It has also been calculated that, at present, humankind has eight times the amount of water necessary for the world's population (Keller 1984; Newson 1996). This is a

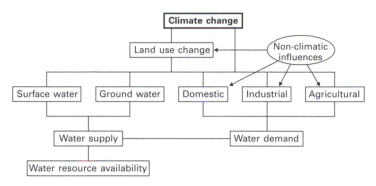

Figure 4.16 The influences of climate on water resources. (Adapted from Arnell *et al.*, 1990.)

somewhat simplistic view to take since it is the distribution of water resources in space and time, and not just the total amount, that is crucial to the human population. The real challenge is to maintain a global equity of water resources and a respect for freshwater ecosystems, while at the same time addressing the problem of the declining overall resource volume, its contamination or pollution, as well as the changes to temporal and spatial distribution caused by climatic change (Newson 1996).

L'vovitch and White (1990) made an assessment of the anthropogenic effect on the global hydrological cycle. They described the major changes in distribution of water on the Earth's surface from 1687 to 1987, listing irrigation as the most extensive influence upon the distribution of water. Impoundment has significantly altered temporal distributions of surface flows; however, the most urgent problem that has been emerging is the question of how to manage the increasing volume of polluted waste water, both now and in the future. The relative proportion of freshwater is extremely small and should not be compromised further through abuse.

Hydrology and climatic change

Dramatic changes of the Earth's climate over the next few decades have been predicted by many climatologists (e.g. World Meteorological Organization 1986). Rises in annual global-mean surface air temperature are expected, owing to the increasing global concentration of radiatively active greenhouse gases in the atmosphere (Houghton *et al.* 1990, 1992). Regional climate changes associated with global warming are usually determined using results from General Circu-

lation Models (GCMs). Analogues may also be used that refer to the identification of warm and cold periods within the climate record of a basin or region and then assume that future trends will mimic those of the past warm spells. Analogues are generally considered to be inferior to GCMs, since they are not derived from 'first principles' and make no allowance for the different nature of radiative forcing mechanisms that produced past climatic change. GCMs are used to predict changes in temperature and in amounts and distribution of precipitation as a result of climatic changes over the forthcoming half-century and beyond. GCMs have a coarse spatial resolution of around 500 km because they are seeking to model the planetary circulation. One of the difficulties associated with GCMs and future climate change scenarios is that many of the atmospheric and oceanic processes that determine regional and local climates are not explicitly simulated by the models. The implication is that less confidence can be placed in model-simulated climates and climatic changes for small regions than for large regions, and that some climatic variables will be better simulated than others (Hulme *et al.* 1993). For instance, greater confidence can be placed in projected monthly mean surface air temperature than in projected daily precipitation rates or mean daily wind velocities.

Hydrology and water management are of key importance when addressing climate change (Figure 4.16). Hydrology therefore plays a central role in the current World Climate Programme and has a leading role in the GEWEX (Global Environment Water Cycle Experiment) scheme and the International Geosphere–Biosphere Program (IGBP), which are attempting to link ground-level hydrological predictions to the global-climate models which are the main basis for the scenarios of climate change that have been publicised (Newson 1996; Vörösmarty *et al.* 1993). The aims of these programmes are to link runoff patterns with catchments or river basins to remotely sensed atmospheric patterns above, often using a grid-square configuration. Climatic changes are expected to have an effect on water resources and water availability. Such impacts are critical, since in many parts of the world, water usage is approaching the limits of water availability. Changes to the availability or distribution of freshwater could have serious consequences (Falkenmark 1989). As a result, a great deal of work has been carried out to develop hydrological models that simulate the effects of predicted climatic change or variation on water resources. In general, these models have assumed that the basin itself would remain unchanged. For example, Kite and Waititu (1981) and Nemec and Schaake (1982) utilised the Sacramento model to study the effects of climatic change in East Africa and the south-eastern United States. Initially, the model was calibrated with recorded precipitation and computed evapotranspiration data, and then run again after making a series of changes to the data (for instance, a 10 per cent increase in precipitation and a 6 per cent decrease in evapotranspiration) to simulate a change in the climate. The results were expressed as per cent changes in the annual average runoff. With these changes in climatological data, an estimated 70 per cent increase in the mean and annual runoff for the Nzoia River in Kenya resulted (Kite and Waititu 1981).

Lettenmaier and Gan (1990) applied soil moisture and snowmelt models to four subcatchments in the Sacramento–San Joaquin Basin. They calibrated the models over a four-year period and verified them over 20 years. The $2 \times CO_2$ climatic outputs from three GCMs were expressed as differences (for temperature) or as ratios (for precipitation) from the $1 \times CO_2$ base-case scenario. Adjustments were made to the recorded climatic data using these differences or ratios. Penman potential evapotranspiration was also adjusted to the scenario temperatures. The models were rerun under the simulated double-carbon-dioxide climate. Annual peak flows were higher and occurred earlier in the year as precipitation fell as rain rather than snow under the projected higher temperatures of the $2 \times CO_2$ scenarios. Nevertheless, the mean monthly flows during the spring were often lower because snowmelt was of less importance. In another study, Troendle (1991) used a detailed hillslope to examine the effects of increased temperature and precipitation on runoff from a subalpine slope. Tsuang and Dracup (1991) created changes in radiation, wind, and humidity in a detailed energy-based snowmelt model to examine changes in runoff for the Emerald Lake Basin, California.

All these models, however, assumed that the catchment or basin would remain constant from one climate regime to the next and did not consider the feedback mechanisms that may be triggered by climatic change. For instance, Kite (1993) points out that climatic change would affect not only precipitation and temperature but also the distribution of land types such as permafrost zones and glaciers, land-surface processes such as weathering, erosion and slope stability, soil processes such as drainage capacity and soil quality, vegetation characteristics such as biomass production, and changes in vegetation zones, population diversity, immigration, and dispersal. In particular, vegetation changes will exert significant influence upon hydrological processes by regulating the temperature, moisture, and wind regime and hence will influence the quantity and quality of water available for infiltration and runoff. Furthermore, changes in vegetation can decrease evapotranspiration, increase erosion, and degrade water quality. The altered evapotranspiration rates then feed back into the altered precipitation and temperature regimes and the concentration of carbon dioxide. As CO_2 concentrations rise, stomatal conductances tend to reduce, although this effect varies with vegetation type and there could be a compensatory increase in leaf area.

To address these implications, Kite (1993) developed another model known as SLURP-GRU. The model parameters were based on land-use classes derived from Landsat satellite data (effectively vegetation type), and the model was calibrated and verified for a watershed in the Rocky Mountains of British Columbia, Canada. The changes in land use associated with a doubled CO_2 scenario were estimated, and the model was rerun incorporating the changed climate and changed land-use. The model showed that the daily peak flows would not increase, but that the frequency of high flows would be significantly raised. Within this model, changes in temperature, precipitation, and evapotranspiration were applied directly from the GCM outputs to the observed data sets, the land use was assumed to vary only with temperature, and the increase in stomatal efficiency was assumed to vary linearly with CO_2 concentration. Kite (1993) stated that further research is required to understand the relationships between plant distribution and climate, to allow development of more physically based methods of evaluating the response of vegetation to climate change, and to address the question of plant response to increasing CO_2 concentrations.

There have been many predictions about the implications of climate changes and the associated hydrological effects. Newson (1996) suggested that the most serious implications of climate change will occur in dryland areas. Development of such areas initially requires irrigated agriculture to support population growth and exportable surpluses followed by urbanisation and industrialisation. Both these processes increase the demand for water but at the same time reduce the supply through pollution, sedimentation of reservoirs, or exploitation of non-renewable groundwater. The ability to forecast drought in these areas will become crucial, but such forecasts will be fruitless without the necessary associated political action and government awareness.

References

Abrahams, A. D. (1984) Channel networks: a geomorphological perspective. *Water Resources Research* **20**, 161–88.

Anderson, M. G. and Burt, T. P. (1978) Toward more detailed field monitoring of variable source areas. *Water Resources Research* **14**, 1123–31.

Arnell, N. W., Brown, R. P. C., and Reynard, N. S. (1990) *Impact of Climatic Variability and Change on River Flow Regimes in the UK* (Institute of Hydrology, Report No. 107). Wallingford, UK: Institute of Hydrology.

Band, L. E. (1986) Topographic partition of watersheds with digital elevation models. *Water Resources Research* **22**, 15–24.

Band, L. E. (1989) A terrain-based watershed information system. *Hydrological Processes* **3**, 151–62.

Bathurst, J. C. and O'Connell, P. E. (1992) Future of distributed modelling: the Système Hydrologique Européen. In K. J. Beven and I. D. Moore (eds) *Terrain Analysis and Distributed Modelling in Hydrology*, pp. 213–26. Chichester: John Wiley & Sons.

Beasley, D. B., Huggins, L. F., and Monke, E. J. (1980) ANSWERS: a model for watershed planning. *Transactions American Society of Agricultural Engineers* **23**, 938–44.

Beckett, P. H. T. and Webster, R. (1962) *The Storage and Collection of Information on Terrain* (Interim Report). Christchurch, Hampshire: Military Engineering Experimental Establishment (mimeo).

Beven, K. and Kirkby, M. J. (1979) A physically-based variable contributing area model of basin hydrology. *Hydrological Science Bulletin* **24**, 1–10.

Beven, K. and Moore, I. D. (eds) (1994) *Terrain Analysis and Distributed Modelling in Hydrology* (Advances in Hydrological Processes). Chichester: John Wiley & Sons.

Beven, K. J. and Wood, E. F. (1983) Catchment geomorphology and the dynamics of runoff contributing areas. *Journal of Hydrology*, **65**, 139–58.

Blackie, J. R. (1987) The Balquhidder catchments, Scotland: the first four years. *Transactions of the Royal Society Edinburgh, Earth Sciences* **78**, 227–9.

Burrough, P. A. and McDonnell, R. A. (1998) *Principles of Geographical Information Systems*. Oxford: Oxford University Press.

Calder, I. R. (1990) *Evaporation in the Uplands*. Chichester: John Wiley & Sons.

Calder, I. R., Harding, R. J., and Rosier, P. T. W. R. (1983) An objective assessment of soil moisture deficit models. *Journal of Hydrology* **60**, 329–55.

Chorley, R. J., Dunn, A. J., and Beckinsale, R. P. (1964) *The History of the Study of Landforms*, Vol. 1. London: Methuen.

Clowes, A. and Comfort, P. (1990) *Process and Landform: Conceptual Frameworks in Geography*, 2nd edn. Harlow: Oliver & Boyd.

Cluis, D., Martz, L. W., Quentin, E., and Rechatin, C. (1996) Coupling GIS and DEM to classify the Hortonian pathways of non-point sources to the hydrographic network. In K. Kovar and H. P. Nachtnebel (eds) *Applications of GIS in Hydrology and Water Resources Management*, pp. 37–45 (IAHS Publication No. 235). Rozendaalselaan, The Netherlands: International Association of Hydrological Sciences.

Costa-Cabral, M. C. and Burges, S. J. (1994) Digital elevation model networks (DEMON): a model of flow over hillslopes for computation of contributing and dispersal areas. *Water Resources Research* **30**, 1681–92.

Costa-Cabral, M. C. and Burges, S. J. (1997) Sensitivity of channel network planform laws and the question of topologic randomness. *Water Resources Research* **33**, 2179–97.

Court, A. and Bare, M. T. (1984) Basin precipitation estimates by Bethlahmy's two-axis method. *Journal of Hydrology* **68**, 149–58.

Crawford, N. H. and Linsley, R. K. (1966) *Digital Simulation in Hydrology: Standard Watershed Model IV*, Technical Report No. 39, Department of Civil Engineering, Stanford University. Stanford, CA: Department of Civil Engineering, Stanford University.

Creutin, J. D. and Obled, C. (1982) Objective analysis and mapping techniques for rainfall fields: an objective comparison. *Water Resources Research* **18**, 413–31.

Daly, C., Neilson, R. P., and Phillips, D. L. (1994) A statistical–topographic model for mapping climatological precipitation over mountainous terrain. *Journal of Applied Meteorology* **33**, 140–58.

De Roo, A. P. J., Wesseling, C. G., Jetten, V. G., and Ritsema, C. J. (1996) LISEM: a physically based hydrological and soil erosion model incorporated into a GIS. *HydroGIS 96: Application of GIS in Hydrology and Water Resources Management (Proceedings of the Vienna Conference, April 1996)* (IAHS Publication No. 235), pp. 395–403. Rozendaalselaan, The Netherlands: International Association of Hydrological Sciences.

Desmet, P. J. J. and Govers, G. (1996) Comparison of routing algorithms for digital elevation models and their implications for predicting ephemeral gullies. *International Journal of Geographical Information Systems* **10**, 3110–31.

Dietrich, W. E. and Dunne, T. (1993) The channel head. In K. Beven and M. J. Kirkby (eds) *Channel Network Hydrology*, pp. 175–219. New York: John Wiley & Sons.

Dietrich, W. E., Wilson, C. J., Montgomery, D. R., McKean, J., and Bauer, R. (1992) Erosion thresholds and land surface morphology. *Geology* **20**, 675–9.

Dietrich, W. E., Wilson, C. J., Montgomery, D. R., and McKean, J. (1993) Analysis of erosion thresholds, channel networks and landscape morphology using a digital terrain model. *Journal of Geology* **101**, 259–78.

Dikau, R. (1989) The application of a digital relief model to landform analysis in geomorphology. In J. Raper (ed.) *Three-dimensional Applications of Geographic Information Systems*, pp. 55–77. London: Taylor & Francis.

Dingman, S. L., Seely-Reynolds, D. M., and Reynolds, R. C. (1988) Application of kriging to estimating mean annual precipitation in a region of orographic influence. *Water Resources Bulletin* **24**, 329–39.

Doan, J. H. (2000) Hydrologic model of Buffalo Bayou using GIS. In D. Maidment and D. Djokic (eds) *Hydrologic and Hydraulic Modeling Support with Geographical Information Systems*, pp. 113–43. Redlands, CA: ESRI Press.

Elsheikh, S. and Guercio, R. (1997) GIS topographic analysis applied to unit hydrograph models: sensitivity to Dem resolution and threshold area. *Remote Sensing and GIS for Design and Operation of Water Resources Systems (Proceedings of Rabat Symposium S3, April 1997)* (IAHS Publication No. 242), pp. 245–53. Rozendaalselaan, The Netherlands: International Association of Hydrological Sciences.

Fairfield, J. and Leymarie, P. (1991) Drainage networks from grid digital elevation models. *Water Resources Research* **30**, 1681–92.

Falkenmark, M. (1989) The massive water scarcity now threatening Africa – Why isn't it being addressed? *Ambio* **18**, 112–18.

Fenneman, N. M. (1914) Physiographic boundaries within the United States. *Annals of the Association of American Geographers* **4**, 84–134.

Freeman, T. G. (1991) Calculating catchment area with divergent flow based on a regular grid. *Computers & Geosciences* **17**, 413–22.

Gallant, J. C. and Wilson, J. P. (2000) Primary topographic attributes. In J. P. Wilson and J. C. Gallant (eds) *Terrain Analysis: Principles and Applications*, pp. 51–85. New York: John Wiley & Sons.

Garbrecht, J. (1988) Determination of the execution sequence of channel flow for cascade routing in a drainage network. *Hydrosoft* **1**, 129–38.

Garbrecht, J. and Martz, L. W. (1997) Automated channel ordering and node indexing for raster channel networks. *Computers & Geosciences* **23**, 961–6.

Garbrecht, J. and Martz, L. W. (2000) Digital elevation model issues in water resource modeling. In D. Maidment and D. Djokic (eds) *Hydrologic and Hydraulic Modeling Support with Geographical Information Systems*, pp. 1–27. Redlands, CA: ESRI Press.

Garbrecht, J. and Starks, P. (1995) Note on the use of USGS Level 1 7.5 minute DEM coverages for landscape drainage analyses. *Photogrammetric Engineering and Remote Sensing* **61**, 519–22.

Garbrecht, J., Starks, P. J., and Martz, L. W. (1996) New digital landscape parameterization methodologies. In C. A. Hallam, J. M. Salisbury, K. J. Lanfear, and W. A. Battaglin (eds) *GIS and Water Resources (Proceedings of AWRA Annual Symposium on GIS and Water Resources, AWRA, September 22–26, 1996, Fort Lauderdale, Florida)*, pp. 357–65. Middleburg, VA: American Water Resources Association.

Gardner, T. W., Sasowsky, K. C., and Day, R. L. (1991) Automated extraction of geomorphic properties from digital elevation data. *Geomorphology Supplements* **80**, 57–68.

Garen, D. C., Johnson, G. L., and Hanson, C. L. (1994) Mean areal precipitation for daily hydrologic modeling in mountainous regions. *Water Resources Bulletin* **30**, 481–91.

Gregory, K. J. and Brown, E. H. (1966) Data processing and the study of landform. *Zeitschrift für Geomorphologie*, NF **10**, 237–63.

Hall, R. L. and Harding, R. J. (1993) The water use of the Balquhidder catchments: a process approach. *Journal of Hydrology* **145**, 283–314.

Horton, R. E. (1945) Erosional development of streams and their drainage basins: hydrophysical approach to quantitative morphology. *Bulletin of the Geological Society of America* **56**, 275–370.

Houghton, J. T., Jenkins, G. J., and Ephranmus, J. J. (eds) (1990) *Climate Change: IPCC Scientific Assessment.* Cambridge: Cambridge University Press.

Houghton, J. T., Callander, B. A., and Varney, S. K. (eds) (1992) *The Supplementary Report to the IPCC Scientific Assessment.* Cambridge: Cambridge University Press.

Hulme, M., Hossell, J. E., and Parry, M. L. (1993) Future climate change and land use in the United Kingdom. *The Geographical Journal* **159**, 131–47.

Hutchinson, M. F. (1989) A new procedure for gridding elevation and stream line data with automatic removal of pits. *Journal of Hydrology* **106**, 211–32.

Jenson, S. K. and Domingue, J. O. (1988) Extracting topographic structure from digital elevation data for Geographical Information System analysis. *Photogrammetric Engineering and Remote Sensing* **54**, 1593–600.

Johnson, R. C. (1985) Mountain and glen climatic contrasts at Balquhidder. *Journal of Meteorology* **10**, 105–8.

Keller, R. (1984) The world's fresh water: yesterday, today, tomorrow. *Applied Geography and Development* **24**, 7–23.

Kirchner, J. W. (1993) Statistical inevitability of Horton's laws and the apparent randomness of stream channel networks. *Geology* **21**, 591–4.

Kirkby, M. J. and Chorley, R. J. (1967) Throughflow, overland flow and erosion. *International Association for Scientific Hydrology Bulletin* **12**, 5–21.

Kite, G. W. (1993) Application of a land class hydrological model to climatic change. *Water Resources Research* **29**, 2377–84.

Kite, G. W. and Waititu, J. K. (1981) *Effects of Changing Precipitation and Evaporation on Nzoia River Flows and Lake Victoria Levels* (Contribution to the World Climate Program, Report 5). Geneva: World Meteorological Organization.

Kotoda, K. (1989) Estimation of river basin evapotranspiration from consideration of topographies and land use conditions. In T. A. Black, D. L. Spittlehouse, M. D. Novak, and D. T. Price (eds) *Estimation of Areal Evapotranspiration* (IAHS Publication No. 177), pp. 271–81. Wallingford, UK: Institute of Hydrology.

Lea, N. J. (1992) An aspect-driven kinematic routing algorithm. In A. J. Parsons and A. D. Abrahams (eds) *Overland Flow: Hydraulics and Erosion Mechanics*, pp. 393–407. London: UCL Press.

Lean, J. and Rowntree, P. R. (1997) Understanding the sensitivity of a GCM simulation of Amazonian deforestation to the specification of vegetation and soil characteristics. *Journal of Climate* **10**, 1216–35.

Lean, J. and Warrilow, D. A. (1989) Simulation of the regional climatic impact of Amazon deforestation. *Nature* **342**, 411–13.

Lee, J., Snyder, P. K., and Fisher, P. F. (1992) Modelling the effect of data errors on feature extraction from digital elevation models. *Photogrammetric Engineering and Remote Sensing* **58**, 1461–7.

Lettenmaier, D. P. and Gan, T. Y. (1990) Hydrologic sensitivities of the Sacramento–San Joaquin River Basin, California, to global warming. *Water Resources Research* **26**, 69–86.

L'vovitch, M. I. and White, G. (1990). Use and transformation of terrestrial water systems. In B. L. Turner II, W. C. Clark, R. W. Kates, J. F. Richards, J. T. Mathews, and W. B. Meyer (eds) *The Earth as Transformed by Human Action: Global and Regional Changes in the Biosphere over the Past 300 Years*, pp. 235–52. Cambridge: Cambridge University Press with Clark University.

Maidment, D. R. (1993) GIS and hydrological modeling. In M. F. Goodchild, B. O. Parks, and L. T. Steyaert (eds) *Environmental Modeling with GIS*, pp. 147–67. New York: Oxford University Press.

Maidment, D. R. (1996) Environmental modeling with GIS. In M. F. Goodchild, L. T. Steyaert, B. O. Parks, C. Johnston, D. Maidment, M. Crane, and S. Glendinning (eds) *GIS and Environmental Modeling: Progress and Research Issues*, pp. 315–24. Fort Collins, CO: GIS World Books.

Maidment, D. R., Olivera, F., Calver, A., Eatherral, A., and Franczek, W. (1996) Unit hydrograph derived from a spatially distributed velocity field. *Hydrological Processes* **10**, 831–44.

Mark, D. M. (1984) Automatic direction of drainage networks from digital elevation models. *Cartographica* **21**, 168–78.

Marks, D., Dozier, J., and Frew, J. (1984) Automated basin delineation from digital elevation data. *GeoProcessing* **2**, 299–311.

Martz, L. W. and Garbrecht, J. (1992) Numerical definition of drainage network and subcatchment areas from digital elevation models. *Computers & Geosciences* **18**, 747–61.

Martz, L. W. and Garbrecht, J. (1995) Automated recognition of valley lines and drainage networks from grid digital elevation models: a review and a new method. Comment. *Journal of Hydrology* **167**, 393–6.

McBean, G. A. (1989) Global energy and water fluxes. *Weather* **44**, 235–52.

McCormack, J. E., Gahegan, M. N., Roberts, S. A., Hogg, J., and Hyle, B. S. (1993) Feature-based derivation of drainage networks. *International Journal of Geographical Information Systems* **7**, 263–79.

Meteorological Office (1981) *The Meteorological Office Rainfall and Evaporation Calculation System, MORECS* (Hydrological Memorandum, No. 45). Bracknell, UK: The Meteorological Office.

Miller, D. R. and Morrice, J. G. (1996) Assessing uncertainty in catchment boundary delimination. *Third International Conference on Integrating Geographical Information Systems and Environmental Modelling, Santa Fe, New Mexico, USA, January 21–25*, unpaginated. No publisher.

Montgomery, D. R. and Dietrich, W. E. (1988) Where do channels begin? *Nature* **336**, 232–4.

Montgomery, D. R. and Dietrich, W. E. (1989) Source areas, drainage density, and channel initiation. *Water Resources Research* **25**, 1907–18.

Montgomery, D. R. and Dietrich, W. E. (1992) Channel initiation and the problem of landscape scale. *Science* **225**, 826–30.

Montgomery, D. R. and Dietrich, W. E. (1994a) A physically based model for the topographic control on shallow landsliding. *Water Resources Research* **30**, 1153–71.

Montgomery, D. R. and Dietrich, W. E. (1994b) Landscape dissection and drainage area-slope thresholds. In M. J. Kirkby (ed.) *Process Models and Theoretical Geomorphology*, pp. 221–46. New York: John Wiley & Sons.

Montgomery, D. R. and Foufoula-Georgiou, E. (1993) [Untitled]. *Water Resources Research* **29**, 3925–34.

Moore, I. D. (1996) Hydrologic modelling and GIS. In M. F. Goodchild, B. O. Parks, and L. T. Steyaert (eds) *Environmental Modeling with GIS*, pp. 143–8. New York: Oxford University Press.

Moore, I. D., Burch, G. J., and MacKenzie, D. H. (1988) Topographic effects on the distribution of surface soil water and the location of ephemeral gullies. *Transactions of the ASAE* **31**, 1098–107.

Moore, I. D., Grayson, R. B., and Ladson, A. R. (1991) Digital terrain modelling: a review of hydrological, geomorphological and biological applications. *Hydrological Processes* **5**, 3–30.

Moore, I. D., Gallant, J. C., Guerra, L., and Kalma, J. D. (1993a) Modelling the spatial variability of hydrological processes using GIS. In K. Kovar and H. P. Nachtnebel (eds) *Application of Geographic Information Systems in Hydrology and Water Resources (Proceedings of the HydroGIS 93 Conference held in Vienna, April 1993, Wallingford, UK)* (IAHS

Publication No. 211), pp. 83–92. Rozendaalselaan, The Netherlands: International Association of Hydrological Sciences.

Moore, I. D., Lewis, A., and Gallant, J. C. (1993b) Terrain attributes: estimation methods and scale effects. In A. J. Jakeman, M. B. Beck, and M. J. McAleer (eds) *Modelling Change in Environmental Systems*, pp. 1989–2014. New York: John Wiley & Sons.

Moore, I. D., Turner, A. K., Wilson, J. P., Jenson, S., and Band, L. (1993c) GIS and land-surface–subsurface process modelling. In M. F. Goodchild, B. O. Parks, and L. T. Steyaert (eds) *Environmental Modeling with GIS*, pp. 196–230. New York: Oxford University Press.

Morris, D. G. and Heerdegen R. G. (1988) Automatically derived catchment boundary and channel networks and their hydrological applications. *Geomorphology* **1**, 131–41.

Nemec, J. and Schaake, J. (1982) Sensitivity of water resources systems to climate variation. *Hydrological Science Journal* **27**, 327–43.

Newson, M. D. (1996) *Hydrology and the River Environment.* Clarendon Press: Oxford.

O'Callaghan, J. F. and Mark, D. M. (1984) The extraction of drainage networks from digital elevation data. *Computer Vision, Graphics and Image Processing* **28**, 323–44.

Olivera, F. and Maidment, D. R. (2000) GIS tools for HMS modeling support. In D. Maidment and D. Djokic (eds) *Hydrologic and Hydraulic Modeling Support with Geographical Information Systems*, pp. 85–111. Redlands, CA: ESRI Press.

O'Loughlin, E. M. (1981) Saturation regions in catchments and their relations to soil and topographic properties. *Journal of Hydrology* **53**, 229–46.

O'Loughlin, E. M. (1986) Prediction of surface saturation zones in natural catchments by topographic analysis. *Water Resources Research* **22**, 794–804.

Penman, H. L. (1948) Natural evaporation from open water, bare soil and grass. *Proceedings of the Royal Society of London* **193**, 120–45.

Perez, A. (2000) Source water protection project: a comparison of watershed delineation methods in ARC/INFO and ArcView GIS. In D. Maidment and D. Djokic (eds) *Hydrologic and Hydraulic Modeling Support with Geographical Information Systems*, pp. 53–64. Redlands, CA: ESRI Press.

Peuker, T. K. and Douglas, D. H. (1975) Detection of surface-specific points by local parallel processing of discrete terrain elevation data. *Computer Graphics Image Processing* **4**, 375–87.

Prosser, I. P. and Abernethy, B. (1996) Predicting the topographic limits to a gully network using a digital terrain model and process thresholds. *Water Resources Research* **32**, 2289–98.

Prosser, I. P. and Dietrich, W. E. (1995) Field experiments on erosion by overland flow and their implication for a digital terrain model of channel initiation. *Water Resources Research* **31**, 2867–76.

Prosser, I. P., Dietrich, W. E., and Stevenson, J. (1995) Flow resistance and sediment transport by concentrated overland flow in a grassland valley. *Geomorphology* **13**, 71–86.

Quinn, P., Beven, K., Chevallier, P., and Planction, O. (1991) The prediction of hillslope flow paths for distributed hydrological modelling using digital terrain models. *Hydrological Processes* **5**, 59–79.

Quinn, P. F., Beven, K. J., and Lamb, R. (1995) The ln(a/tan$b$) index: how to calculate it and how to use it within the TOPMODEL framework. *Hydrological Processes* **9**, 161–82.

Robinson, M. (1986) Changes in catchment runoff following drainage and afforestation. *Journal of Hydrology* **86**, 71–84.

Saulnier, G.-M., Beven, K., and Obled, C. (1997) Digital elevation analysis for distributed hydrological modeling: reducing scale dependence in effective hydraulic conductivity values. *Water Resources Research* **33**, 2097–101.

Saunders, W. (2000) Preparation of DEMs for use in environmental modelling analysis. In D. Maidment and D. Djokic (eds) *Hydrologic and Hydraulic Modeling Support with Geographical Information Systems*, pp. 29–51. Redlands, CA: ESRI Press.

Savabi, M. R., Klik, A., Grulich, K., Mitchell, J. K., and Nearing, M. A. (1996) Application of WEPP and GIS on small watersheds in USA and Austria. In K. Kovar and H. P. Nachtnebel (eds) *HydroGIS 96: Application of Geographic Information Systems in Hydrology and Water Resources Management* (Proceedings of the Vienna Conference, IAHS Publication No. 235), pp. 469–76. Rozendaalselaan, The Netherlands: International Association of Hydrological Sciences.

Savigear, R. A. G. (1965) A technique for morphological mapping. *Annals of the Association of American Geographers* **55**, 514–38.

Shaw, E. M. (1994) *Hydrology in Practice.* London: Chapman & Hall.

Shreve, R. L. (1975) The probabilistic–topologic approach to drainage basin geomorphology. *Geology* **3**, 527–9.

Smart, J. S. (1978) The analysis of drainage network composition. *Earth Surface Processes* **3**, 129–70.

Speight, J. C. (1980) The role of topography in controlling throughflow generation – a discussion. *Earth Surface Processes* **5**, 187–91.

Strahler, A. N. (1957) Quantitative analysis of watershed geomorphology. *Transactions of the American Geophysical Union* **38**, 913–20.

Strahler, A. N. (1964) Quantitative geomorphology of drainage basins and channel networks. In V. T. Chow (ed.) *Handbook of Applied Hydrology*, Section 4–11. New York: McGraw-Hill.

Summerfield, M. A. (1991) *Global Geomorphology: An Introduction to the Study of Landforms.* Harlow: Longman Scientific & Technical.

Tabios, G. Q. and Salas, J. D. (1985) A comparative analysis of techniques for spatial interpolation of precipitation. *Water Resources Bulletin* **21**, 365–80.

Tarboton, D. G., Bras, R. L., and Rodrigues-Iturbe, I. (1991) On the extraction of channel networks from digital elevation data. *Water Resources Research* **5**, 81–100.

Tribe, A. (1992) Automated recognition of valley heads from digital elevation models. *Earth Surface Processes and Landforms* **16**, 33–49.

Troendle, C. A. (1991) Global change: can we detect its effect on subalpine hydrographs? In B. Shafer (ed.) *Proceedings 59th Western Snow Conference, Juneau, Alaska, 12–15 April 1991*, pp. 1–7. Fort Collins, CO: Colorado State University.

Tsuang, B. J. and Dracup, J. A. (1991) Effect of global warming on Sierra Nevada mountain snow storage. In B. Shafer (ed.) *Proceedings 59th Western Snow Conference, Juneau, Alaska, 12–15 April 1991*, pp. 17–28. Fort Collins, CO: Colorado State University.

United States Geological Survey, Department of the Interior (1990) *Digital Elevation Models: Data Users' Guide* (National Mapping Program, Technical Instructions, Data Users Guide 5). Reston, VA: United States Geological Survey.

Vörösmarty, C. J., Gutowski, W. J., Person, M., Chien, T.-C., and Case, D. (1993) Linked atmosphere–hydrology models at the macroscale. In W. B. Wilkinson (ed.) *Macroscale Modelling of the Hydrosphere* (Proceedings of the Yokohama Symposium, July) (IAHS Publication No. 214), pp. 3–27. Rozendaalselaan, The Netherlands: International Association of Hydrological Sciences.

Ward, R. C. (1975) *Principles of Hydrology*, 1st edn. Maidenhead: McGraw-Hill.

Ward, R. C. (1981) River systems and river regimes. In J. Lewin (ed.) *British Rivers*, pp. 1–53. London: Allen & Unwin.

Watson, R. A. (1969) Explanation and prediction in geology. *Journal of Geology* **77**, 488–94.

Wilson, J. P., Repetto, P. L., and Snyder, R. D. (2000) Effect of data source, grid resolution, and flow-routing method on computed topographic attributes. In J. P. Wilson and J. C. Gallant (eds) *Terrain Analysis: Principles and Applications*, pp. 133–61. New York: John Wiley & Sons.

Wischmeier, W. H. and Smith, D. D. (1978) *Predicting Rainfall Erosion Losses: A Guide to Conservation Planning* (USDA Agricultural Handbook 537). Washington DC: United States Department of Agriculture, Science and Education Administration.

Wise, S. M. (1998) The effect of GIS interpolation errors on the use of digital elevation models in geomorphology. In S. Lane, K. Richards, and J. Chandler (eds) *Landform Monitoring, Modelling and Analysis*, pp. 139–64. Chichester: John Wiley & Sons.

Wise, S. M. (2000) Assessing the quality for hydrological applications of digital elevation models derived from contours. *Hydrological Processes* **14**, 1909–29.

Wolcock, D. M. and McCabe, G. J. (1995) Comparison of single and multiple flow direction algorithms for computing topographic parameters in TOPMODEL. *Water Resources Research* **31**, 1315–24.

Wolcock, D. M. and Price, C. V. (1994) Effects of digital elevation model and map scale and data resolution on a topography-based watershed model. *Water Resources Research*, **30**, 3041–52.

Woods, R. A., Sivapalan, M., and Robinson, J. S. (1997) Modeling the spatial variability of subsurface runoff using a topographical index. *Water Resources Research* **33**, 1061–73.

Wooldridge, S. W. (1932) The cycle of erosion and the representation of relief. *Scottish Geographical Magazine* **48**, 30–6.

World Meteorological Organization (1986) *Report of the International Conference on the Assessment of the Role of Carbon Dioxide and of Other Greenhouse Gases in Climate Variations and Associated Impacts (Villach, October)* (WMO Publication No. 661). Geneva: World Meteorological Organization.

Wright, I. R. and Harding, R. J. (1993) Evaporation from natural mountain grassland. *Journal of Hydrology* **145**, 267–83.

Zhang, W. H. and Montgomery, D. R. (1994) Digital elevation model grid size, landscape representation and hydrologic simulations. *Water Resources Research* **30**, 1019–28.

CHAPTER 5

Soil and sediments

Topography influences soils and soil processes in several ways. As latitudinal climatic belts tend to produce zonal soils and latitudinal (and longitudinal) climosequences, so altitudinal belts tend to produce different soil types and altitudinal climosequences. Local factors, such as slope exposure and substrate, often modify altitudinal soils and soil properties, just as they do zonal soils. Likewise, as climatic zones foster different geomorphic zones (e.g. glacial, humid, arid), so altitudinal climatic zones normally nurture altitudinal geomorphic zones. Soil, sediments, and slopes are under many circumstances sensitive to slope exposure. Soil on north-facing and south-facing slopes can differ markedly, while sediments on the same slopes commonly reflect the differing geomorphic process regimes, and the hillslopes themselves, owing to the different process regimes, are often asymmetrical.

Soils and sediments are affected by slope inclination and often vary in a constant way along slope sequences (catenas or toposequences). Catenary soil and sediment processes are intimately linked to hillslope hydrology. Soils and sediments are part of three-dimensional landscapes. Studies of soil–landscapes incorporate the third dimension when considering soil distribution and the spatial variability of soil properties and processes. Some studies consider the broad relationships between soil types and landscape elements, while other studies explore more refined connections between soils and landform elements using DEMs and GIS technology. Yet other work focuses on soils and sediments in the context of the wider landscape, including river networks.

Soil and altitude

Altitudinal soil sequences

Some altitudinal trends

As a rule of thumb, temperature decreases and rainfall increases with increasing elevation, often within short distances. Altitudinal climatic gradients are normally more compressed than latitudinal and longitudinal gradients. This climatic compression sometimes makes it easier to investigate 'vertical' soil climosequences than the more extensive 'horizontal' ones. Studies conducted in a range of the world's mountainous regions report the effects of altitudinal temperature and moisture changes on soils and soil properties. Generalisations from these studies capture the basic relationships between soils and altitude, though, as with all generalisations, notable exceptions exist. Ordinarily, increasing elevation is associated with an increasing content of organic matter and nitrogen, an increasing ratio of carbon to nitrogen (C/N ratio), and increasing acidity (pH), and usually with a decreasing content of calcium, magnesium, and potassium.

However, latitude, aspect, and substrate distort broad altitudinal trends. For instance, forested soils gain humus with increasing altitude, primarily because the drop in temperature slows the rate of humus decomposition. The nearer a mountain is to the equator, the faster is the rate of humus gain. A doubling of soil nitrogen in the Sierra Nevada, USA, requires an ascent of 2760 m, in India 1350 m, and in Colombia 890 m (Jenny 1980, 320). These increases in altitude are equivalent to mean annual temperature reductions of 14.6, 7.6, and 5.0°C, respectively. Moreover, the type of humus formed on mountains

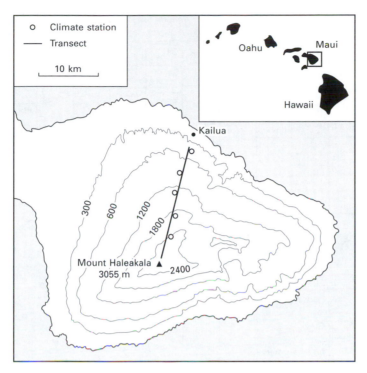

Figure 5.1 Topography of Mount Haleakala, Maui, Hawaii, showing the transect and climate stations. (Adapted from Kitayama and Mueller-Dombois, 1994.)

depends upon latitude, aspect, and parent material, as well as altitude (Duchaufour 1982). In temperate regions, middle and lower montane zones support mull humus and acid brown soils. In the upper montane zone, biological activity is still sufficient to create mull or thick mull–moder, providing some deciduous trees are mixed in with conifers. This zone encourages the genesis of humic ochric brown soils. Above the montane zone, in the subalpine zone, the vegetation is dominated by species, such as conifers and Ericaceae, that promote acidification. Soils evolved in this more acidic environment are brown podzolics and podzols. On the alpine meadows, which lie above the tree-line, humus is a moderately active moder or mull–moder, but the subsoil is not well developed. Rankers form on silicate rocks and humic lithocalcic soils on limestones. This altitudinal climosequence is best expressed on north-facing slopes. On south-facing slopes, the extra intensity of solar radiation with altitude offsets the lowering of temperature. The result is that the speed of organic matter decomposition is increased and the climatic belts become blurred.

Many altitudinal soil climosequences bear out these trends in soil properties, but they also bring out the exceptions and local effects due to aspect, parent material, and other factors. Case studies from mountains in Hawaii, the Spanish Pyrenees, and Borneo will illustrate the general relationships, the exceptions, and the significance of local factors.

A soil climosequence on Mount Haleakala

Mount Haleakala is a volcanic peak that rises to 3055 m on Maui, in the Hawaiian Islands. The windward slope receives persistent trade winds for nearly three-quarters of the year and is wet in its lower and middle sections. It affords complex temperature and moisture gradients on a uniform substrate of volcanic rocks, mainly alkalitic basalt, and layered volcanic ashes. The bottom of the slope has a warm-tropical and perhumid year-round climate, whereas the top has a cool-tropical climate with a summer dry season.

An altitudinal transect on the windward side of Mount Haleakala was used to inspect altitudinal variations in climate and soils (Kitayama and Mueller-Dombois 1994). The transect ran from 350 m to 3055 m (Figure 5.1). Climatic conditions along the transect displayed two points of significant change associated

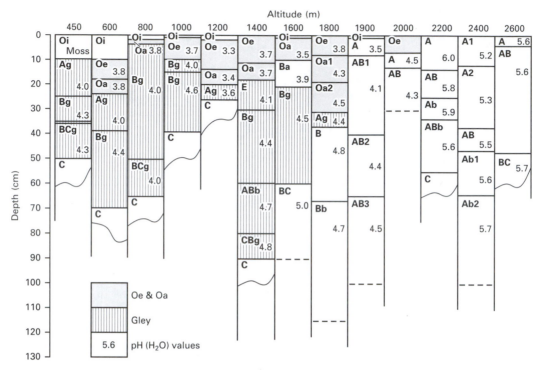

Figure 5.2 Altitudinal sequence of soil profiles along the Mount Haleakala transect, Maui. The figures are pH values (determined for a 1:1 mixture of free soil and deionised water). (Adapted from Kitayama and Mueller-Dombois, 1994.)

with the lower and with the upper cloud limits. The precise cloud limits vary, but in summer they normally sit at about 1000 m and 1900 m respectively. The lower limit corresponds to the lifting level of condensation and the upper limit to the base of the trade-wind inversion. The two limits yield three fairly distinct moisture zones on the mountain: the moist lowlands, which lie below about 1000 m; the wet montane cloud zone, which spans 1000 to 1900 m; and the arid high-altitude zone, which lies above 1900 m. The high-altitude zone comprises three subzones: the frost-free zone (1900–2400 m); the winter ground-frost zone (above 2700 m); and the ecotone between the other two.

Two broad soil zones exist on the mountain, the dividing line between them coinciding with the most frequent trade-wind inversion base at 1900 m (Figure 5.2). Soils in the lower zone are wet with accumulations of surface organic material (histic horizons) and strongly reduced (gleyed) subsurface horizons. These soils are highly acidic with pH values in the range 3 to 4 in the uppermost horizons. Reducing conditions even occur on flat interfluves at 450 and

600 m, despite the drier atmosphere there. This gleying on lower-altitude interfluves results from poor lateral drainage on the flat terrain and from the addition of water by downslope run-on. The montane soils between 1000 and 1800 m have thick peaty organic horizons (Oa) and are moderately gleyed, but the gleying is not so strong as in the lowland soils. Some montane profiles on better-drained sites have eluviated horizons indicative of stronger leaching and lower water-tables. Montane-zone soils become better drained towards the trade-wind inversion, but thin gleyed horizons persist to 1800 m. No signs of waterlogging are found above 1800 m, at which elevations the soils have a friable crumb structure that contrasts with the massive, coherent, and structureless soils at lower elevations. Soils in the upper zone, above 2000 m, are arid. They are better drained, drier, and less acidic with pH values above 5. They have diffuse horizon boundaries and lack thick organic horizons. Between 2000 and 2700 m, the soils are dominantly sandy-clay loams to sandy loams. Above 2700 m, the soils are sandy and the terrain is stony.

Table 5.1 Selected chemical properties of soil profiles along the Mount Haleakala transect, Maui, Harvaii

Horizon	Depth (cm)	Organic carbon (%)	Total nitrogen (%)	C/N ratio	Exchangeable potassium (milliequivalents per 100 g)	Ca/Mg ratio (mol)
600 m						
Oe[a]	10–18	47.2	1.35	35	1.5	0.6
Oa	18–24	37.8	2.16	18	1.1	0.6
Ag	24–39	7.1	0.24	29	0.15	1.0
Bg	39–70	2.7	0.10	27	0.10	2.1
1000 m						
Oe[b]	1–10	41.3	2.65	16	1.84	0.1
Bg	10–15	2.4	0.16	15	0.07	1.2
Bg	15–39	1.1	0.10	11	0.11	2.7
1400 m						
Oe	0–11	51.4	2.56	20	1.54	2.4
Oa	11–18	46.0	2.69	17	0.49	1.5
E	18–30	4.5	0.25	18	0.22	2.3
Bg[c]	30–60	11.5	0.35	33	0.12	6.0
Abg	60–80	18.9	0.76	25	0.06	8.7
1800 m						
Oe	0–8	29.7	2.48	12	0.82	2.3
Oa1	8–19	20.6	1.45	14	0.29	2.2
Oa2	19–31	23.0	1.33	17	0.10	5.3
Ag	31–37	11.2	0.52	22	0.08	5.0
B	37–67	9.8	0.48	20	0.06	5.2
Bb	67–>115	8.7	0.44	20	0.04	6.0
2200 m						
A	0–14	9.3	0.82	11	0.21	7.7
AB	14–25	10.0	0.97	10	0.15	8.2
Ab	25–34	11.9	1.03	12	0.09	8.5
ABb	34–55	13.4	0.88	15	0.09	7.1

[a] A 10-cm thick Oi horizon was not measured.
[b] A 1-cm thick Oi horizon was not measured.
[c] Figures for a grey subsample (gleyed).
Source: Adapted from Kitayama and Mueller-Dombois (1994).

The physical and chemical properties of the soils change up the mountainside in accordance with the changing soil-water regime, which becomes drier with increasing elevation (Table 5.1). Organic carbon in the surface horizons ranges from more than 40 per cent below 1800 m to less than 10 per cent at 2200 m. Total nitrogen follows the same trend as organic carbon. It is high (2.16–2.48 per cent) in organic horizons below 1800 m, and low (0.82 per cent) in the surface soil at 2200 m. Carbon–nitrogen (C/N) ratios in surface horizons decrease upslope in response to increasing rates of nitrogen mineralisation and nitrification. Exchangeable potassium in the uppermost horizon decreases with increasing elevation. The calcium–magnesium (Ca/Mg) ratio, which is indicative of ex-changeable cation leaching, increases with elevation – 0.63 at 600 m and 7.73 at 2200 m – suggesting that leaching effectiveness decreases upslope.

Soil climosequences in the Spanish Pyrenees

A study of two altitudinal soil climosequences in the Alta Garrotxa region, south-eastern Pyrenees, Spain, shows how parent material modifies altitudinal variations in soils and soil properties (Martí and Badia 1995). One transect ranged from 1020 to 1605 m on siliceous parent material (schists and mica-schists, with outcrops of sandstones, conglomerates, and limolites) in Montfalgars. The second ranged from 730 to 1120 m on calcareous parent material (marls and limestones)

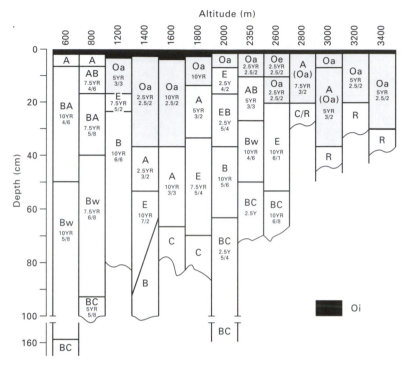

Figure 5.3 Altitudinal sequence of soil profiles along the Mount Kinabalu transect. Soil colours follow the Munsel colour-chart system. (Adapted from Kitayama, 1992.)

on the Mare de Déu del Mont Massif, with one soil profile coming from Camanegre peak (1500 m). Vegetation on the siliceous transect was temperate deciduous. Grassland dominated higher elevations and white oak (*Quercus humilis*) and beech (*Fagus sylvatica*) middle and lower elevations. Beech favoured the wetter, colder, and moister sites; white oak was common on warmer and drier sites. The grassland areas were produced by forest clearance, and supported such shrubs as common broom (*Sarothamnus scoparius*) and blackthorn (*Prunus spinosa*). Hooked pine (*Pinus uncinata*) was also present, as was snow woodrush (*Luzula nivea*) in the herb layer. On the calcareous transect, the vegetation was Mediterranean. The lowest two soil profiles carried a mixed holm oak (*Quercus ilex*) – white oak forest, with wild privet (*Ligustrum vulgare*) the dominant shrub. The next higher soil profile carried the same mixed forest, but with box (*Buxus sempervirens*) as the dominant shrub. A pure white-oak forest was present at the next soil profile, and this gave way to grassland at the highest site.

Acidity in the siliceous soils was acid or slightly acid, lying in the pH range 6.6–4.6. It decreased with increasing altitude, primarily owing to the increased water content of soils at higher elevations. In the calcareous soils, pH ranged from basic, through neutral, to acidic in the surface horizons at higher elevations. Soil carbonate was present only in the profiles at lower elevation in the calcareous soil transect, having been leached from profiles at higher elevations. The organic carbon content of surface horizons increased significantly with altitude in both transects. The C/N ratio also increased with altitude in both transects, though the increase was greater in the siliceous transect (9.3 to 14.4 under forest and 9.3 to 13.4 under grassland) than in the calcareous transect (8.4 to 9.7). These trends reflected the lowering of the organic-matter decomposition rate at higher elevations. The humus in all profiles consisted of O horizons lying upon thick A horizons, with a sharp boundary between the two. The humus on the calcareous transect was in OL/A horizons of variable thickness, and was classified as eumull (standard mull), though its acidity ranged from basic, through neutral, to slightly acidic with increasing altitude. In the siliceous transect, organic horizons and humus types were more variable. The range was from eutrophic mull at the lowest site, through mull–moder at middle elevations, to oligotrophic mull at the highest site. Of the exchangeable cations, calcium concentrations decreased with increasing altitude along both transects. Exchangeable sodium concentrations also decreased with increasing altitude. Calcium is more mobile an ion than sodium, potassium, or magnesium so it should suffer a greater net loss from profiles at higher altitudes. This was borne out by the ratios of calcium to magnesium and calcium to potassium, both of which decreased with increasing altitude along the calcareous transect and, less significantly, the siliceous transect.

A climosequence on Mount Kinabalu

At 4101 m, Kinabalu is the highest mountain in southeast Asia between the Himalayas and New Guinea. It lies in the north of Borneo, in Sabah, and is isolated from the central Bornean ranges. Formed by the intrusion of an adamellite pluton into Tertiary sedimentary rocks some 1.5 million years ago, Kinabalu

has steep slopes with granitic bedrock above about 2700 m. Ultrabasic intrusions outcrop around the massif. The mountain has a humid tropical climate.

Several conspicuous changes of soil are noticeable on Mount Kinabalu (Figure 5.3) (Kitayama 1992). Soil depth tends to decrease with increasing elevation from more than 1.5 m at 600 m to about 0.3 m at 3400 m. Organic (Oa) horizons form above 1200 m. Between 1200 m and 2600 m the organic horizons are fibrous mors. Those at 2800 and 3000 m are mixed with much mineral matter and designated as A horizons. Oa and A horizons above 2800 m lie directly on consolidated granitic rock, suggesting either a short time of evolution following glacial stripping or a sluggish rate of soil evolution. Eluviated (E) horizons are common in mid-slope soils between 1200 and 2600 m.

Soil properties also change with altitude (Figure 5.4). Midslope (1200–2600 m) topsoils have

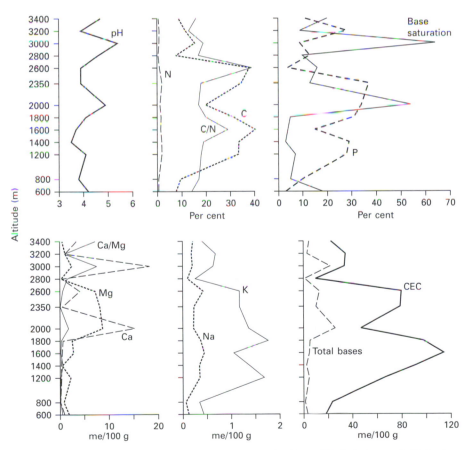

Figure 5.4 Altitudinal changes in soil properties along the Mount Kinabalu transect. (Data from Kitayama, 1992.)

higher amounts of organic carbon, total nitrogen, extractable phosphorus, exchangeable magnesium and potassium, and cation exchange capacity. Increased humus levels on midslope sites, which are reflected in high organic-carbon levels, are responsible for the higher values of the other soil properties. C/N ratios are high in some midslope soils, which points to a lower nitrogen-mineralisation rate there than in other soils. The ratio of calcium to magnesium, which is an index of leaching, is usually lower in soils below 2600 m, suggesting a more intense leaching regime below that height. Soil acidity should be most acidic in midslope sites as leaching is most intense there, but measurements show that it bears no consistent relationship with altitude. Two soil profiles at 2000 and 3000 m have higher pH values than the rest, one at pH 4.9 and the other at pH 5.4. These high values explain the remarkably high base-saturation levels (54 per cent and 64 per cent). The soil at 2000 m has a high base saturation and great depth, possibly because the substrate there might be derived from an early Quaternary till.

Altitudinal geomorphic zones

Different land-surface forms and processes are, to varying degrees, associated with different zonal climates. They are also associated with different altitudinal climatic belts, which, like altitudinal vegetation belts, are much compressed compared to their latitudinal equals. This compression leads to a quintessential property of mountain environments – the possession of precipitous ecological gradients that drive an elevated flux of water, sediments, and energy. So, most mountain environments boast highly dynamic geomorphic systems that are inherently unstable and prone to such high-magnitude events as debris slides and rock avalanches. A stability of sorts emerges in young mountain landscapes where a rapid sediment flux is episodically counterbalanced by tectonic rejuvenation – the geomorphic systems have a quick turnover of materials and a fairly constant form (cf. O'Connor 1984).

High-mountain landform zones

Of all the landform zones on mountains, the high-mountain zone is perhaps the most individual. It is delimited in three ways: by the upper tree-line, by landforms, and by the lower limit of denudation by snow and ice, also known as cryonival denudation (Gerrard 1990a, 40). The upper tree-line is of enormous

geomorphic, and ecological, significance. It usually marks a major change of process regime as the forested zone gives way to the alpine zone. The forested zone's regime is dominated by fluvial processes, soil creep, and throughflow, while the alpine zone's regime is dominated by nival, cryonival, and mass movement processes (Gerrard 1990a, 49). High-mountain landforms are multifarious yet similar on all the world's mountains. They were all to varying degrees fashioned by glacial processes, and especially glacial erosion, during Pleistocene glaciations. The Pleistocene snowline, which is ordinarily taken as the lowest level at which cirques occur, sets the lower limit of high-mountain landforms. Interestingly, Pleistocene snow-lines and upper tree-lines often rise or fall together (Troll 1973). In temperate latitudes they almost coincide, though in more arid mountains they drift apart by 400–800 m because a lack of water and freezing conditions preclude tree growth at high elevations. And on temperate mountains with extreme oceanic climates, such as the Atlas Mountains of North Africa, the Pleistocene snow-line was lower than the current tree-line (cf. p. 174). The cryonival belt lies between the tree-line and the snowline. It is a zone of intense frost action and patterned ground (stone stripes, polygons, and so on). In addition, its lower limit – the tree-line – is the lower limit of solifluction in temperate regions.

On the basis of these features, high-mountain environments may be split into three zones (Gerrard 1990a, 43). First, there is an uppermost or nival zone that sits above the limit of year-round snow. Second, there is an intermediate zone of frost-shattered debris, patterned ground, and a very patchy cover of pioneer vegetation, that is sandwiched between the snowline and the line where a dense plant cover and soil layer begin. Third, the lowermost zone extends down to the tree-line as an almost unbroken carpet of alpine meadow with dwarf shrubs, tussocks, rosette plants, and alpine herbs, all of which struggle to survive as frost action churns the soil. However, the existence of altitudinal geomorphic zones within arctic–alpine mountains is questionable, and the zones that exist are primarily topographic zones rather than altitudinal climatic zones. The topographic zones themselves have different climatic characteristics, but these pertain to local climates rather than broad altitudinal belts. Take the schema devised for a typical toposequence in arctic–alpine environments in Greenland (Stäblein 1984). Four geomorphodynamic zones are recognised, each associated with a particular prevailing process-regime (Figure 5.5). First is the zone of peaks,

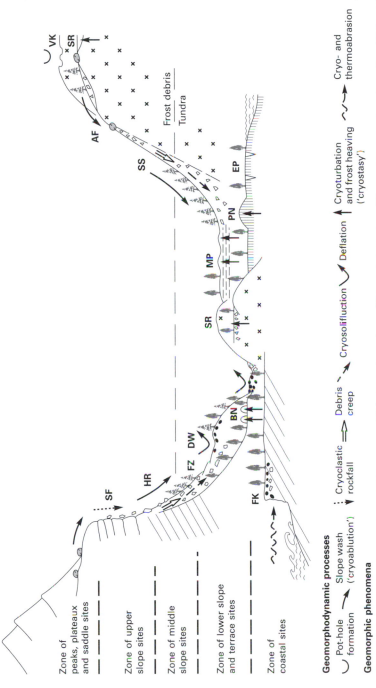

Zone of peaks, plateaux and saddle sites

Zone of upper slope sites

Zone of middle slope sites

Zone of lower slope and terrace sites

Zone of coastal sites

Geomorphodynamic processes

⌣ Pot-hole formation ('cryoablution')   → Slope wash ('cryoablution')   ⟹ Cryoclastic rockfall   ⟹ Debris creep   ⟿ Cryosolifluction   ⤵ Deflation   ↑ Cryoturbation and frost heaving ('cryostasy')   ⤳ Cryo- and thermoabrasion

Geomorphic phenomena

AF Denudational compensating frost slope;   **BN** Hummocks;   **DW** Deflation depressions;   **EP** Ice wedge polygons;   **FK** Cryoclastic and thawing frost cliffs;   **FZ** Cryosolifluction lobes;   **HR** Slope rills and gullies;   **MP** Mud-pits, unsorted circles;   **PN** Palsas;   **SF** Cryogenic rockfall slope;   **SR** Sorted stone circles;   **SS** Sorted stone stripes;   **VK** Pot-holes.

Substrata of the surface

Sedimentary rocks   Crystalline rocks   Moraines   Frost debris   Sandy terrace sediments with pebbles   Silty-clay marine terrace sediments   Turf and peat

Figure 5.5 A typical geomorphic toposequence in arctic–alpine environments, Greenland. (Adapted from Stäblein, 1984.)

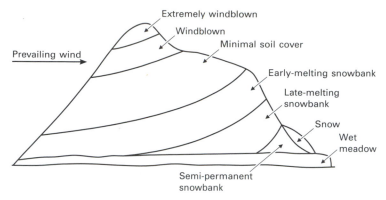

Figure 5.6 Schematic diagram of the ridge-top tundra geomorphic province according to the synthetic alpine model. (Adapted from Burns and Tonkin, 1982.)

plateaux, and saddles. The chief processes in this zone are frost weathering, solution weathering, slope wash, cryoturbation, and wind deflation. Landforms created here include weathering pans in exposed bedrock, solution hollows, and tafoni (hemispherical hollows in rock surfaces). In more sheltered sites, where enough frost-weathered debris and moisture accumulate, patterned ground is formed, sorted stone circles being a typical landform. Slope wash on steep (>25°) slopes produces rills and gullies. The second zone is the upper slope zone, where frost weathering, cryogenic rockfall, and debris creep are the main processes. Slopes are steep, the lower portions sometimes carrying stone stripes. The third, or midslope, zone is characterised by slope wash, solifluction, nivation, and slope dissection. Landforms include gelifluction and solifluction lobes and sorted stone stripes. At the foot of the sequence is the lower slope zone. Cryoturbation, frost heaving, and wind deflation dominate this valley-bottom zone. Gelifluction and solifluction feed valley-side sediment into the valley bottom. Landforms found in the lower-slope zone include terraces and bare rock outcrops.

Alpine soil and landform zones

A model of alpine slopes was constructed to clarify soil distribution and soil evolution in alpine geomorphic provinces (Burns and Tonkin 1982). This model had a historical, as well as a geographical, dimension and recognised periods of stability alternating with periods of instability in different parts of the alpine landscape. Although the model was devised for the southern Rocky Mountains in the USA,

it has a universal applicability in similar environments. It distinguished three geomorphic provinces: ridge-top tundra, valley-side tundra, and valley-bottom tundra. Broad interfluves, untouched by glaciation in late Pleistocene times, are the stable landforms of the ridge-top tundra province. They carry deep, well-evolved soils. The valley-side tundra province consists of steep slopes of the valley-side walls. During the Holocene, these slopes have experienced cycles of erosion and deposition producing thin soils intermingled with rock outcrops. In the valley-bottom province, which is located in cirque floors (though may be present further downvalley), deposition during phases of erosion upslope produces a mix of glacial, fluvial, and slope-derived sediments. The three geomorphic provinces are divided again into smaller 'zones'. Take the ridge-top zone in the Rocky Mountains. Using relationships between soils, topography, and snow cover as criteria, the ridge-top province is divisible into seven microenvironmental sites (Figure 5.6).

1. Extremely windblown sites sit on the crests of drainage divides. About 300 snow-free days occur each year. Slope angles vary from 0° to 8°. Mean soil temperature at 50 cm is 0.5°C. Vegetation is sparse and consists mainly of cushion plants. Rocky Mountain nailwort (*Paronychia pulvinata*) and moss campion (*Silene acaulis*) are indicator species. Soils are little evolved and well drained, with thin and sandy A horizons over thin cambic B horizons. They comprise 90 per cent Dystic Cryochrepts and 10 per cent Typic Cryumbrepts.
2. Windblown sites run from the tops to 30 per cent of the way down slopes. They are dry sites with

Colour Plate 1   A three-dimensional render of a digital elevation model (DEM) of Longdendale, north Derbyshire, England. (© Crown Copyright).

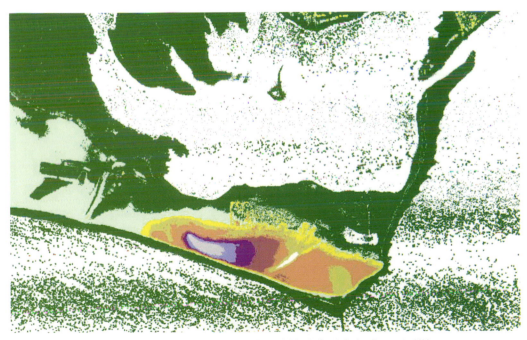

Colour Plate 2   Environment Agency LIDAR DEM of Mudeford Spit, Dorset, UK.

Colour Plate 3 The western slopes of Mount Hauhungatahi, New Zealand, with the forest sequence studied by Druitt *et al.* (1990) forming a band across the escarpment, but extending upwards beyond it on the right, where tongues of forest interdigitate with the subalpine shrubbery. Above the alpine tree-line, the subalpine scrub grades into tussock grasslands near the summit. The lower tree-line abuts a swampy plain and is determined in part by anthropogenic fire. The large macrophyte in the foreground is the New Zealand flax (*Phormium tenax*). (Photograph: John Ogden)

Colour Plate 4 Subalpine forest on Mount Hauhungatahi, New Zealand. The dominant tree with the coarse striated bark is mountain cedar (*Libocedrus bidwillii*). The dense shrub understorey is predominantly *Coprosma parviflora*, stinkwood (*C. foetidissima*), and rohutu (*Neomyrtus pedunculata*). The person is Dr Mark Horrocks. (Photograph: John Ogden)

Colour Plate 5 Hummock and hollow microrelief in an Atlantic white-cedar (*Chamaecyparis thyoides*) swamp on Shinn Branch, a first-order stream in the New Jersey Pinelands, USA. (Photograph: Joan G. Ehrenfeld)

Colour Plate 6 Landscape in Nahuel Huapi National Park, Argentina. (Photograph: Marcela Ferreyra)

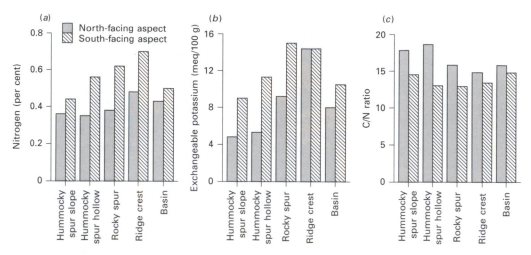

Figure 5.7 Variations of (a) soil nitrogen, (b) exchangeable potassium (in milliequivalents per 100 grams), and (c) carbon–nitrogen (C/N) ratio, in pasture soils on different slope units and aspects, Coopers Creek, near Oxford, North Canterbury, New Zealand. (Adapted from Radcliffe, 1982.)

limited inclusions of rock material from underlying schist bedrock (Figure 5.8). The occurrence of soil series within the landscape was found to relate to a combination of aspect and slope angle (Figure 5.9). Soils in Twin Stream, a transverse valley of the Ben Ohau Range, South Canterbury, New Zealand, are greatly affected by aspect (Archer and Cutler 1983). Their variability and distribution are different on warmer north and north-east aspects compared with cooler south and east aspects. Soils on the warmer aspects are more disturbed by periodic phases of erosion and deposition producing composite profiles with buried horizons. Soil development on the cooler slopes is more progressive and involves gleying and podzolisation.

Soil and slopes

Several methods are used for investigating relationships between soils and hillslopes. One method examines correlations between soil types or soil properties and hillslope elements (summit, shoulder, backslope, and so forth) or hillslope topographic variables (usually slope gradient, slope curvature, or aspect). A second method tries to establish causal connections between hillslope soils and soil processes and hillslope hydrology. This method may involve the field measurements of soil water and solute movements, the assessment of long-term translocation soils material using elemental ratios or index minerals, or mathematical modelling.

Soils and slope form

Sundry studies correlate soil properties with particular topographic (hillslope) variables. The simplest cases measure a few soil properties and single topographic attributes at sites along toposequences, while more complicated cases investigate many soil properties and several topographic properties, and sometimes vegetation, to boot. Examples from toposequences in Wyoming, New Zealand, and Taiwan will typify the nature of such soil–slope-form surveys.

Morainic soil toposequences in Wyoming

In two toposequences formed on coarse-grained tills near Willow Lake, in the Wind River Mountains, Wyoming, a plot of several soil properties (including free iron oxides content and free aluminium content) against hillslope curvature showed a consistent relationship (Swanson 1985). Soils on convex slopes differed from soils on concave slopes along a catena, the convex-slope soils containing less free iron and

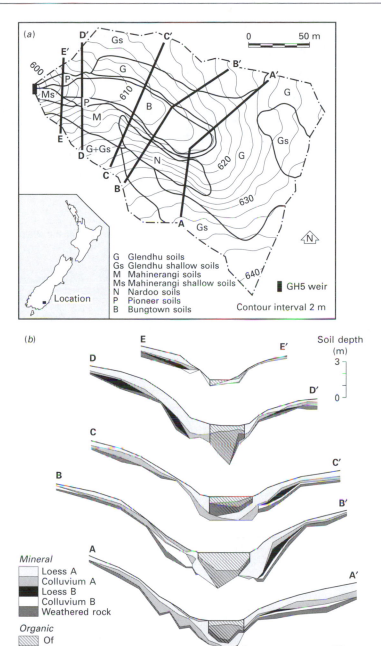

Figure 5.8 Soils and regolith in a catchment of the Waipori River, New Zealand. (a) Soil map of subcatchment with transects marked. (b) Regolith stratigraphy within the subcatchment along the five transects marked in (a). (Adapted from Webb *et al.*, 1999.)

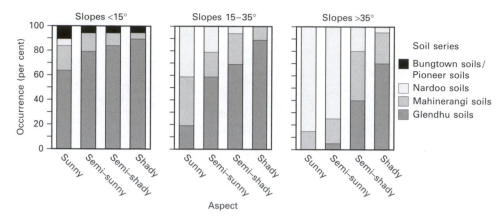

Figure 5.9 Relationship between soil series and aspect in a catchment of the Waipori River, New Zealand. (Adapted from Webb *et al.*, 1999.)

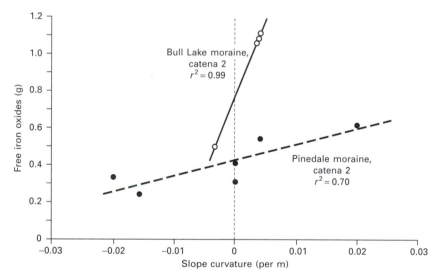

Figure 5.10 Free iron oxides (in a 1 × 1 × 130 cm soil column) versus hillslope curvature for catenas Pinedale 2 and Bull Lake 2, Wind River Mountains, Wyoming. (Adapted from Swanson, 1985.)

aluminium than the concave-slope soils (Figure 5.10). Interestingly, the two moraines differ in age – the Bull Lake moraine is 140 000 years old and the Pinedale moraine is 20 000 years old. The effect of slope curvature on soil properties was stronger in the older moraine than in the younger moraine, owing presumably to the extra 120 000 years' worth of soil evolution.

Downland soil toposequences in New Zealand

A study of soil–landscape relationships of downland soils on eastern South Island, New Zealand, which considered the connections between soils and several topographic variables, including aspect, is quite revealing (Webb and Burgham 1997). The study

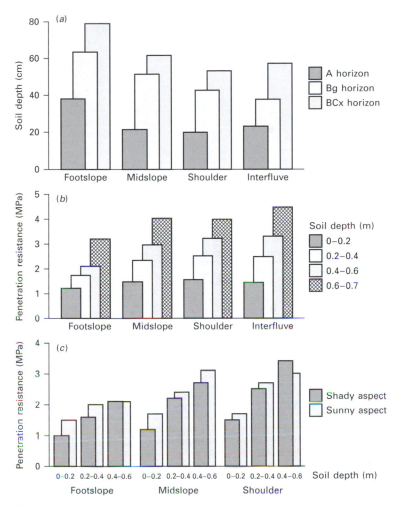

Figure 5.11 Properties of Timaru and Claremont soil series along transects over downland of eastern South Island, New Zealand. (a) Soil depth to base of A horizon and depths to Bg and BCx horizons versus landform element for Timaru and Claremont soils combined. (b) Penetration resistance versus soil depth and landform element for Timaru and Claremont soils combined. (c) Penetration resistance versus soil depth, landform element, and aspect for Timaru and Claremont soils combined. (Adapted from Webb and Burgham, 1997.)

assessed variability in the Timaru and Claremont soil series along transects at 12 sites, covering a range of aspects (sunny and shady), topographies (easy-rolling to strongly rolling terrain), and land uses (non-ploughed, pastoral, and mixed cropping). Both soil series displayed consistent trends with landscape elements. Topsoil thickness, depth to reducing conditions, and depth to fragipans were least on shoulder slopes and generally increased to footslopes. Data for all transects indicated that topsoil horizons on

footslopes were double the thickness of topsoil horizons on other landform elements (Figure 5.11(a)). Horizons quickly thickened at the junction between straight midslopes and concave lower slopes. In cultivated land, footslope sites had over-thickened topsoils that resulted from the relocation of topsoil material from upper to lower slopes. Cultivation practices were the chief cause of the topsoil movement, the mechanical movement of soil material associated with cultivation operations and the overall promotion

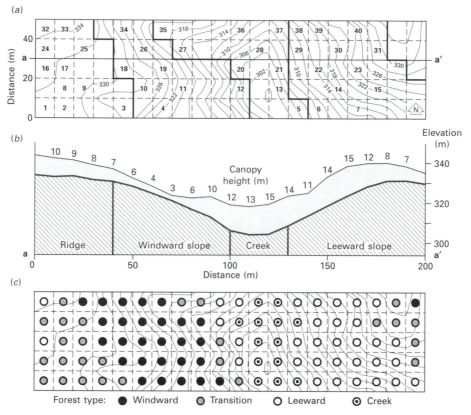

Figure 5.12 Taiwan soil–vegetation study. (a) Topography of the 1-ha plot with the location of the 40 sampling sites. Landforms (ridge, windward slope, creek, and leeward slope) are delimited by solid lines. (b) Cross-sectional profile of the plot with canopy height shown. (c) The distribution of four vegetation types based on quadrat groups delimited along the DCA axis 1. (Adapted from Chen *et al.*, 1997.)

of soil erosion under cultivation both playing a role. No significant correlations were found between topsoil depth and slope gradient. Soil strength was measured as penetration resistance. Footslopes bore the weakest soils and shoulders the strongest, while lower horizons were stronger than upper horizons for all landform elements (Figure 5.11(b)). Aspect significantly affected soil penetration resistance (Figure 5.11(c)). Soils on sunny aspects were weaker than soils on shaded aspects, probably because evaporation rates are greater on sunny aspects so sunny-aspect soils dry faster than shaded-aspect soils.

A soil and vegetation toposequence in Taiwan

A more complicated investigation probed the relations between soil properties and topography and vegeta-

tion in southernmost Taiwan (Chen *et al.* 1997). Soil chemical properties for a subtropical rain-forest in the Nanjenshan Reserve, Kenting National Park, were examined during 1989 and 1990. The forest is a constituent of the evergreen oak–laurel forest in Taiwan. A previous study had set up a permanent 3-ha plot, measuring 100 m (north–south) × 300 m (east–west), between two ridge crests. The south-western hectare, divided into 10 m × 10 m parcels, was used in the soil–topography–vegetation survey (Figure 5.12). The dominant woody species were long-leaf evergreen chinkapin (*Castanopsis carlesii*), rosy anise (*Illicium arborescens*), Oldham's daphniphyllum (*Daphniphyllum glaucescens* subsp. *oldhamii*), leather-leaf holly (*Ilex cochinchinensis*), and common schefflera (*Schefflera octophylla*). Soils were sampled at three depths (0–20 cm, layer 1; 20–40 cm, layer 2; and 40–60 cm, layer 3) in each of four landform

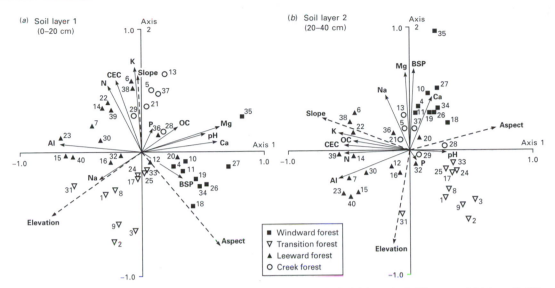

Figure 5.13 Biplots based on redundancy analysis (RDA) of soil data in (a) layer 1, 0–20 cm, and (b) layer 2, 20–40 cm, in Taiwan soil–vegetation study. Soil samples are classed by symbols representing the four vegetation types. Abbreviations are: N is available nitrogen, P is available phosphorus, K, Ca, Na, Mg, and Al are, respectively, exchangeable potassium, calcium, sodium, magnesium, and aluminium, BS is base saturation, OC is organic carbon, and CEC is cation exchange capacity. (Adapted from Chen *et al.*, 1997.)

Table 5.2 A summary of redundancy analysis (first two axes) of soil data from the Nanjenshan forest site, Taiwan

	Axis 1			Axis 2		
	Soil layer 1 (0–20 cm)	Soil layer 2 (20–40 cm)	Soil layer 3 (40–60 cm)	Soil layer 1 (0–20 cm)	Soil layer 2 (20–40 cm)	Soil layer 3 (40–60 cm)
Eigenvalue, λ	0.18	0.14	0.10	0.12	0.11	0.07
Cumulative variance (%)	18.20	13.90	10.40	29.7	24.80	17.40
Correlation coefficient, r	0.72	0.77	0.74	0.72	0.66	0.57
Significance level, p	0.01	0.01	0.16	0.01	0.02	0.51

Source: Adapted from Chen *et al.* (1997).

zones – ridge, windward slope, creek, and leeward slope. Vegetation in the plot consisted of 10 894 stems of 104 woody species. Correspondence analysis (CA) and detrended correspondence analysis (DCA) of importance values for all species revealed four vegetation types based on quadrat groups – windward-slope vegetation (windward slopes bearing the brunt of the north-east monsoons), leeward-slope vegetation, creek vegetation, and transitional vegetation. The forest structure and floristic composition changed radically along the wind-stress gradient. To display soil–landscape and soil–vegetation relationships, redundancy analysis (RDA) was applied to soil data

with topographic attributes (aspect, slope, and altitude) and vegetation types (windward, transition, leeward, and creek) as external variables. RDA is a linear multivariate technique for relating two sets of variables. Figure 5.13 shows RDA biplots of 40 soil samples (from layers 1 and 2) on the first two ordination axes. Soil and topographic variables are shown by arrows. A Monte Carlo permutation test was used to find out which variables best explained the soil data. It showed that the first two axes are significant in soil layers 1 and 2 (Table 5.2).

To see whether topographic variables alone were sufficient to explain the variation in the soil properties,

Table 5.3   The correlation of external variables (topography and vegetation-type) with the first two axes of a redundancy analysis for the Nanjenshan forest site, Taiwan

	Axis 1		Axis 2	
	Soil layer 1 (0–20 cm)	Soil layer 2 (20–40 cm)	Soil layer 1 (0–20 cm)	Soil layer 2 (20–40 cm)
Topographic variable				
Altitude	−0.56**	−0.08	−0.35*	−0.50**
Aspect	0.44**	−0.55**	−0.55**	0.11
Slope	−0.02	−0.54**	0.43**	0.17
Vegetation-type variable				
Windward	0.54**	0.21	−0.15	0.45**
Transition	−0.18	0.45**	−0.45**	−0.38*
Leeward	−0.36*	−0.63**	0.27	−0.11
Creek	0.08	0.05	0.35*	0.08

**$p = 0.01$; *$p = 0.05$.
Source: Adapted from Chen *et al.* (1997).

a partial RDA was carried out with topographic variables as covariables and vegetation types as variables of interest. Additionally, vegetation types were then treated as covariables to see how much soil variation could be explained by vegetation. The partial RDA tests indicated that the topographic variables were more significant than the vegetation variables in explaining soil variations (Table 5.3). For soil layer 1, topographic variables alone were sufficient to explain the soil variation, but in soil layer 2, vegetation type contributed significantly to soil variation. ANOVA results of soil and landform data revealed that soil properties differing significantly among landforms were pH, available nitrogen, cation exchange capacity, and exchangeable aluminium, potassium, calcium, and magnesium. Levels of pH and exchangeable Ca and Mg increased downslope in windward and leeward soils. Exchangeable Al was consistently higher on the leeward slopes where it increases with increasing altitude. Exchangeable K and cation exchange capacity were the only soil properties to vary significantly with slope, perhaps due to less leaching on gentler slopes. For soil layer 2, aspect is the major determinant of soil properties, with slope, but not altitude, making a significant contribution. The primary difference between steep-slope soils and gentle-slope soils lies in the higher concentrations of available N, exchangeable Al and K, and CEC on gentle portions of slopes. Analysis of variance shows leeward soils to have higher concentrations of available N and exchangeable K than

windward soils. On the second biplot axis, elevation is the main factor. Exchangeable Ca and Mg increase on a downslope direction, as in soil layer 1. In soil layer 3, the only soil properties to vary significantly between landforms are Al content and CEC, both of which are higher in leeward soils. These trends in soil properties, and particularly the downslope increase of pH and exchangeable Ca and Mg, testified to the importance of slope processes in redistributing soluble soil materials from valley-side slopes and their accumulation in the valley floor, near the creek site. The downslope trends may also mirror an underlying gradient of increasing soil moisture in a downslope direction. Vegetation appeared to affect soil properties through aspect. The contents of available N, exchangeable K, and CEC in soil layers 1 and 2 (0–40 cm) under the windward, low-stature (mostly sclerophyllous) forest were consistently lower than those under the leeward forest. For a given toposequence, however, soil variability associated with vegetation differences seemed to be muddled by slope processes.

Semi-arid grassland toposequences, Colorado

It would be wrong to give the impression that all soil properties show predictable relationships with topography along all catenas. Evidence for considerable systematic variation of soil properties with topography is overwhelming, and yet there are exceptions. Take a study of four toposequences in a short-grass steppe area in north-central Colorado (Singh *et al.* 1998).

Each toposequence comprised upland, midslope, and lowland positions. There was a sandy loam toposequence, a clay loam toposequence, and two sandy clay loam toposequences. Soil water dynamics was measured in these toposequences between 1985 and 1992. According to the classic catena model, lowland sites would normally be wetter than upland sites as water tends to drain downslope. Indeed, a study of subhumid tallgrass prairie in eastern Kansas showed that soil water content was significantly lower in upland sites than in lowland sites (Knapp *et al.* 1993). In the semi-arid shortgrass steppe, no regular variation of soil water content was observed. The lowland position was wettest in one toposequence, and driest in the other three. The classical catena model further predicts that the proportion of fine particles should increase from the convex upland to the concave lowland in accordance with water movement. In the semi-arid shortgrass steppe toposequences, total fines (clay and silt) were highest in the lowland positions in two toposequences, and lowest in one. Nor did vegetation vary systematically over any of the toposequences. The apparent failure of the catena concept at these sites, and at 24 toposequences within the Central Plains Experimental Range (Yonker *et al.* 1988), may result from the rarity of runoff and deep drainage episodes. It is possible, therefore, that the catena concept does not fully apply to semi-arid and other regions where runoff and throughflow are uncommon events, at least in the short term.

Soils and hillslope hydrology: current processes

Some of the toposequence work mentioned in the previous section suggests that mobile soil constituents are moved downslope. The sideways connectivity of soil profiles was first proposed by Geoffrey Milne (1935) in his catena concept and enlarged upon by Cecil Morison and his colleagues (1948), who underscored the connection between hillslope soils and hillslope hydrology. During the 1960s, researchers started to take up Milne's and Morison's seminal ideas and investigate soil evolution in the context of hillslopes. A consensus grew that soils on lower slopes are potential sumps for the drainage of soils upslope of them (Hallsworth 1965); that, on hilly terrain, water movement connects soils with one another and differentiates their properties (Blume 1968); and that adjacent soils at different elevations are linked by a lateral migration of chemical elements to form a single geochemical landscape (Glazovskaya 1968). A key point in this work is that solution and water transport act selectively, so that, as vertical movement in soil profiles produces A and B horizons, the lateral concatenation of soils leads to a differentiation of soil materials along a slope, with hilltop soils the analogues of A horizons, and valley-bottom soils the analogues of B horizons (Blume and Schlichting 1965). Subsequently, the burgeoning sophistication of hillslope hydrological investigations has prompted increasingly detailed and revealing examinations of slope soils (see Huggett 1995, 171–2).

Soil–hillslope hydrology studies fall into two broad groups: those that measure the current movements of soil materials and water in the field; and those that combine techniques for reconstructing the past movements of materials with an understanding of hillslope hydrology. This section will cover the first group of studies, the next section the second group.

Measuring lateral movements of soil materials

Studies of current soil and hydrological processes involve either field measurements of material translocation or a more general consideration, still using field observations, of soils and soil characteristics within a hydrological setting.

Direct monitoring of soil translocation processes often confirms the downslope movement of mobile soil constituents. A study in south-eastern Saskatchewan revealed that salts coming from summits moved downslope during periods of abnormally high precipitation and accumulated in toeslope soils (Ballantyne 1963). In Hettinger County, North Dakota, water exceeding crop use leached soluble salts from the root zone and transported them, by overland flow and throughflow, to lower landscape positions where they collected as saline seeps. Evaporation of water from the seeps caused the dissolved salts to rise through the soil, resulting in salt-crust formation at the surface (Timpson *et al.* 1986). Near Cape Thompson, Alaska, very poorly drained soils occupying low ground had a higher burden of strontium-90 than better drained soils on high ground, probably because the strontium-90 had been washed downhill (Holowaychuk *et al.* 1969).

Soils contaminated by radionuclides offer an undesired but convenient opportunity to gauge soil processes in landscapes. Radionuclide contamination of soils in Rocky Flats, Colorado, resulted from leaking steel drums of plutonium-contaminated oil stored at an outdoor area. The mobility of plutonium (Pu-239 and Pu-240) and americium (Am-241) in soils and groundwater is not well known. To evaluate the mechanisms of radionuclide transport from the contaminated soils to groundwater, an extensive monitoring system was installed across a toposequence at Rocky Flats Plant, 25 km north-west of Denver, Colorado (Litaor *et al.* 1998). The study investigated the impact of natural rain, snowmelt, and large-scale rain simulations on the mobility and distribution of the radionuclides in soil interstitial water. Combined plutonium-239 and plutonium-240 activity in the top 20 cm of soil ranged from 2220 to 11 460 Bq/kg (becquerels per kilogram), and the americium-214 activity ranged from 1840 to 8840 Bq/kg. These radionuclide activities accounted for over 90 per cent of activity in the top metre of soil, except in the pit at the top of the catena (pit 5), in which significant vertical translocation of all radionuclides had taken place due to the coarse texture of this soil. Radionuclide activity in soil solutions was highest in the upper layer of the top pit and decreased down the toposequence and with soil depth. The distribution of radionuclide activity in soil and soil water was the diametrical opposite of patterns of pH, alkalinity, and specific conductance in the soil solutions. The distribution of radionuclides during the monitoring period from 1993 to 1995 suggested that plutonium-239 with plutonium-240 and americium-241 are largely immobile in semi-arid soils, and that, contrary to findings in some other studies, americium-241 does not move faster in soils than plutonium. The vertical distribution of radionuclides in interstitial soil water indicated minimal transport of these contaminants after rainfall events and snowmelt. As none of the recorded snowmelt and natural rainfall during the period exceeded a 25-year recurrence interval, large-scale rain simulations with a 100-year recurrence interval were conducted. The rain simulation did not mobilise the radionuclides down the soil column in large amounts, but occasional 'outliers' and extreme fluxes, sometimes recorded to a depth of 70 cm, suggested that the radionuclides may be carried to greater depths than during natural events of small rain simulations. At 40–70 cm, the flux of radionuclides is between 1 and 3.3 per cent of the stored plutonium and americium in these soils. In consequence, it is possible that isolated but large rainfall events are significant in translocating radionuclides to groundwater, where the level of radionuclide activity rises to well above the drinking standard of 0.002 Bq/l. Fractionation of the radionuclides to different particle sizes in the soil interstitial water suggested that most of the radionuclides (83–97 per cent) were associated with suspended particles, whereas the level of radionuclides associated with colloidal and non-filterable fractions ranged from 1.5 to 15 per cent. This finding has implications for computer models of radionuclide movement in soils, which usually assume that most of the radionuclides are in 'dissolved' form.

A study of exchangeable base cations on a three-dimensional, acid hillslope in Bicknoller Combe, Quantock Hills, Somerset, England, aimed to map the spatial distribution of exchangeable base cations – calcium, magnesium, potassium, and sodium – and to investigate possible cation release processes from slope soils to the stream (Burt and Park 1999). The vegetation and parent materials across the slope are fairly uniform, so the concentration of exchangeable cations in the hillslope soils should reflect the nutrient stores and cation leaching processes along the toposequence. The parent material is deep (>2 m) stratified regolith of Devonian Hangman Grits, which are sandstones with subordinate shales, siltstones, and mudstones. The vegetation is characteristic of sheep pasture. A two-way analysis of variance (ANOVA) exposed soil depth as the principal variable in explaining the total variance of exchangeable bases, despite the steep slope gradient and clear podzolic catena development. Major nutrient base cations displayed uniform topsoil storage right along the toposequence, a pattern that might suggest that the spatial distribution of major nutrient cations is tightly controlled by the soil–vegetation system in nutrient-poor heathland environments. Sodium does not follow this putative vegetation-controlled spatial distribution, presumably because it is not particularly involved with soil–vegetation and soil exchangeable systems. In subsurface soils, cations liberated from the soil–vegetation system are redistributed along hydrological flowlines that run along the toposequence. Some cations are released into the stream. A saturated wedge develops at the slope base and plays a central role in the storage and release processes of base cations from slope soils to the stream. The saturated wedge stores calcium, magnesium, and sodium carried by throughflow, and slowly releases them into the stream at times of high flow. Potassium, on the other hand, displays a different spatial behaviour, being scarce in the saturated wedge.

A related toposequence study looked at aluminium in Bicknoller Combe, and its release to the stream (Park and Burt 1999). The spatial distribution of acid ammonium oxalate extractable aluminium, Al_o, and

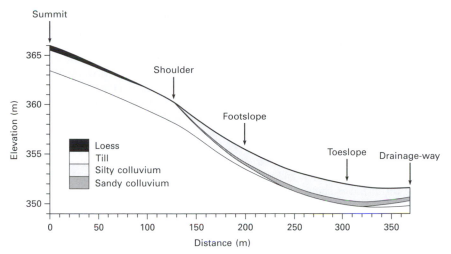

Figure 5.14 Catena east of Faribault, south-eastern Minnesota, USA. Topography, parent materials, and monitoring stations are shown. (After Thompson and Bell, 1998.)

exchangeable aluminium was measured. Eluviated aluminium from topsoils was mainly deposited in the lower soil horizons, forming podzolic B horizons. Some aluminium moved downslope through the medium of throughflow. Aluminium oxides may provide the main source of exchangeable Al^{3+} on the study slope due to high soil acidity. Examination of the spatial distribution of exchangeable Al^{3+} suggested that the slope hollow, where active convergent throughflow occurs, and the saturation wedge at the base of the slope are the main delivery routes of dissolved Al^{3+} to the stream. Divalent base cations (Ca^{2+} and Mg^{2+}), supplied from atmospheric input and organic decomposition and carried by throughflow, exchange Al^{3+} via cation-exchange reactions under high water content. Laterally illuviated aluminium oxides in the lower hollow adjacent to the saturation wedge probably provide a pool for continuous delivery of aluminium as either soluble or complexed forms to the stream via the saturated wedge.

Soils and their properties in a hydrological context

Soil hydrology is a key factor influencing soil processes and soil types along toposequences. Several studies single out the effect of hillslope hydrology on catenary soil development. For instance, the status of water tables, groundwater hydrology, and saturation zones were examined on selected Aquolls and Hapludolls in a Mollisol catena in Iowa, USA (Khan and Fenton 1994). Water table depths and precipita-tion were measured for a 10-year period on five soils formed in glacial till or till-derived sediments. Water table depth, duration of saturation, morphological features, and recharge and discharge to groundwater varied with geomorphic position in the catena. Most soils in summit and shoulder positions (Hapludolls) were not saturated, had deeper water tables, maximum water-table fluctuations, and rich chromas without redoximorphic features in the B horizons. Soils on toeslopes and depressions (Aquolls) had the shallowest water tables, longest saturation times, and B horizons with grey matrices, bright mottles, and iron and manganese concretions. Soils on backslope positions (Aquic Hapludolls) showed intermediate characteristics with redoximorphic features relating to the fluctuating water-table depths, which were influenced by artificial tile drainage.

A study of a Mollisol catena in south-eastern Minnesota, east of Faribault, looked at the distribution of hydric soils in relation to hillslope hydrology (Thompson and Bell 1998; Thompson *et al.* 1998). Part of the study evaluated the presence of wetland hydrology and hydric soil conditions, and examined associations between observed soil hydrology and soil morphologic properties, in five soils along a summit to valley-bottom transect (Figure 5.14) (Thompson and Bell 1998). Saturated conditions were measured with wells, piezometers, and tensiometers, anaerobic and reducing conditions with platinum electrodes. Saturated conditions occurred within 25 cm of the soil surface at all five sites, at least for a few days each year, between 1992 and 1995. They occurred mainly in the early spring and the late autumn. Only in the valley bottom (the drainage-way) were soils saturated and anaerobic during the microbial activity season in more than one year. Thick and black surface horizons formed where conditions were saturated and anaerobic. Another part of the study assessed soil and geomorphic properties controlling hillslope water movement, monitored the water movement, and related the water movement to the distribution of hydric soils (Thompson *et al.* 1998). Local subsurface flow occurred laterally along the catena, largely under the guidance of a sandy colluvium subsoil layer lying upon a dense till in lower landscape positions (Figure 5.14). Water accumulation in the landscape was predicted using three terrain attributes: slope gradient, specific catchment area, and a compound topographic index. Low-order drainage basins running at right-angles to the main drainage-way, largely owing to increased specific catchment areas, tend to focus surface and near-surface flow in lower slope positions, forming wetter areas that extend upslope of the valley floor. Moreover, the boundary between hydric and non-hydric soils extends farther upslope in the vicinity of these same low-order drainage basins.

Hydrological conditions along catenas greatly affect the formation of clay minerals in soils. The causal mechanism involved lies in the local water balance, which affects the opportunity that soluble and colloidal weathering products have to interact and create clay crystals. A 100-m-long soil catena formed in serpentinite in the Ligurian Apennines, north-west Italy, exemplifies the effect of drainage on clay mineralogy (Bonifacio *et al.* 1997). Seven soil profiles were studied that, from summit to toeslope, were classed as Lithic Ustorthents (2 summit pits and shoulder pit), Lithic Ustorchrept (backslope pit), Typic Haplustalf (upper footslope pit), Typic Ochraqualf (lower footslope pit), and Typic Haplustoll (toeslope pit). The main mineral in all pits was serpentine, which weathers to low-charge vermiculite or to smectite, depending on soil drainage. Traces of interstratified, low-charge vermiculites were present in the surface horizons of the well-drained summit and backslope soils, as well as the surface horizons of the upper footslope and toeslope soils. Smectites were found in horizons of lower porosity and restricted drainage. Such horizons occurred in the two footslope pits. The upper footslope soil had vermiculites in the surface horizons and smectites in the deepest ones; in the lower footslope soil, smectites occurred throughout the profile. These two profiles had different drainage regimes. It seems likely that the smectite in the base

of the upper footslope soil is formed from low-charge vermiculite under the restricted drainage regime. The vermiculite was probably translocated from the upper horizons and upslope. In the toeslope soils, where conditions were very wet, due to poor drainage, smectite, with a composition similar to saponite, appeared to form directly from serpentine. In short, different 2:1 phyllosilicates form along the catena, depending on porosity and hence drainage. Low-charge vermiculites form in upper and drier horizons, and smectites in poorly drained conditions. And, where conditions are wet, vermiculite in turn may be transformed to smectite.

Non-destructive surveys of toposequences

A problem with all soil investigations is that they are destructive – the digging of soil pits demolishes the features being studied. In an effort to side-step this problem, ground-penetrating radar (GPR) can be used to delineate subsurface soil features along toposequences. This method was applied to map continuously and non-destructively shallow subsurface features along a small wetland catena in south-eastern Newfoundland, Canada (Lapen *et al.* 1996). Detailed profiles of soil dielectric constant and common midpoint velocity surveys were used to determine radar pulse velocities through subsurface features. Major reflectors identified in the study included (Figure 5.15): (a) organic soil–mineral soil contact, (b) placic horizon (saturated mineral soil–unsaturated mineral soil contact), (c) water table (unsaturated mineral soil–saturated mineral soil contact), and (e) mineral soil–bedrock contact. Thicknesses of major soil features were estimated from radar profiles and compared with thicknesses determined from soil core-auger data. The relationship between estimated and observed thicknesses was strong ($r = 0.99$). Spatial relationships between placic horizons and wetland community types were also identified.

Soils and hillslope hydrology: reconstruction techniques

Radioactive tracers and mobile salts do not normally reveal long-term changes in catenas. Geochemists have used heavy metal concentration patterns in landscapes to trace veins of ore, and their investigations plainly show that soil material moves downslope (e.g. Rose *et al.* 1979; Thomas *et al.* 1985). However, the geochemical work appears to have proceeded independently of pedological work. Changes in soil

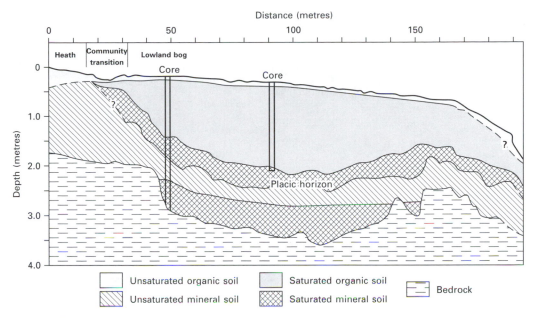

Figure 5.15 Interpretation of a radar profile through a lowland bog, Cape Race, Newfoundland, Canada. (Adapted from Lapen *et al.*, 1996.)

properties during soil catena evolution are indicated by the depth functions (vertical distribution) of sensitive soil constituents, such as iron and manganese oxides, and are quantified by reconstructing gains and losses of soil constituents relative to their concentrations in parent materials (e.g. Evans 1978). They may also be modelled mathematically.

Depth functions

A growing number of studies investigate the lateral movement of soil materials by establishing depth functions. The examples of soil properties in slickspots, a vernal pool, and in a tropical toposequence will bear out the value of the method.

Slickspots are barren and shallow depressions that are common features in sodic soil-landscapes. To investigate slickspot formation in the Carrizo Plain, San Luis Obispo County, California, USA, 16 pairs of pedons, one of each in a slickspot and the other under nearby vegetation (Figure 5.16(a)), were sampled (Reid *et al.* 1993). Field and laboratory measurements of soil properties enabled an interpretation of the conditions and processes leading to slickspot growth (Figure 5.16(b)). Slickspot soils appear to start from vegetated soils where erosion causes a bare patch. Once a potential slickspot is initiated, a key factor in its further development appears to be the

impedance of infiltration on the slickspot soil surface. This leads to leaching occurring in the vegetated pedons but not in the slickspot pedons. Water and dissolved salts are leached through the vegetated soils and move along a water-potential gradient into the drier slickspot subsoils, and then move upwards owing to a gradient induced by evaporation at the slickspot soil surface. This pattern of movement causes the salinisation of slickspot soils. As salinisation increases, the surrounding vegetation is stressed, producing a shortgrass zone with bare areas that is subject to wind and water erosion. Thus, the slickspots expand irreversibly under the present climate of the area.

Vernal pools are shallow surface depressions that fill with water in winter and spring and dry out during dry summer months. To form, they require a Mediterranean climate, almost level topography, and a microrelief of shallow basins with restricted drainage from an impermeable soil horizon or bedrock. These requirements are found in California. The basalt-capped mesas of the Santa Rosa Plateau Ecological Preserve, Riverside County, California, support 13 vernal pools that range in size from 0.4 to 10.2 ha (Weitkamp *et al.* 1996). These pools normally form ponds up to 30 cm deep between February and April. An inspection of a 1.43-ha vernal pool catena revealed the relationships between soil morphology, soil chemical properties, and water movement. This

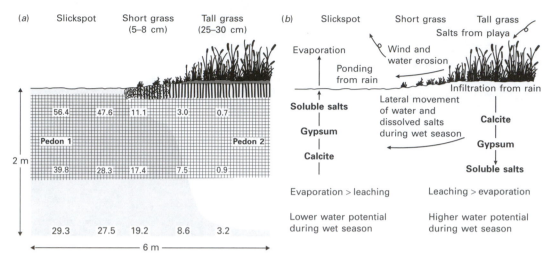

Figure 5.16 Microcatenary relationships in slickspot soils on the Carrizo Plain, California, USA. (a) The transition between a slickspot soil, Pedon 1, and a vegetated soil, Pedon 2. Structural features are shown above 125 cm depth, below which soils are massive. Stippled area are zones of visible salt efflorescence. Numbers are electrical conductivity values (dS/m) of saturation extracts. (b) Schematic representation of processes contributing to salinisation and expansion of slickspots. (After Reid *et al.*, 1993.)

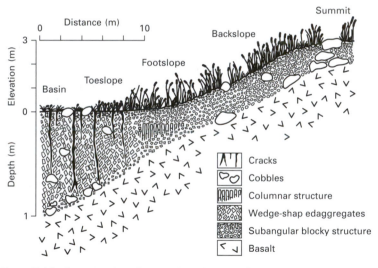

Figure 5.17 Cross-section of a vernal pool catena, Riverside County, California, showing soil morphology and slope elements. (After Weitkamp *et al.*, 1996.)

vernal pool receives drainage from roughly 4 ha and sits on the Mesa de Burro at an elevation of 730 m. The catena was a transect on a north-facing slope and had a maximum gradient of 9 per cent (Figure 5.17). Grasses and forbs, mainly soft brome (*Bromus bordeaceus*) and slender oat (*Avena barbata*), dominated the summit and backslope, toad rush (*Juncus sphaerocarpus*) the toeslope, and button celery (*Eryngium aristulatum*) the basin. Soils on the

summit and backslope have a weak and moderate subangular blocky structure. They are shallow with a coarse-loamy texture and are maintained as Entisols by extensive burrowing of rodents and their predators. Footslope soils are clayey Alfisols with a moderate subangular blocky structure above horizons with a strong prismatic structure. The basin and toeslope soils are Vertisols with a strong, coarse and very coarse angular blocky structure, and wedge-shaped

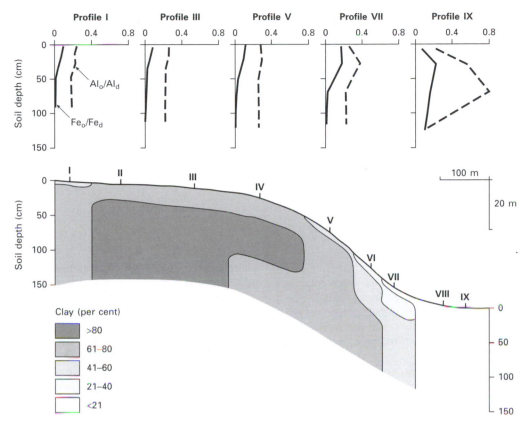

Figure 5.18   Soil toposequence with clay distribution and depth functions for Fe_o/Fe_d and Al_o/Al_d ratios near Itacoatiara, Amazonas, Brazil. (Adapted from Botschek *et al.*, 1996.)

aggregates, slickensides (smooth surfaces with parallel striae and grooves characteristic of Vertisols), and shrinkage cracks. Of the soil properties in the vernal pool catena, iron and manganese display interesting variations. Conditions in much of the catena are not conducive to iron movement. Iron oxides are concentrated in the upper slope soils where they impart 5YR hues and high chromas to soil material. However, the form of iron does vary along the catena. Amorphous iron oxides, Fe_o, as measured in an ammonium oxalate extract, are more important, relative to crystalline iron oxides, in lower-slope soils, whereas crystalline iron oxides, Fe_d, as measured in a citrate–bicarbonate–dithionate extract, predominate in the well-drained upper-slope soils. Manganese oxides are soluble at higher redox potentials than iron oxides and manganese is preferentially solubilised and carried down the catena in throughflow. The throughflow is produced by water infiltrating the upper-slope soils and moving downslope along the soil–basalt bedrock contact and within the vesicular basalt. It bears man-

ganese from upper-slope sites and transports it to footslope, toeslope, and basin sites, where it is oxidised and precipitated as manganese oxides when the vernal pool dries out. In the basin, the Vertisols are wetted unevenly as water moves downwards through soil cracks and upwards from throughflow in the underlying vesicular basalt. Much of the manganese in the lower slope sites is found in manganese stains and nodules, features that are most abundant in the basin and decrease in number towards the footslope. Manganese stains are found below 5 cm, and manganese nodules below 10 cm. The nodules tend to lie just above the Bss1 horizons of the toeslope and basin soils, and in the Bt1 horizon of the footslope soil.

A larger toposequence under primary rain forest, near Itacoatiara, Amazonas, Brazil, used depth functions to investigate lateral movement of soil materials (Botschek *et al.* 1996). Nine soil profiles were studied along an 840-m transect with a relief of 50 m, although five were selected for presentation of results (Figure 5.18). The soils of the area are formed in

Table 5.4   Total gains and losses of phosphorus in soil from an Illinois catena

Soil	Area (ha)	Net change in total phosphorus content to a depth of 230 cm	
		(g/m^2)	(kg/soil area)
Site 1	0.0376	+493.49	+188.0
Site 2	0.1596	+284.96	+460.5
Site 3	0.1808	+58.86	+107.7
Site 4	0.1708	−103.19	−178.3
Site 5	0.2184	−99.94	−221.0
Site 6	0.2184	−151.34	−545.6
Landscape balance	0.3564	−16.60	−188.7

Source: Adapted from Smeck and Runge (1971).

Tertiary fluviolacustrine deposits derived from deeply weathered Precambrian rocks. Concretions and crusts, colluvium and alluvium formed during the Pleistocene, and fluviomarine sediments were laid down in valley bottoms during the Holocene. In the toposequence studied, soil texture changes from predominantly clay in upper slope positions to sand at the slope base. The ratios between oxalate-soluble and dithionate-soluble iron and aluminium (Fe_o/Fe_d and Al_o/Al_d) measure the relative proportions of more mobile forms of those elements. The ratios for both iron and aluminium increase down the toposequence, which might be indicative of lateral translocation, although differences in the weathering environment might also account for the patterns.

Quantifying downslope movement

An early attempt to quantify the downslope movement of soil material was made in a catena in Cass County, Illinois (Smeck and Runge 1971). In some soils of the catena, more phosphorus had accumulated in the B horizons than could be accounted for by eluviation from the overlying A horizons, and in other soils more phosphorus had been lost from the A horizons than had accumulated in the B horizons. Net gains and losses in each profile were used to investigate phosphorus dynamics. Using zirconium oxide as an index against which to gauge changes in phosphorus content, and estimating the area represented in the landscape by each profile, absolute gains and losses for each soil unit were calculated (Table 5.4). In the entire catenary sequence, a minimum of 944.9 kg of phosphorus had been translocated laterally within an area of 1.1 ha. Summit soils had lost 151.34 g/m^2 of phosphorus; footslope soils had gained 493.49 g/m^2 of phosphorus.

Sommer and Stahr (1996) devised a method for inferring gains and losses of iron, manganese, and phosphorus in landscapes developed on sedimentary rocks. To do this, they introduced an element–clay ratio (see also Sommer 1992). The amounts of pedogenic iron, manganese, and phosphorus (in kg) were computed on a pedon basis, that is, for the volume under 1 m^2 down to the C horizon, or to an arbitrary depth in soil lacking a C horizon. These amounts were then divided by the amounts of clay in the same volume. By comparing these ratios to the ratio of elements to clay in the parent material (or mixture of parent materials), semi-quantitative statements about elemental gains and losses due to solutional processes can be given. If only transformation processes were operative, the ratios in the soil material should equal the ratios in the parent material. Any depletion or (relative or absolute) accumulation of the elements is suggested from values departing from this ratio. For phosphorus, the ratio in soil material should decline with time in comparison with the ratio in the parent material. Minor differences between the ratios in soil and parent material for phosphorus, or cases where the soil material ratios exceed the parent material ratio, register relative and absolute gains. Applying the ratio method to 27 pedons in south-west Germany produced encouraging results. Figure 5.19 shows two toposequences with element–clay ratios calculated at different sites. Along the clayey-sandstone plateau toposequence (Figure 5.19(a)), the top pedon (a Typic Dystrochrept) reveals neither losses nor gains of manganese but slight losses of iron. The middle pedon (a Histic Humaquept)

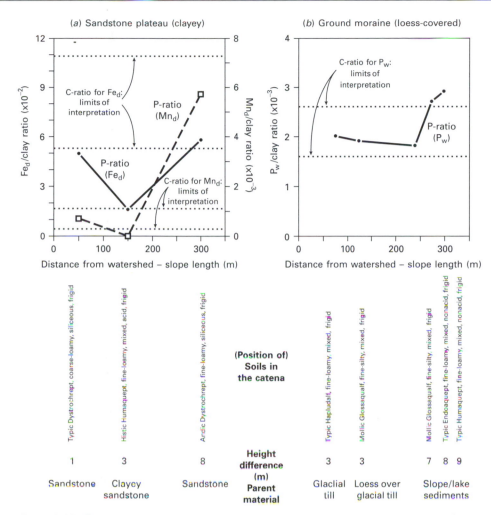

Figure 5.19 Two toposequences in south-west Germany: (a) toposequence on a clayey sandstone plateau; (b) toposequence on a loess-covered ground moraine. In both toposequences, areas within the dotted lines define areas with non-significant changes in the P ratios. (Adapted from Sommer and Stahr, 1996.)

has lost nearly all its manganese and about 75 per cent of its iron. The iron and manganese from this pedon appear to have been translocated downslope to the bottom pedon (an Andic Dystrochrept), where a higher Eh (redox potential) has led to their being immobilised. Along the loess-covered ground moraine toposequence, the top three pedons display neither gains nor losses of phosphorus (Figure 5.19(b)). The bottom two pedons show an accumulation of phosphorus. The first three pedons are covered by non-fertilised forest, while the bottom pedons are influenced by groundwater and are under fertilised pasture, hence the accumulation of phosphorus. This

catena is an example of an absolute accumulation in a landscape without any attendant areas of depletion.

Other reconstructions of catenary development, like the loess-covered ground moraine catena in Germany, suggest that lateral movement of soil materials is not always important in pedogenesis, though the profiles nonetheless vary according to their topographic setting. In Texas, reconstruction techniques were used to explore catenary influences on the evolution of carbonate-rich horizons (West *et al.* 1988). The conclusion was that gains and losses of carbonates in summit soils were best explained by deeper leaching in this stable landscape position. Backslope soils were

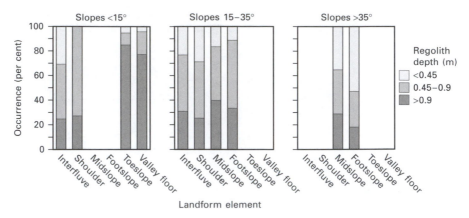

Figure 5.20 Relationship between regolith depth, slope angle, and landform element in a catchment of the Waipori River, New Zealand. (Adapted from Webb *et al.*, 1999.)

unstable, and carbonate distribution with depth was affected more by erosion than by downslope enrichment with carbonates. Soil reconstruction techniques were employed to examine relationships between landscape position and soil genesis in the Piedmont and Blue Ridge Highlands region in Virginia (Stolt *et al.* 1993a, 1993b). Four toposequences were selected for detailed study. Two were on mica gneiss, one on augen gneiss, and the fourth on gneissic schist. The chief processes that could be quantified using reconstruction analysis were sand and silt weathering, subsequent transport and leaching of weathering products, clay illuviation, and the accumulation of free iron oxides. Results showed that summit and backslope soils undergo the same process of soil evolution. Footslope soil genesis is, in part, dependent on the type and composition of parent material: horizons formed in substantially weathered local alluvium displayed minimal sand and silt weathering and minimal leaching; subjacent horizons evolved in parent rock yielded signs of weathering and clay eluviation. There was evidence that some material in footslope soils was derived from upslope. However, differences between the soils on summits and backslopes, which are morphologically very similar, appeared to have resulted from soil disturbance by tree-throw or natural hillslope erosion, from accelerated erosion due to cultural practices, or from differences in parent materials, and not simply from catenary position.

Sediments and hillslopes

The transfer of the clastic debris of weathering is greatly influenced by landform and position. In a catchment of the Waipori River, New Zealand (see Figure 5.8), the depth of regolith was found to relate to slope angle and landform element (Figure 5.20). Topographic surveys of small plots (7–10 ha) in fields about 80 km north-west of Saskatoon, Canada, were carried out along transects using an average density of 50 observations per hectare (Pennock and de Jong 1987). Slope curvature, contour curvature, and slope gradient were calculated at all points. Soil samples were taken at 50-m intersections of a grid, and analysed for caesium-137, the redistribution of which compared to native or control sites was used as a measure of soil erosion. Distinct differences in soil gains and losses were associated with landform elements according to the vergency pattern of flowlines: on shoulders, convergence caused enhanced erosion; deposition was associated mainly with footslopes. This pattern is clear for the data for the entire data set (Figure 5.21). A study in a 10.5- ha first-order drainage basin in the south-western part of Bureau County, north-west Illinois, disclosed that erosion varies with landscape position (Kreznor *et al.* 1989). Transect data of all geomorphic units at a cultivated site showed that shoulders were slightly or moderately eroded, while lower backslopes and upper footslopes were either severely or very severely eroded, hinting that slope length was the key determinant of erosion. Landscape form also exerted an influence on erosion: geomorphic units with positive contour curvature (hollows) were less eroded than those with negative contour curvature (spurs).

Soil erosion depends also on vegetation cover. In a toposequence in south-west Niger, erosional and depositional sites were identified by comparing [137]Cs

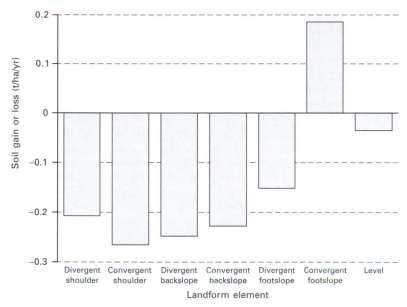

Figure 5.21 Soil gains and losses in different landform elements in small plots in fields north-east of Saskatoon, Canada. (From data in Pennock and de Jong, 1987.)

inventories with a reference site (Chappell *et al.* 1998). It was found that the vegetation canopy protects accumulated dust from water erosion on steep slopes and from wind erosion on gentle slopes.

Soil landscapes

The term soil–landscape bears many meanings. In traditional soil survey, soil–landscape models relate landscape units to soil classes or to soil properties. Such soil–landscape relationships are mostly used as aids to pedological mapping and either go unreported or are summarised in a general diagram (cf. Rijkse and Trangmar 1995; Bui *et al.* 1999). These soil–landscape models commonly have utilitarian goals of national survey programmes and serve a valuable purpose in that context (cf. McSweeney *et al.* 1994). However, they portray soil and landform data in a form that is of little use when considering the generative processes of soil–landscape evolution. An alternative is a dynamic view of soil–landscapes that focuses on soil and geomorphic processes and the roles that they play in shaping the soil–landscape system. At least two process-based models of soil–

landscape systems exist. The first was promulgated before sophisticated tools and techniques of geographical information systems were available. The second arose from the burgeoning fields of remote sensing and digital terrain modelling.

Dynamic soil–landscape models

Soil–landscape systems

A close association between soils, sediments, water, and topography is seen in landscapes. Several frameworks for linking pedological, hydrological, and geomorphic processes within landscapes have been proposed, most of them concerned with two-dimensional catenas (e.g. the nine-unit land-surface model). The idea of soil–landscape systems was an early attempt at an integrated, three-dimensional model (Huggett 1975). The argument was that dispersion of all the debris of weathering – solids, colloids, and solutes – is, in a general and fundamental way, influenced by land-surface form, and is organised in three dimensions within a framework dictated by the drainage network. In moving down slopes, weathering products tend to move at right-angles to land-surface contours. Flowlines of material converge and diverge according to contour curvature. The pattern of vergency influences the storage of water, solutes, colloids, and clastic sediments at different landscape positions. Naturally, the movement of weathering products alters the topography, which in turn influences the movement of the weathering products – there is feedback between the two systems.

If soil evolution involves the change of a three-dimensional mantle of material, it is reasonable to propose that the spatial pattern of many soil properties will reflect the three-dimensional topography of the land surface. This hypothesis can be examined empirically by observation and statistical analysis, and theoretically using mathematical models. Three-dimensional topographic influences on soil properties were originally investigated in small drainage basins (Huggett 1973, 1975; Vreeken 1973) and larger ones (Roy *et al.* 1980). Later work has confirmed that a three-dimensional topographic influence does exist, and that some soil properties are very sensitive to minor variations in topographic factors. A case in point is an investigation into natural nitrogen-15 abundance in plants and soils within the Bîrsay study area, southern Saskatchewan, Canada (Sutherland *et al.* 1993). Two sampling grids were used, each involving 144 points, in an irrigated field. The large grid was 110×110 m and the small grid 11×11 m (Figure 5.22). Samples of soils from both grids, and samples of durum wheat (*Triticum durum*) from the large grid, were analysed for nitrogen-15 (Figure 5.22(c) and (d)). Spatial statistical analysis indicated that the distribution of nitrogen-15 was random in the small grid, but in the large grid it was concentrated in depressions and followed the same pattern as denitrification activity and related soil properties (Eh, soil water content, bulk density, and total respiration). Spatial variability of nitrogen-15 in plants was greater than that in soils. Extreme outliers of nitrogen-15 in plants were associated with the landscape elements with highest denitrification activity and lowest Eh values. Elevation was the single most important variable for both plant and soil nitrogen-15 abundances, accounting for 26 per cent of the variation in soil nitrogen-15 and 31 per cent of the variation in plant nitrogen-15. Overall, the analysis suggests that topography had a significant influence on landscape patterns of nitrogen-15 in soil and plants.

The soil–landscape continuum

A new model of the soil–landscape continuum (McSweeney *et al.* 1994) builds on technical advances made through linking hydrology and land-surface form for terrain-based modelling of hydrological processes. Digital terrain modelling came into its own with the development of geographical information system (GIS) technology. The value of digital terrain modelling to hydrology, geomorphology, and ecology was recognised by the early 1990s (e.g. Moore *et al.* 1991) (p. 20). In the new model of soil–landscape systems, various sources of spatial data (e.g. vegetation and geological substrate) and attribute data (e.g. soil organic-matter content and particle-size distribution) are integrated through GIS technology. Four inter-related and iterative stages are applied in the model. The stages are designed to be clear-cut and applicable to any scale of accuracy specified by the user. They are as follows.

1. *Physiographic domain characterisation*   The first stage is the gathering and analysis of pertinent data to characterise a physiographic domain. It involves bringing together available data on geology, climate, and vegetation, as well as remotely sensed data, to define and characterise the physical geography of an area.

2. *Geomorphometric characterisation*   The second stage uses digital terrain models to perform a geomorphometric analysis and characterisation of the landscape. This stage derives primary and secondary terrain attributes from elevation data. Primary attributes are flow direction and first and second derivatives of the surface topographic surface – slope gradient, slope curvature, slope aspect, and contour curvature. Secondary attributes are simple or complex combinations of primary attributes that provide indices representing specific landscape processes (see Table 2.2). Stage 2 provides a land surface representation, to which other data are referenced, and a division of the land surface into areas that may correspond with soil patterns.

3. *Soil horizon characterisation*   The third stage uses georeferenced sampling as a basis for defining soil horizons, their attributes, and their spatial arrangement in the landscape. Soil horizons are used instead of soil profiles as a basic unit for field investigation. This practice departs from traditional methods of soil survey, but is central to developing a three-dimensional approximation of soil diversity in the landscape.

4. *Soil property characterisation*   The fourth stage is the laboratory and statistical analyses of soil-horizon attributes collected during the third stage.

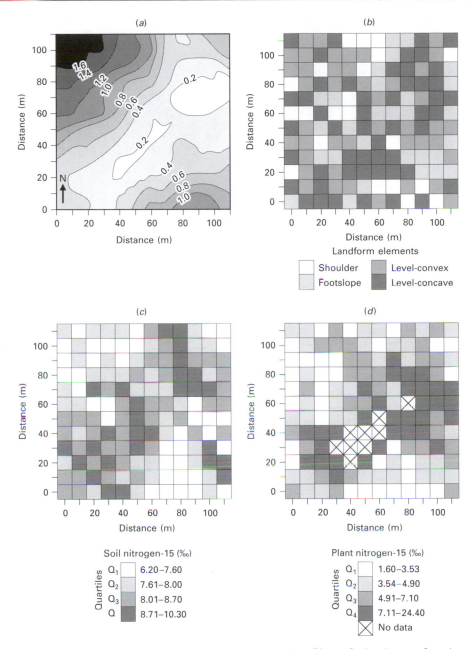

Figure 5.22 Study site and quartile maps of soil properties, Birsay, Saskatchewan, Canada: (a) topography; (b) landform elements in grid cells; (c) soil nitrogen-15; (d) plant nitrogen-15. (Adapted from Sutherland *et al.*, 1993.)

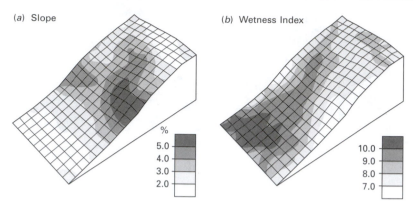

(a) Slope

(b) Wetness Index

Figure 5.23 A catena at a site near Stirling, Colorado, USA: (a) slope; (b) wetness index. The grid is 15.24 × 15.24 m. (Adapted from Moore *et al.*, 1993.)

It focuses on the basic structure of the model to provide insight about the range, variance, and correlation of soil and associated landform attributes, and the soil and geomorphic processes that shaped the landscape.

An underlying assumption in this model is that soil and geomorphic patterns are related and linked to the same suite of processes. If this assumption should be correct – and much evidence suggests that it is – then in many parts of landscapes, landforms should correlate strongly with the underlying nature and arrangement of soil horizons. The challenge is to establish where landform–soil horizon correlations are strong; to determine the feasibility of using such relationships to extrapolate soil patterns across the landscape; and to interpret the relationships through processes and events that shape soil–landscape evolution (McSweeney *et al.* 1994). However, some landscapes contain very complex landform–soil horizon patterns at small scales that require special treatment if they are to be portrayed in a three-dimensional model. Landscapes with cradle–knoll or gilgai microtopography are cases in question. Additionally, deciphering landform–soil horizon relationships in landscapes where subsurface features, such as varied bedrock, or processes, such as saline groundwater, exert a powerful influence on soil patterns may prove especially difficult.

The new model uses GIS, image processing, and statistical analysis software. No single system meets all requirements, and users mix and match software and hardware platforms (McSweeney *et al.* 1994).

Nonetheless, most research adopting this kind of model follows the same basic stages that in turn characterise the physiographic domain, geomorphometry, soil horizons, and soil properties (e.g. Bornand *et al.* 1994, 1997; see also Basher 1997). The rest of this section will describe and discuss several examples of soil–landscape models of this sort.

Soil processes and soil types in landscapes

Soil processes

Relationships between soil attributes and terrain attributes were revealed in a landscape at Sterling, Logan County, Colorado (Moore *et al.* 1993). The area of the site is 5.4 ha. Relative elevation and A horizon thickness were measured at, and soil samples were taken from, 231 locations on a 15.4 × 15.4 m grid. Several primary and secondary topographic attributes were derived from the elevation data. Primary attributes were slope (per cent), aspect (degrees clockwise from north), specific catchment area (m²/ m), maximum flowpath length (m), profile curvature (/m), and plan curvature (/m). Secondary attributes were a wetness index, a stream-power index, and a sediment-transport index. The 'best' combination of terrain variables for explaining variation in soil attributes was explored using stepwise linear regression. Slope and a wetness index (Figure 5.23) were the topographic variables most highly correlated with soil properties, accounting individually for about 50 per cent of the variability in A horizon thickness, organic matter content, pH, extractable phosphorus,

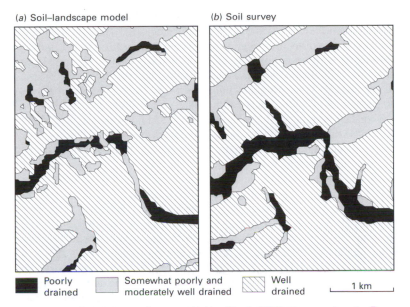

(a) Soil–landscape model   (b) Soil survey

■ Poorly drained   ▨ Somewhat poorly and moderately well drained   ▧ Well drained   1 km

Figure 5.24   Soil drainage classes at Licking Creek, Mifflintown quadrangle, Pennsylvania, USA. (a) Soil drainage class predicted by a soil–landscape model. (b) Soil drainage class derived from the Juniata–Mifflin County Soil Survey. (Adapted from Bell *et al.*, 1994.)

and silt and sand contents. The regression equations were used to predict the spatial distribution of soil attributes. Correlation among the three terrain attributes and soil attributes suggests that pedogenesis in this landscape has been influenced by the way in which water flows through and over the soil.

Soil mapping

During the last 10 years or so, technological advances in GIS and remote sensing have allowed quantitative soil–landscape models to be applied directly to soil mapping. The models are used for various purposes, including the delimitation of soil drainage classes, which is discussed in this section, and the discrimination of terrain objects relevant to soil distribution (e.g. de Bruin and Stein 1998; de Bruin *et al.* 1999).

For example, one study combined a statistically based soil–landscape model with GIS technology to create soil drainage class maps (Bell *et al.* 1994). The starting point was an existing soil–landscape model that predicted soil drainage class from topographic variables, proximity to surface drainage features, and parent material (Bell *et al.* 1992). A digital geographic database of parent material, terrain, and surface-drainage feature proximity variables of the Mifflintown 7.5′ topographic quadrangle covering 14 448 ha of south-central Pennsylvania, USA, was used as a source of inputs to the model. Soil drainage class probabilities were estimated from unique combinations of the landscape variables in each raster. A map of the most likely drainage class was then constructed by assigning to each raster the drainage class with the highest probability of occurrence (Figure 5.24(a)). The modelled soil drainage class map agreed with a 1:20 000-scale soil survey for 67 per cent of the study area (Figure 5.24(b)). The majority of the disagreement – 54 per cent – was attributed to areas where the soil survey predicts well-drained soils and the model predicts somewhat poorly drained to moderately well-drained soils. This technique provides a consistent means of assigning soil drainage class according to landscape attributes, documents the data and decision criteria used for drainage class assignment, estimates the uncertainty associated with drainage class assignment, and generates a digital map for GIS applications.

Another study developed a colour index – the Profile Darkness Index (PDI) – to aid in delineating the change from upland to wetland soils (Thompson *et al.* 1997). The PDI is well correlated with the duration of saturated and reducing conditions in specific Mollisol catenas in humid regions of the north-central USA. A 10-m DEM was created from a ground survey using an infrared laser surveyor. A universal kriging procedure within ARC/INFO was used to interpolate the approximately 1500 irregular grid points to a regular grid DEM of 1978 points (43 × 46). Primary and secondary terrain attributes were computed from this DEM. Primary terrain attributes were slope gradient, profile curvature, plan curvature, tangential curvature, specific catchment area, upslope length, distance to a local depression, and elevation above a local depression. Secondary terrain attributes were a compound topographic index, a stream power index, a sediment transport capacity index, and a depression proximity index. Regression models were used to quantify the relationships between terrain attributes and the PDI. It was found that variations in slope gradient, profile curvature, and elevation above local depression explained up to 65 per cent of the variation in the PDI. The approach is helpful in delineating hydric soils, which are associated with wetlands.

Skidmore *et al.* (1991) developed an expert system to map forest soil–landscape units formed on granite. They considered three thematic maps to be important in the distribution of the soils. The first layer showed the distribution of nine classes of native eucalypt forests, and the second and third were derived from a DEM and represented slope gradient and a soil wetness index combined with topographic position. These layers were combined in a raster GIS. From a knowledge of soil distributions, the relationships between the soil–landscape units and the three data layers were quantified by an experienced soil scientist and used as rules in a rule-based expert system. The thematic layers in the GIS provided data for the expert system to infer the forest soil–landscape unit most likely to occur at any given pixel. When compared with a conventional soil–landscape map generated using interpretation of aerial photographs, the soil–landscape map output was considered to be good.

Soils and stream order

Position within a drainage network should have a bearing on soil type and soil properties (Bunting 1965, 75). This relationship would be expected if, as seems reasonable, drainage basin expansion and integration produce a sequence of valley development in which each component drainage basin in the network has a characteristic combination of landscape elements.

In turn, the landscape elements influence the slope and soil processes. Thus the evolution of a soil profile should be influenced by the landscape at several scales: microscale landforms, hillslopes, single drainage basins, and the entire landscape or drainage basin network. The relationship between soil, soil catenas, and stream order has not been much explored, but the available evidence suggests that it does exist. In the Rocksberg Basin, Queensland, soil type is related to catenary position and stream order (Arnett and Conacher 1973). In west Essex, England, consistent relationships exist between soil series and stream order (Huggett 1973; 1995, 194–5). A study conducted on soils and landscapes in drainage basins in central Spain suggests a structural correlation between the spatial organisation of the fluvial systems and the soil landscape (Ibáñez *et al.* 1990). An investigation of soils on Dartmoor, England, found that relationships between soil type and slope position were modulated by location within a drainage basin network (stream order) (Gerrard 1988, 1990b).

Far more work on relationships between soil types, soil properties, and stream order needs to be carried out before any firm conclusions can be drawn. However, there does appear to be a relationship between these variables, so vindicating the view that soil should be viewed as a three-dimensional body interacting with the topography of the landscape in which it evolves.

Sediments in landscapes

Processes affecting regolith thickness operate in three dimensions and should, therefore, be influenced by landscape form as well as landscape position. It was mentioned earlier (p. 148) that, in a 10.5-ha first-order drainage basin in south-west Bureau County, north-west Illinois, the three-dimensional land-surface form influences soil erosion: geomorphic units with positive contour curvature (hollows) were less eroded than those with negative contour curvature (spurs) (Kreznor *et al.* 1989). Three-dimensional landscape influences on regolith thickness are seen in steepland hillslopes of Taranaki, New Zealand (DeRose *et al.* 1991). Regolith depth varies greatly on both spur and swale sites, but the mean depth for swales (122 cm) is significantly deeper than for spurs (59 cm), as is the variation about the mean – the standard deviations are 79 cm for swales and 43 cm for spurs. These differences point to major dissimilarities in the processes influencing regolith depth on convex and concave sites.

A main thrust of research into sediments in landscapes has focused on soil erosion. As this is such a large topic, it is given its own section.

Soil erosion modelling

Background to soil erosion models

Soil erosion has become a global issue because of its environmental consequences, including pollution and sedimentation. Major pollution problems may occur from relatively moderate and frequent erosion events in both temperate and tropical climates. The control and prevention of erosion are necessities in virtually every country of the world under almost all land-cover types (Morgan 1995). Prevention of soil erosion is the reduction of the rate of soil loss to approximately the rate that would occur under natural conditions. It is of critical importance and depends upon implementation of suitable soil conservation strategies (Morgan 1995). This, in turn, requires a thorough understanding of the processes of erosion and the ability to provide some sort of prediction. The factors that influence the rate of soil erosion include rainfall, runoff, wind, soil, slope, land cover, and the presence or absence of conservation strategies.

Before implementing any conservation strategies, it is useful to be able to make an estimate of how fast soil is being eroded. Such estimates can be compared with what is considered acceptable and will enable evaluation of the effectiveness of conservation strategies. Methods of prediction of soil loss under a wide range of conditions are required, soil erosion models fitting the bill. There is a range of different types of soil erosion model, including physical, analogue, digital, physically based, stochastic, and empirical. Early models within this range are well described by Gregory and Walling (1973). A great many types of soil erosion model, which are applicable to different temporal and spatial scales, have been developed. Since slope gradient is one of the factors influencing soil erosion, most of these models include some form of topographic parameter. For instance, one of the first attempts to develop a soil-loss equation for hillslopes related erosion to slope steepness and slope length (Zingg 1940):

$$E \propto (\tan^m \theta) L^n \qquad (5.1)$$

where E is soil loss per unit area, θ is the slope angle, and L is slope length. Further refinements have involved the addition of a climatic factor (Musgrave 1947), a crop factor (Smith 1958), a conservation factor, and a soil erodibility factor.

Empirical soil erosion models

The Universal Soil Loss Equation

The Universal Soil Loss Equation (USLE) is a popular empirical soil erosion model. It is based upon statistical analysis of soil erosion data collected from small erosion plots (sometimes referred to as 'Wischmeier plots') in the USA. The USLE was developed by changing the climatic factor of Zingg's modified equation to a rainfall erosivity index (R), and then subsequently modified further and updated by Wischmeier and Smith (1978):

$$E = R \times K \times L \times S \times C \times P \qquad (5.2)$$

where E is the mean annual soil loss, R is the rainfall erosivity factor, K is the soil erodibility factor, L is the slope length factor, S is the slope steepness factor, C is the crop management factor, and P is the erosion control practice factor.

The slope length, L, and slope steepness, S, factors are combined to produce a single index that represents the ratio of soil loss under a given slope steepness and slope length to the soil loss from the standard condition of a 5° slope, 22 m long, for which $LS = 1.0$. The value can be produced from nomographs (Wischmeier and Smith 1978) or from the equation

$$LS = (x/22.13)^n (0.065 + 0.045x + 0.0065s^2) \qquad (5.3)$$

where x is the slope length (m) and s is the slope gradient (per cent). The value of n is varied according to the slope steepness. Morgan (1995) provides the details for the other factors (i.e. R, K, C, and P) for the USLE.

The USLE has been widely used, especially in the USA, for estimating sheet and rill erosion in national assessments of soil erosion. However, there are a number of problems in using the USLE, including limitations of the database to model slopes where cultivation is possible, normally 0° to 7°, and there is considerable interdependence between the variables. A revised USLE (RUSLE) has been developed (Renard et al. 1991) that includes a number of modifications, including a seasonally variable K factor that allows the LS factor to take account of the susceptibility of the soils to rill erosion, and a new method for computing the C factor value through the multiplication of various subfactor values. The USLE is also of limited use in European sites because it became apparent that, owing to different climatic conditions and soil properties, the rainfall factor R and the K-nomographs developed in the USA to determine the erodibility factor K are not applicable to the European loess soils without some fundamental modifications. Other disadvantages of the USLE are that the process of sedimentation is not represented in the equation and that other soil losses and gains over neighbouring areas are not accounted for (De Roo *et al.* 1989).

Soil Loss Estimator for Southern Africa

Another example of an empirical soil erosion model is the Soil Loss Estimator for Southern Africa (SLEMSA) (Elwell 1978), which was developed mainly from data from the Zimbabwe highveld to evaluate the erosion resulting from different farming systems so that conservation approaches could be installed. The equation is

$$Z = K \times X \times C \qquad (5.4)$$

where Z is the mean annual soil loss (t/ha), K is the mean annual soil loss (t/ha) from a standard field plot (30 × 10 m, with a 2.5° slope for a soil of known erodibility, F, under a weed-free bare fallow), X is a dimensionless combined slope length and steepness factor, and C is a dimensionless crop management factor.

Non-spatial, physically based soil erosion models

During the 1980s there was a shift from empirical models, such as the USLE, that calculate soil erosion only on a single slope, towards more analytical, physically based models such as CREAMS (Knisel 1980) and ANSWERS (Beasley and Huggins 1982). Empirical models are based upon statistical analysis of important factors in the soil erosion process and provide only approximate and probable outcomes. Physically based models describe the process of erosion with physical mathematical relationships and should provide more exact results. Physically based models can be divided into 'lumped' and 'distributed' models. 'Lumped' models, such as CREAMS, describe an overall or average response for the watershed and are non-spatial. 'Distributed' models, such as ANSWERS, are spatial models and predict the spatial distribution of runoff and sediment over the land surface during

individual storm events, as well as the total runoff and soil loss.

Physically based models use the laws of conservation of mass and energy. The majority use a particular differential equation called the continuity equation that is a statement of the conservation of matter as it moves through space over time (Huggett 1985; Morgan 1995). This equation can be applied to soil erosion of a small area of a hillslope, with an input to material to the area (or segment) from the upslope segment and an output of material to the downslope segment. A difference between the input and output relates to either erosion or deposition on the segment. If ∂ denotes a change in the value, the continuity equation can be expressed as a mass balance (Bennett 1974; Kirkby 1980):

$$\frac{\partial(AC)}{\partial t} + \frac{\partial(QC)}{\partial x} - e(x, t) = q_s(x, t) \qquad (5.5)$$

where A is the cross-sectional area of the flow, C is the sediment concentration in the flow, Q is the discharge, t is time, x is horizontal distance downslope, e is the net pick-up rate or erosion of sediment on the slope segment, and q_s is the rate of input or extraction of sediment per unit length of flow from land outside the segment (which would be zero on a plane slope segment).

CREAMS

CREAMS (Chemicals, Runoff and Erosion from Agricultural Management Systems) is a field-scale model developed to assess non-point-source pollution and to quantitatively investigate the environmental consequences of different agricultural practices (Knisel 1980). CREAMS comprises three components including hydrology, erosion, and chemistry. The hydrology component estimates runoff volume and peak runoff rate as well as infiltration, evapotranspiration, soil water content, and percolation on a daily basis. The erosion component produces daily estimates of erosion and sediment yield as well as particle-size distribution of the eroded material at the edge of a field. The chemistry section provides estimates of storm loads, and average concentrations of adsorbed and dissolved chemicals, in the runoff, sediment, and percolated water. CREAMS has become one of the most widely applied and tested process-based models. Trials have been carried out in the USA (Foster and Lane 1981), Czech Republic (Holý *et al.* 1982), and Lithuania (Kairiukstis and Golubev 1982).

WEPP

Another process-based model is WEPP (Water Erosion Prediction Project). It was designed to replace the USLE for routine assessment of soil erosion by organisations responsible for soil and water conservation and environmental planning (Nearing *et al.* 1989). There are three computer models, including a profile version, a watershed version, and a grid version. The profile model estimates soil detachment and deposition on a hillslope profile and the net total soil loss from the end of the slope. It can be used for modelling areas up to about 260 ha in size. The watershed and grid models estimate net soil loss and deposition over small catchments and can handle ephemeral gullies on valley floors. The models have parameters for climate, soils, topography, management, and any conservation practices. Although they are designed to run on a continuous simulation, they can be run for a single storm event.

EUROSEM

The European Soil Erosion Model (EUROSEM) is an event-based model which computes the sediment transport, erosion, and deposition over the land surface throughout a storm event. It can be applied to small catchments or individual fields (Morgan 1994; Morgan *et al.* 1994). Unlike other models, EUROSEM simulates rill and inter-rill erosion explicitly, including the transport of water and sediment from internal areas to rills, so allowing for the deposition of material *en route*. This is thought to be more realistic than assigning all or a set proportion of the detached material to the rills. In addition, a more physically based approach to simulating the effect of vegetation or crop cover is utilised that takes account of the influence of leaf drainage. Soil conservation measures can also be accounted for by choosing appropriate values of the soil, microtopography and plant cover parameters that describe the condition associated with the measures. EUROSEM does not, however, describe the eroded sediment in terms of its particle size.

Quinton (1994) validated the EUROSEM model on a data set from the Rothamsted Experimental Station's Woburn Experimental Farm in England. The soil here is predominantly Cottenham Series, a Brown Sand on Lower Greensand, with some patches of Stackyard Series, a Brown Earth on sandy colluvium. The results showed that, as a predictive model, EUROSEM is able to predict soil loss to within 1 t/ha in 90 per cent of simulations, and runoff to within 0.8 mm in 50 per cent of simulations. Observed and predicted sedigraphs showed good agreement. Nonetheless, the EUROSEM hydrograph was sharper than the observed hydrograph.

Spatial, physically based soil erosion models

Lumped, physically based models such as CREAMS have been criticised because they do not allow for the spatial variability of the erosion process (De Roo *et al.* 1989). Distributed-parameter models attempt to overcome this limitation and to increase the accuracy of the model predictions by preserving and using information concerning the areal distribution of all spatially variable, non-uniform processes being modelled. Potentially, these models should produce a more accurate simulation of natural catchment behaviour. They are capable of simulating conditions at all points within a watershed simultaneously and of forecasting the spatial patterns of hydrological conditions within a catchment, as well as simple outflows and bulk storage volumes (Beven 1985).

ANSWERS

One of the first examples of a distributed-parameter model is ANSWERS (Areal Nonpoint Source Watershed Environment Response Simulation), which was presented by Beasley *et al.* (1980) and can be used to model surface runoff and soil erosion. When first developed, ANSWERS was considered to be one of the most complete erosion models (Foster 1982). However, more process-based models have been developed since. ANSWERS is designed to simulate the behaviour of watersheds that are primarily agricultural in use, during, and immediately after, a rainfall event. Its main purpose is in planning and evaluating various strategies for controlling pollution from intensively cropped areas. The catchment being modelled is subdivided into square elements. Values of such factors as slope, aspect, soil variables (porosity, moisture content, field capacity, infiltration capacity, erodibility factor), crop variables (coverage, interception capacity, USLE *C/P* factors), surface variables (roughness and surface retention), and channel variables (width and roughness) are defined for each element.

A rainfall event is simulated with time increments of one minute, with spatial and temporal variability of rainfall accounted for. The continuity equation is employed to compute the composite response of the single elements. The input of downslope elements is provided by the output of upslope elements. Physically based mathematical relationships are used to describe interception, infiltration, surface retention, drainage, overland flow, channel flow, subsurface flow, detachment by rainfall, and/or overland flow, and sediment transport by overland flow (inter-rill erosion). When the water and sediment reach an element that has a channel, they are transported to the catchment outlet. Sedimentation within a channel occurs when the transport capacity has been exceeded.

Spatial soil erosion models and GIS

Many of the physically based models have been adapted to, or developed to use, raster-based GIS. These GIS-based models usually incorporate a DEM from which slope and slope length can be calculated and used as parameters in the predictive models. Many existing erosion models have been modified and interfaced with GIS, including a range of modifications of the empirical USLE equation (e.g. Warren *et al.* 1989; Flacke *et al.* 1990; Huang and Ferng 1990), the WEPP model (Savabi *et al.* 1996), and watershed models for non-point-source pollution such as AGNPS or ANSWERS (De Roo *et al.* 1989; Rewerts and Engel 1991; Srinivasan and Engel 1991).

A number of studies have looked at the automation of the USLE and RUSLE using GIS and a DEM (Desmet and Govers 1996; Mitasova *et al.* 1996). The *LS* topographic factor is calculated automatically. The advantage of the computer procedure is that if data on land use and soils are available specific *K*, *C*, and *P* values can be assigned to each land unit so that predicted soil losses can be calculated using simple overlay procedures. The automated procedure results have been compared to manual results. Desmet and Govers (1996) found that the manual method led to significantly lower *LS* values for plan-concave areas, because the effect of flow concentration on rill development cannot be accounted for. Both the manual and automated methods yield results in agreement with soil map information. However, the automated method was advantageous in terms of speed of execution and objectivity. Furthermore, the ease of linking the topographic module with a GIS facilitated the application of the Revised USLE to complex land-units, thus extending the applicability and flexibility of the USLE in land resources management. Additionally, Desmet and Govers (1996) stated that the applicability of the algorithm could be significantly improved by including procedures to predict the location and intensity of ephemeral gully erosion that

is not accounted for by the USLE. However, the USLE was originally developed for agricultural fields and its application at a landscape scale to model erosion has been considered inappropriate (Mitasova *et al.* 1996). Models based upon the unit stream-power theory (Moore and Burch 1986; Moore and Wilson 1992; Mitasova and Iverson 1992) incorporate the influence of terrain shape and are considered more appropriate for complex topographic conditions. They require a high-resolution DEM and reliable tools for topographic analysis within the GIS. Mitasova *et al.* (1996) demonstrated how the topographic factors for both the standard USLE and the unit stream-power based model could be computed using a GIS and high-resolution DEM.

ANSWERS is another model that has been integrated with a GIS by writing interfacing programs to input data from the GIS to the model and to display the results. Originally, the element files for the input of spatially varying parameters to ANSWERS developed by Beasley *et al.* (1980) were manually created. GIS allows many of the parameters to be stored in digital format and extracted for erosion modelling. In particular, the information about terrain form could be derived from a raster-based DEM. The advantages of a GIS-based ANSWERS model over older empirical models such as the USLE have been identified by De Roo *et al.* (1989) as: (1) the potential increased accuracy for predicting runoff and erosion; (2) the use of physically based mathematical relationships; (3) the capacity to incorporate newly developed relationships; (4) the incorporation of information about the spatial variability of land characteristics; (5) the potential ease of validation because it deals with individual rainfall events; and (6) the detailed, spatially displayed output of the models, which is useful for planners because the effectiveness of potential control measures can be determined. However, De Roo *et al.* (1989) have also pointed out the main disadvantages of ANSWERS as being theoretical weaknesses (processes such as subsurface flow, gully erosion, and infiltration), the quantity and required quality of necessary input data, and the possible costs of data acquisition.

De Roo *et al.* (1989) ran a set of experiments with the GIS-based ANSWERS model to identify to which state variables the model is most sensitive. The work was carried out on two watersheds – Catsop and Etzenrade – in the Dutch province of Limburg, which is located on the loess soils of the Netherlands. They found that the sensitivity of the model to the infiltration, soil moisture content, and soil roughness variables is a major problem. They recommended that field data for these variables should be collected very carefully. Insufficient data will mean that thousands of the square grid elements have to be 'filled' and there is a serious risk of substantial errors occurring. The problem can be partly addressed by using geostatistical methods such as block kriging to interpolate point data to blocks of the same size as the elements. De Roo *et al.* (1989) also noted that a major problem when attempting accurate modelling is the degree to which reliable location-specific estimates of the attributes can be made. However, today this should not be such an important issue with the development of accurate GPS equipment.

In another study (De Roo and Walling 1994), the GIS-based ANSWERS model was tested under different conditions at the Yendacott catchment in Devon in England. At this catchment, surface runoff and soil erosion have been measured and simulated and field observation and hydrograph analysis have confirmed that saturation overland flow processes are of major importance. As a consequence, it was inevitable that the ANSWERS model would produce poor results, because it is based on the Horton theory of storm runoff generation. The variable contributing area concept (Beven and Kirkby 1979) was introduced into the model, using a GIS to calculate the $\ln(a/\tan b)$ parameter and the initial lambda, λ, value. The modified results, with λ calibrated for each storm, were improved, but the hydrograph was not closely simulated. De Roo and Walling suggested that this may be because other important sources of water, such as subsurface lateral flow caused by perched water tables or macropores, are not represented and the fact that groundwater flow is poorly simulated. They validated the model in a distributed manner using caesium-137 (^{137}Cs). They found correlations of simulated soil loss with observed ^{137}Cs values to be low, but statistically significant. This was not unexpected because it was difficult to simulate the hydrograph properly. They concluded that quantitative spatial modelling of soil erosion still involves larger percentage errors, especially when the model is used in areas other than those for which it was originally developed and tested.

LISEM

LISEM (the Limburg Soil Erosion Model) developed by De Roo *et al.* (1996) is also a physically based hydrological and soil erosion model which can be used for planning and conservation purposes. LISEM

is one of the first examples of a physically based model that is entirely incorporated in a raster GIS. LISEM simulates hydrological and soil erosion processes during single rainfall events on a catchment scale. Using LISEM, it is possible to calculate the impacts of land-use change and explore soil conservation scenarios. A number of processes are incorporated into the model, including rainfall, interception, surface storage in micro-depressions, infiltration, vertical movement of water in the soil, overland flow, channel flow, detachment by rainfall and throughfall, detachment by overland flow, and transport capacity of the flow. Special attention is also given to the influence of tractor wheelings, small roads, and surface sealing, with different surfaces being defined at a sub-grid level. Vertical movement of water in soil is simulated using Richards' equation and storage of surface water in micro-depressions is modelled using the concept of random roughness. The detachment and transport equations are similar to those used in EUROSEM. For the distributed flow routing, a four-point finite-difference solution of the kinematic wave is used along with the Manning equation. LISEM has been found to be very useful for planners owing to the ease with which newly developed relationships can be added, the inclusion of information about the spatial variability of land characteristics, and the spatially detailed displayed output of the model (De Roo *et al.* 1996).

References

Archer, A. C. and Cutler, E. J. B. (1983) Pedogenesis and vegetation trends in the alpine and upper subalpine zones of northeast Ben Ohau Range, New Zealand. 1. Site description, soil classification, and pedogenesis. *New Zealand Journal of Soil Science* **26**, 127–50.

Arnett, R. R. and Conacher, A. J. (1973) Drainage basin expansion and the nine unit landsurface model. *Australian Geographer* **12**, 237–49.

Ballantyne, A. K. (1963) Recent accumulation of salts in the soils of south-eastern Saskatchewan. *Canadian Journal of Soil Science* **43**, 52–8.

Basher, L. R. (1997) Is pedology dead and buried? *Australian Journal of Soil Research* **35**, 979–94.

Beasley, D. B. and Huggins, L. F. (1982) *ANSWERS User's Manual*. West Lafayette, IN: Agricultural Engineering Department, Purdue University.

Beasley, D. B., Huggins, L. F., and Monke, E. J. (1980) ANSWERS: a model for watershed planning. *Transactions of the American Society of Agricultural Engineers* **23**, 938–44.

Bell, J. C., Cunningham, R. L., and Havens, M. W. (1992) Calibration and validation of a soil–landscape model for predicting soil drainage class. *Soil Science Society of America Journal* **56**, 1860–6.

Bell, J. C., Cunningham, R. L., and Havens, M. W. (1994) Soil drainage class probability mapping using a soil–landscape model. *Soil Science Society of America Journal* **58**, 464–70.

Bennett, J. P. (1974) Concepts of mathematical modeling of sediment yield. *Water Resources Research* **10**, 485–92.

Beven, K. (1985) Distributed models. In M. G. Anderson and T. P. Burt (eds) *Hydrological Forecasting*, pp. 405–35. Chichester: John Wiley & Sons.

Beven, K. J. and Kirkby, M. J. (1979) A physically based, variable contributing area model of basin hydrology. *Hydrological Sciences Bulletin* **24**, 43–69.

Blume, H.-P. (1968) Die pedogenetische Deutung einer Catena durch die Untersuchung der Bodendynamik. *Transactions of the Ninth International Congress of Soil Science, Adelaide* **4**, 441–9.

Blume, H.-P. and Schlichting, E. (1965) The relationships between historical and experimental pedology. In E. G. Hallsworth and D. V. Crawford (eds) *Experimental Pedology*, pp. 340–53. London: Butterworths.

Bonifacio, E., Zanini, E., Boero, V., and Franchini-Angela, M. (1997) Pedogenesis in a soil catena on serpentinite in northwestern Italy. *Geoderma* **75**, 33–51.

Bornand, M., Legros, J.-P., and Rouzet, C. (1994) Les banques régionales de données-sols. Exemple du Languedoc–Roussillon. *Étude et Gestion des Sols* **1**, 64–82.

Bornand, M., Robbez-Masson, J. M., Donnet, A., and Lacaze, B. (1997) Caractérisation des sols et paysages des garrigues méditerranéennes. *Étude et Gestion des Sols* **4**, 27–42.

Botschek, J., Ferraz, J., Jahnel, M., and Skowronek, A. (1996) Soil chemical properties of a toposequence under primary rain forest in the Itacoatiara vicinity (Amazonas, Brazil). *Geoderma* **72**, 119–32.

Bui, E. N., Loughhead, A., and Corner, R. (1999) Extracting soil–landscape rules from previous soil surveys. *Australian Journal of Soil Research* **37**, 495–508.

Bunting, B. T. (1965) *The Geography of Soil*. London: Hutchinson.

Burns, S. F. and Tonkin, P. J. (1982) Soil–geomorphic models and the spatial distribution and development of alpine soils. In C. E. Thorn (ed.) *Space and Time in Geomorphology* (Binghamton Symposium in Geomorphology 12), pp. 25–43. London: George Allen & Unwin.

Burt, T. P. and Park, S. J. (1999) The distribution of solute processes on an acid hillslope and the delivery of solutes to a stream: I. Exchangeable bases. *Earth Surface Processes and Landforms* **24**, 781–97.

Carter, B. J. and Ciolkosz, E. J. (1991) Slope gradient and aspect effects on soils developed from sandstone in Pennsylvania. *Geoderma* **49**, 199–213.

Chappell, A., Warren, A., Oliver, M. A., and Charlton, M. (1998) The utility of ^{137}Cs for measuring soil redistribution rates in southwest Niger. *Geoderma* **81**, 313–37.

Chen, Z.-S., Hsieh, C.-F., Jiang, F.-Y., Hsieh, T.-H., and Sun, I.-F. (1997) Relations of soil properties to topography and vegetation in a subtropical rain forest in southern Taiwan. *Plant Ecology* **132**, 229–41.

De Bruin, S. and Stein, A. (1998) Soil–landscape modelling using fuzzy *c*-means clustering of attribute data derived from a digital elevation model (DEM). *Geoderma* **83**, 17–33.

De Bruin, S., Wielemaker, W. G., and Molenaar, M. (1999) Formalisation of soil–landscape knowledge through interactive hierarchical disaggregation. *Geoderma* **91**, 151–72.

De Roo, A. P. J. and Walling, D. E. (1994) Validating the ANSWERS soil erosion model using ^{137}Cs. In R. J. Rickson (ed.) *Conserving Soil Resources: European Perspectives*, pp. 246–63. Wallingford: CAB International.

De Roo, A. P. J., Hazelhoft, L., and Burrough, P. A. (1989) Soil erosion modelling using 'ANSWERS' and Geographical Information Systems. *Earth Surface Processes and Landforms* **14**, 517–32.

De Roo, A. P. J., Wesseling, C. G., Jetten, V. G., and Ritsema, C. J. (1996) LISEM: a physically-based hydrological and soil erosion model incorporated in a GIS. In *HydroGIS 96: Applications of Geographic Information Systems in Hydrology and Water Resources Management (Proceedings of the Vienna Conference, April 1996)* (IAHS Publication No. 235), pp. 395–403. Rozendaalselaan, The Netherlands: International Association of Hydrological Sciences.

DeRose, R. C., Trustrum, N. A., and Blaschke, P. M. (1991) Geomorphic change implied by regolith–slope relationships on steepland hillslopes, Taranaki, New Zealand. *Catena* **18**, 489–514.

Desmet, P. J. J. and Govers, G. (1996) A GIS procedure for automatically calculating the USLE *LS* factor on topographically complex landscape units. *Journal of Soil and Water Conservation* **51**, 427–33.

Duchaufour, P. (1982) *Pedology: Pedogenesis and Classification*. Translated by T. R. Paton. London: George Allen & Unwin.

Elwell, H. A. (1978) Modelling soil losses in Southern Africa. *Journal of Agricultural Engineering Research* **23**, 117–27.

Evans, L. J. (1978) Quantification and pedological processes. In W. C. Mahaney (ed.) *Quaternary Soils*, pp. 361–78. Norwich: Geo Abstracts.

Flacke, W., Auerswald, K., and Neufang, L. (1990) Combining a modified USLE with a digital terrain model for computing high resolution maps of soil loss resulting from rain wash. *Catena* **17**, 383–97.

Foster, G. R. (1982) Modelling the erosion process. In C. T. Haan, M. D. Johnson, and D. L. Brakensiek (eds) *Hydrologic Modelling of Small Watersheds* (ASAE Monograph 5), pp. 328–47. St Joseph, MO: American Society of Agricultural Engineers.

Foster, G. R. and Lane L. J. (1981) Simulation of erosion and sediment yield from field-sized areas. In R. Lal and E. W. Russell (eds) *Tropical Agricultural Hydrology*, pp. 375–94. Chichester: John Wiley & Sons.

Gerrard, A. J. (1988) *Soil–Slope Relationships: A Dartmoor Example* (University of Birmingham, Occasional Paper

No. 26). Birmingham: School of Geography, University of Birmingham.

Gerrard, A. J. (1990a) *Mountain Environments: An Examination of the Physical Geography of Mountains.* London: Belhaven Press.

Gerrard, A. J. (1990b) Soil variations on hillslopes in humid temperate climates. *Geomorphology* 3, 225–44.

Glazovskaya, M. A. (1968) Geochemical landscapes and geochemical soil sequences. *Transactions of the Ninth International Congress of Soil Science, Adelaide* 4, 303–12.

Gregory, K. J. and Walling, D. E. (1973) *Drainage Basin: Form and Process.* London: Edward Arnold.

Hallsworth, E. G. (1965) The relationship between experimental pedology and soil classification. In E. G. Hallsworth and D. V. Crawford (eds) *Experimental Pedology*, pp. 354–74. London: Butterworths.

Holowaychuk, N., Gersper, P. L., and Wilding, L. P. (1969) Strontium-90 content of soils near Cape Thompson, Alaska. *Soil Science* 107, 137–44.

Holý, M., Svetlosanov, V., Hándova, Z., Kos, Z., Váska, J., and Vrána, K. (1982) *Procedures, Numerical Parameters and Coefficients of the CREAMS Model: Application and Verification in Czechoslovakia* (International Institute for Applied Systems Analysis Collaborative Paper CP-82-23). Laxenburg, Austria: International Institute for Applied Systems Analysis.

Huang, S. L. and Ferng, J. J. (1990) Applied land classification for surface water quality management. II. Land process classification. *Journal of Environmental Management* 31, 127–41.

Huggett, R. J. (1973) *Soil Landscape Systems: Theory and Field Evidence.* Unpublished Ph.D. Thesis, University of London.

Huggett, R. J. (1975) Soil landscape systems: a model of soil genesis. *Geoderma* 13, 1–22.

Huggett, R. J. (1985) *Earth Surface Systems* (Springer Series in Physical Environment, Vol. 1). Heidelberg: Springer.

Huggett, R. J. (1995) *Geoecology: An Evolutionary Approach.* London: Routledge.

Ibáñez, J. J., Ballestra, R. J., and Alvarez, A. G. (1990) Soil landscapes and drainage basins in Mediterranean mountain areas. *Catena* 17, 573–83.

Jenny, H. (1980) *The Soil Resource: Origin and Behavior* (Ecological Studies 37). New York: Springer.

Kairiukstis, L. and Golubev, G. (1982) Application of the CREAMS model as part of an overall system for optimizing environmental management in Lithuania, USSR: first experiments. In V. Svetlosanov and W. G. Knisel (eds) *European and United States Case Studies in Application of the CREAMS Model* (International Institute for Applied System Analysis Collaborative Proceedings Series CP-82-511), pp. 99–119. Laxenburg, Austria: International Institute for Applied Systems Analysis.

Khan, F. A. and Fenton, T. E. (1994) Saturated zones and soil morphology in a Mollisol catena of central Iowa. *Soil Science Society of America Journal* 58, 1457–64.

Kirkby, M. J. (1980) Modelling water erosion processes. In M. J. Kirkby and R. P. C. Morgan (eds) *Soil Erosion*, pp. 183–216. Chichester: John Wiley & Sons.

Kitayama, K. (1992) An altitudinal transect study of the vegetation on Mount Kinabalu, Borneo. *Vegetatio* 102, 149–71.

Kitayama, K. and Mueller-Dombois, D. (1994) An altitudinal transect analysis of the windward vegetation on Haleakala island mountain: (1) climate and soils. *Phytocoenologia* 24, 111–33.

Knapp, A. K., Fahnestock, J. T., Hamburg, S. P., Statland, L. B., Seastedt, T. R., and Schimel, D. S. (1993). Landscape patterns in soil–plant water relations and primary production in tallgrass prairie. *Ecology* 74, 549–60.

Knisel, W. G. (ed.) (1980) *CREAMS: A Field Scale Model for Chemicals, Runoff and Erosion from Agricultural Management Systems* (USDA, Conservation Research Report, No. 26). Washington DC: United States Department of Agriculture.

Kreznor, W. R., Olson, K. R., Banwart, W. L., and Johnson, D. L. (1989) Soil, landscape, and erosion relationships in a northwestern Illinois watershed. *Soil Science Society of America Journal* 53, 1763–71.

Lapen, D. R., Moorman, B. J., and Price, J. S. (1996) Using ground-penetrating radar to delineate subsurface features along a wetland catena. *Soil Science Society of America Journal* 60, 923–31.

Litaor, M. I., Barth, G., Zika, E. M., Litus, G., Moffitt, J., and Daniels, H. (1998) The behavior of radionuclides in the soils of Rocky Flats, Colorado. *Journal of Environmental Radioactivity* 38, 17–46.

Martí, C. and Badia, D. (1995) Characterization and classification of soils along two altitudinal transects in the Eastern Pyrenees, Spain. *Arid Soil Research and Rehabilitation* 9, 367–83.

McSweeney, K., Slater, B. K., Hammer, R. D., Bell, J. C., Gessler, P. E., and Petersen, G. W. (1994) Towards a new framework for modeling the soil–landscape continuum. In R. Amundson, J. Harden, and M. Singer (eds) *Factors of Soil Formation: A Fiftieth Anniversary Retrospective* (Soil Science Society of America Special Publication Number 33), pp. 127–45. Madison, WI: Soil Science Society of America.

Milne, G. (1935) Some suggested units of classification and mapping, particularly for East African soils. *Soil Research* 4, 183–98.

Mitasova, H. and Iverson, L. R. (1992) Erosion and sedimentation potential analysis for Hunter lake. In W. U. Brigham and A. R. Brigham (eds) *An Environmental Assessment of the Hunter Lake Project Area* (Project Completion Report submitted to City Water, Light and Power, Springfield, IL), pp. 1.2.01–08. Champaign, IL: Illinois Natural History Survey.

Mitasova, H., Hofierk, J., Zlocha, M., and Iverson, L. R. (1996) Modelling topographic potential for erosion and deposition using Geographical Information Systems. *International Journal of Geographical Information Systems* 10, 629–41.

Moore, I. D. and Burch, G. J. (1986) Physical basis of the length–slope factor in the universal soil loss equation. *Soil Science Society of America Journal* 50, 1294–8.

Moore, I. D. and Wilson, J. P. (1992) Length–slope factors for the Revised Universal Soil Loss Equation: simplified method of estimation. *Journal of Soil and Water Conservation* 423–8.

Moore, I. D., Gessler, P. E., Nielsen, G. A., and Peterson, G. A. (1993) Soil attribute prediction using terrain analysis. *Soil Science Society of America Journal* **57**, 443–52.

Moore, I. G., Grayson, R. B., and Ladson, A. R. (1991) Digital terrain modelling: a review of hydrological, geomorphological, and biological applications. *Hydrological Processes* **5**, 3–30.

Morgan, R. P. C. (1994) The European Soil Erosion Model: an up-date on its structure and research base. In R. J. Rickson (ed.) *Conserving Soil Resources: European Perspectives*, pp. 286–99. Wallingford: CAB International.

Morgan, R. P. C. (1995) *Soil Erosion and Conservation*, 2nd edn. Harlow: Longman.

Morgan, R. P. C., Quinton, J. N., and Rickson, R. J. (1994) Modelling methodology for soil erosion assessment and soil conservation design: the EUROSEM approach. *Outlook in Agriculture* **23**, 5–9.

Morison, C. G. T., Hoyle, A. C., and Hope-Simpson, J. F. (1948) Tropical soil–vegetation catenas and mosaics: a study in the south-western part of Anglo-Egyptian Sudan. *Journal of Ecology* **36**, 1–84.

Musgrave, G. W. (1947) The quantitative evaluation of factors in water erosion: a first approximation. *Journal of Soil and Water Conservation* **2**, 133–8.

Nearing, M. A., Foster, G. R., Lane, L. J., and Finkner, S. C. (1989) A process-based soil erosion model for USDA-Water Erosion Prediction Project technology. *Transactions of the American Society of Agricultural Engineers* **32**, 1587–93.

O'Connor, K. F. (1984) Stability and instability of ecological systems in New Zealand mountains. *Mountain Research and Development* **4**, 15–29.

Park, S. J. and Burt, T. P. (1999) The distribution of solute processes on an acid hillslope and the delivery of solutes to a stream: II. Exchangeable Al^{3+}. *Earth Surface Processes and Landforms* **24**, 851–65.

Pennock, D. J. and de Jong, E. (1987) The influence of slope curvature on soil erosion and deposition in hummock terrain. *Soil Science* **144**, 209–17.

Quinton, J. N. (1994) Validation of physically based erosion models, with particular reference to EUROSEM. In R. J. Rickson (ed.) *Conserving Soil Resources: European Perspectives*, pp. 300–13. Wallingford: CAB International.

Radcliffe, J. E. (1982) Effects of aspect and topography on pasture production in hill country. *New Zealand Journal of Agricultural Research* **25**, 485–96.

Reid, D. A., Graham, R. C., Southard, R. J., and Amrhein, C. (1993) Slickspot soil genesis in the Carrizo Plain, California. *Soil Science Society of America Journal* **57**, 162–8.

Renard, K. G., Foster, G. R., Weesies, G. A., and Porter, J. P. (1991) RUSLE: Revised Universal Soil Loss Equation. *Journal of Soil and Water Conservation* **46**, 30–3.

Rewerts, C. C. and Engel, B. A. (1991) *ANSWERS on GRASS: Integrating a Watershed Simulation with a GIS* (ASAE Paper No. 91-2621). St Joseph, MO: American Society of Agricultural Engineers.

Rijkse, W. C. and Trangmar, B. B. (1995) Soil–landscape models and soils of Eastern Highlands, Papua New Guinea. *Australian Journal of Soil Research* **33**, 735–55.

Rose, A. W., Hawkes, H. E., and Webb, J. S. (1979) *Geochemistry in Mineral Exploration*, 2nd edn. London: Academic Press.

Roy, A. G., Jarvis, R. S., and Arnett, R. R. (1980) Soil–slope relationships within a drainage basin. *Annals of the Association of American Geographers* **70**, 397–412.

Savabi, M. R., Klik, A., Grulich, K., Mitchell, J. K., and Nearing, M. A. (1996) Application of WEPP in GIS on small watersheds in USA and Austria. In K. Kovar and H. P. Nachtnebel (eds) *HydroGIS 96: Applications of Geographic Information Systems in Hydrology and Water Resources Management (Proceedings of the Vienna Conference, April 1996)* (IAHS Publication No. 235), pp. 469–76. Rozendaalselaan, The Netherlands: International Association of Hydrological Sciences.

Singh, J. S., Milchunas, D. G., and Lauenroth, W. K. (1998) Soil water dynamics and vegetation patterns in a semiarid grassland. *Plant Ecology* **134**, 77–89.

Skidmore, A. K., Ryan, D. J., Dawes, W., Short, D., and O'Loughlin, E. (1991) Use of an expert system to map forest soils from a Geographical Information System. *International Journal of Geographical Information Systems* **5**, 431–45.

Smeck, N. E. and Runge, E. C. A. (1971) Phosphorus availability and redistribution in relation to profile development in an Illinois landscape segment. *Soil Science Society of America Proceedings* **35**, 952–9.

Smith, D. D. (1958) Factors affecting rainfall erosion and their evaluation. *International Association of Scientific Hydrology Publication* **43**, 97–107.

Sommer, M. (1992) *Musterbildung und Stofftransporte in Bodengesellschaften Baden–Württembergs* (Hohenheimer Bodenkundliche Heft 4). Stuttgart: Institut für Bodenkunde und Standortslehre, Universität Hohenheim.

Sommer, M. and Stahr, K. (1996) The use of element:clay-ratios assessing gains and losses of iron, manganese and phosphorus in soils of sedimentary rocks on a landscape scale. *Geoderma* **71**, 173–200.

Srinivasan, R. and Engel, B. A. (1991) *A Knowledge Based Approach to Extract Input Data from GIS* (ASAE Paper No. 91-7045). St Joseph, MO: American Society of Agricultural Engineers.

Stäblein, G. (1984) Geomorphic altitudinal zonation in the Arctic–alpine mountains of Greenland. *Mountain Research and Development* **4**, 319–30.

Stolt, M. H., Baker, J. C., and Simpson, T. W. (1993a) Soil–landscape relationships in Virginia: I. Soil variability and parent material uniformity. *Soil Science Society of America Journal* **57**, 414–21.

Stolt, M. H., Baker, J. C., and Simpson, T. W. (1993b) Soil–landscape relationships in Virginia: II. Reconstruction analysis and soil genesis. *Soil Science Society of America Journal* **57**, 422–8.

Sutherland, R. A., van Kessel, C., Farrell, R. E., and Pennock, D. J. (1993) Landscape-scale variations in soil nitrogen-15 natural abundance. *Soil Science Society of America Journal* **57**, 169–78.

Swanson, D. K. (1985) Soil catenas on Pinedale and Bull Lake moraines, Willow Lake, Wind River Mountains, Wyoming. *Catena* **12**, 329–42.

Swanson, D. K. (1996) Soil geomorphology on bedrock and colluvial terrain with permafrost in central Alaska, USA. *Geoderma* **71**, 157–72.

Thomas, M. F., Thorp, M. B., and Teeuw, R. M. (1985) Palaeogeomorphology and the occurrence of diamondiferous placer deposits in Koidu, Sierra Leone. *Journal of the Geological Society of London* **142**, 789–802.

Thompson, J. A. and Bell, J. C. (1998) Hydric conditions and hydromorphic properties within a Mollisol catena in south-eastern Minnesota. *Soil Science Society of America Journal* **62**, 1116–25.

Thompson, J. A., Bell, J. C., and Butler, C. A. (1997) Quantitative soil–landscape modeling for estimating the areal extent of hydromorphic soils. *Soil Science Society of America Journal* **61**, 971–80.

Thompson, J. A., Bell, J. C., and Zanner, C. W. (1998) Hydrology and hydric soil extent within a Mollisol catena in south-eastern Minnesota. *Soil Science Society of America Journal* **62**, 1126–33.

Timpson, M. E., Richardson, J. L., Keller, L. P., and McCarthy, G. J. (1986) Evaporite mineralogy associated with saline seeps in southwestern North Dakota. *Soil Science Society of America Journal* **50**, 490–3.

Troll, C. (1973) High mountain belts between the polar caps and the equator: their definition and lower limit. *Arctic and Alpine Research* **5**, 19–28.

Vreeken, W. J. (1973) Soil variability in small loess watersheds: clay and organic matter content. *Catena* **2**, 321–36.

Warren, S. D., Diersing, V. E., Thompson, P. J., and Goran, W. D. (1989) An erosion-based land classification system for military installations. *Environmental Management* **13**, 251–7.

Webb, T. H. and Burgham, S. J. (1997) Soil–landscape relationships of downlands soils formed from loess, eastern South Island, New Zealand. *Australian Journal of Soil Research* **35**, 827–42.

Webb, T. H., Fahey, B. D., Giddens, K. M., Harris, S., Pruden, C. C., and Whitton, J. S. (1999) Soil–landscape and soil–hydrological relationships in the Glendhu Experimental Catchments, East Otago Uplands, New Zealand. *Australian Journal of Soil Research* **37**, 761–85.

Weitkamp, W. A., Graham, R. C., Anderson, M. A., and Amrhein, C. (1996) Pedogenesis of a vernal pool Entisol–Alfisol–Vertisol catena in Southern California. *Soil Science Society of America Journal* **60**, 316–23.

West, L. T., Wilding, L. P., Stahnke, C. R., and Hallmark, C. T. (1988) Calciustolls in central Texas: I. Parent material uniformity and hillslope effects on carbonate enriched horizons. *Soil Science Society of America Journal* **52**, 1722–31.

Wischmeier, W. H. and Smith, D. D. (1978) *Predicting Rainfall Erosion Losses: A Guide to Conservation Planning* (USDA Agricultural Handbook 537). Washington, DC: United States Department of Agriculture, Science and Education Administration.

Yonker, C. M., Schimel, D. S., Paroussis, E., and Heil, R. D. (1988) Patterns of organic carbon accumulation in a semiarid grassland steppe, Colorado. *Soil Science Society of America Journal* **52**, 478–83.

Zingg, A. W. (1940) Degree and length of land slope as it affects soil loss in runoff. *Agricultural Engineering* **21**, 59–64.

CHAPTER 6

Animals and plants

Topography influences animals and plants in several fundamental and many subtle ways. Altitude, like latitude, acts through climatic conditions to exert a major influence upon the distribution and abundance of living things. Altitudinal vegetation zones mirror altitudinal climatic zones. Moreover, the animal and plant species that inhabit mountain environments are partly under a strong climatic influence. The altitudinal tree-line is a striking feature, marking the transition from forest to alpine vegetation. Animal species may manifest altitudinal zoning. Landform, including such attributes as slope exposure (aspect), slope inclination, and slope sequences, influences living things. Slope exposure leads to differences of plants and, sometimes, animals on sunny and shady aspects of mountains, hills, and such smaller-scale features as sand dunes. Slope inclination affects life through steepness and through the agency of linked slope elements in a toposequence or catena. Microtopography is highly influential in determining some basic vegetation processes, such as germination, and may affect the distribution and life history of some animal species. Three-dimensional landform (surfaces) affects animals and plants. Vegetation, for instance, commonly varies in coves, sideslopes, and spurs. Landscape elements – patches, corridors, networks, and mosaics – exert major constraints on animal and plant distributions, abundance, diversity, movement, and population dynamics.

Life and altitude

Altitudinal floral zones

Carl Linnaeus, the eighteenth-century Swedish botanist, was aware that, on ascending mountainsides, a traveller would pass though different zones of life. Alexander von Humboldt, the German philosopher and geographer, recognised that a mountain was a miniature version of a hemisphere, carrying altitudinal replicates of tropical, temperate, boreal, and frigid floral zones. Clinton Hart Merriam (1894) came to a similar conclusion and estimated that each mile of altitude was equivalent to about 800 miles of latitude. Today, altitudinal vegetation zones are well established, the basic zones in ascending order being submontane, montane, upper montane, subalpine, alpine, subnival, and nival (Figure 6.1). However, the causes of this zonary arrangement are not wholly resolved and are the subject of continuing research.

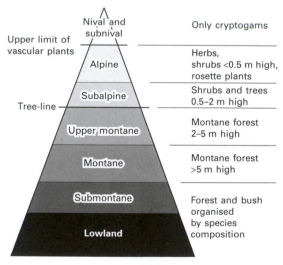

Figure 6.1 Altitudinal vegetation zones. The physiognomic characters of the vegetation refer to tropical and warm-temperate oceanic islands, but would be typical of many other mountains. (Adapted from Leuschner, 1998.)

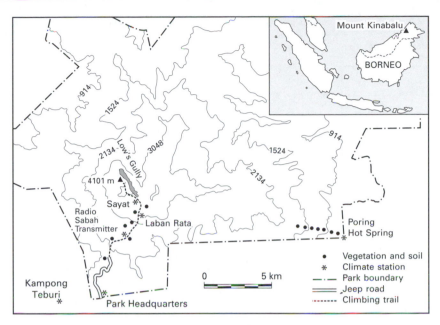

Figure 6.2 Location map of Mount Kinabalu, north Borneo, showing study area, location of transects with vegetation and soil sampling sites, and climate stations. (Adapted from Kitayama, 1992.)

The example of Mount Kinabalu will illustrate the research challenges posed by a single mountain, and consideration of the Andes will show how altitude and latitude act together to help shape montane vegetation zones.

Mount Kinabalu: a case study

Plant communities on Mount Kinabalu change with increasing altitude, and also to a lesser extent with aspect (Kitayama 1992). On the south face, rice and vegetable cultivation lead to a fragmented forest below about 1600 m, but on the east and north slopes the forest is still continuous and meets the lowland forest at about 500 m. There are four altitudinal vegetation belts composed of mutually exclusive species groups: lowland (less than 1200 m), lower montane (1200 to 2000–2350 m), upper montane (2000–2350 to 2800 m), and subalpine (2800 to the forest line at 3400 m). Above 3400 m the subalpine community grades into an alpine community, which sits between the tree-line at 3700 m and the summit. The alpine community consists of closed-canopy scrub to about 4000 m, where it is replaced by a vast alpine rock-desert with small scrub patches in pockets of accumulated soil. A few extremely stunted trees eke out a

living near the peak, which suggests that the tree-line lies below the climatic shrub line.

A transect study of vegetation on the mountain demonstrated the distinctiveness of each altitudinal vegetation belt. Three research questions were posed. First, are canopy species parcelled into separate altitudinal groups? Second, are floristic changes reflected in forest structure? Third, how are changes in vegetation (and soils) related to altitudinal changes in climate? These questions have global applicability and are relevant to all mountains. To answer the questions on Mount Kinabalu, a transect was made on the south face and east ridge, both of which are accessible. The transect comprised two discontinuous segments in undisturbed primary forest (Figure 6.2). Fourteen vegetation sampling plots were located at 200-m altitudinal intervals. Plots were chosen to avoid geological anomalies, such as ultrabasic rocks, and topographical extremes, such as ridge tops and valley bottoms. No stands were sampled at 2200 m owing to the steepness of the terrain. The results for vegetation structure show that, with increasing altitude, species richness decreases, total basal area tends to increase, tree density increases (with sharp increases at 1400 and 2800 m), and canopy height decreases. Two-way indicator species analysis (TWINSPAN)

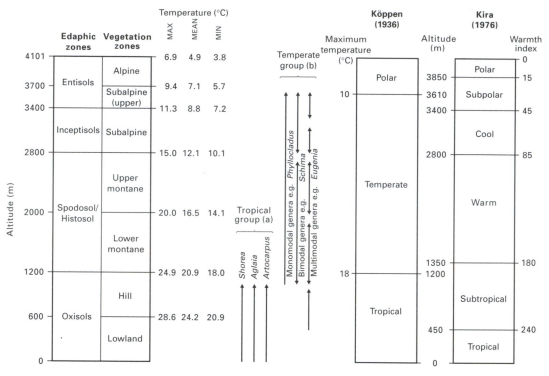

Figure 6.3 Vegetation and climatic zones on Mount Kinabalu. (Adapted from Kitayama, 1992.)

revealed four altitudinal floristic zones, each with indicator species. The lowland group contains the greatest number of indicator species at 36, which include pantropical and Malesian elements, seraya tembaga (*Shorea leprosula*) being a Malesian example. The lower montane zone contains 25 indicator species. They are members of the Fagaceae, Lauraceae, Theaceae, Myrtaceae, and Elaeocarpaceae. An example is gatal (*Schima wallichii*). The upper montane zone is indicated by four associated species: *Ascarina philippinensis*, which is called parukanak in the Philippines, obar (*Eugenia punctilimba*), an olive, *Olea rubrovenia*, and a holly, *Ilex zygophylla*. The subalpine zone is indicated by 10 associated species. Most of them are sclerophyllous or micro-nanophyllous. An example is the tea-tree, locally called sayat-sayat (*Leptospermum recurvum*).

The structure of the forest was assessed from air photographs and field observation. Air photographs revealed emergent trees below 1200 m, suggesting a structural transition from the lowland zone to the montane zone. Field observations showed a sharp change in forest structure at 2800 m on the south face, which marks the upper limit of the upper montane

zone. No sharp change was observed for the upper limit of the lower montane zone at 2000–2350 m.

The combined evidence of abrupt change in species composition in altitudinal zones, and changes in forest structure at the same altitudes, indicate that 'critical altitudes' exist on Mount Kinabalu. These critical altitudes should relate to climatic factors. Climatic control is suggested by bioclimatic parameters. Correlation between the floristic vegetation boundaries and the bioclimatic thermal thresholds is very close (Figure 6.3; see also p. 55). The lowland forest is limited altitudinally by the threshold of Köppen's tropical climate, while the upper limits of the upper montane and subalpine zones are marked by Kira's warm temperate and cool temperate climates, respectively. Ground-frost is another bioclimatically significant thermal factor. On Kinabalu, the daily ground-frost line lies at 3680 m and coincides with the tree-line (the upper limit of open-canopy subalpine forest). The boundary of Köppen's temperate climate also coincides with the tree-line.

Kitayama (1992) explains the disagreement between the Köppen and Kira zones on Mount Kinabalu. The two most important turnover points are the upper limit ·

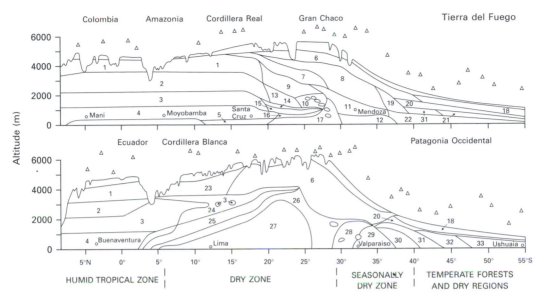

Figure 6.4 Altitudinal and latitudinal vegetation belts from Colombia to Tierra del Fuego. The top section is through the eastern Andes and the bottom section is through the western cordilleras. Triangles signify mountain peaks. Vegetation belts are identified by numbers: **1** Microthermal tropical mountain grasslands (páramo). **2** Mountain cloud forest. **3** Tropical montane forest. **4** Lowland wet tropical forest. **5** Tropical moist savanna. **6** High mountain vegetation (puna). **7** Mountain pastures. **8** Mountain steppe. **9** Tall grassland. **10** Dry prairie. **11** Patchy woodland. **12** Subtropical thorn steppe. **13** Subtropical aliso forest. **14** Subtropical mirtaceas forest. **15** Subtropical laurel forest. **16** Subtropical pacara forest ecotone. **17** Subtropical thorn savanna. **18** Patagonian mountain bushland. **19** Mountain steppe with some timber. **20** Mixed subantarctic forest. **21** Subantarctic forest. **22** Patagonian steppe and bushland. **23** Moist puna belt. **24** Mesophyte bush vegetation. **25** Succulent vegetation. **26** Low thorn bush vegetation. **27** Peruvian desert and semi-desert with loma vegetation. **28** Thorn bush and succulent vegetation. **29** Wet temperate Chilean hardwood forest. **30** Deciduous forest. **31** Valdivian forest. **32** Patagonian inland forest. **33** Subantarctic evergreen forest. (Adapted from Ives, 1992.)

of lowland forest and the tree-line, both of which are set by minimum temperatures. The latitudinal lowland tropical floristic elements, represented by seraya tembaga (*Shorea leprosula*), and the altitudinal lowland temperate elements, represented by gatal (*Schima wallichii*), drop out respectively at the two turnover points. These two floristic groups occur over a wide altitudinal range owing to Kinabalu's year-round equable temperature regime and correspond to Köppen's tropical and temperate zones. As a result, heat deficiency constrains some widespread floristic elements near their upper limits, and splits them into altitudinal parapatric congeners, for example as obar (*Eugenia*) of group b in Figure 6.3 is split. This evolutionary adaptation to altitudinal temperature regimes leads to the subdivision of the two broad vegetation zones into finer altitudinal vegetation zones that correlate with Kira's climatic zones.

The Andes

Mountain ranges, like individual mountains, display altitudinal life zones. In addition, especially if they run north to south, they exhibit latitudinal and local longitudinal effects. The vegetation of the Andes sports this mix of elevational, latitudinal gradient, and longitudinal effects (Figure 6.4) (Ives 1992). The northern Andes lie in the humid tropics where the high humidity supports luxuriant vegetation. Tropical rainforests and other kinds of evergreen and deciduous forests are dominant. Altitudinal belts of vegetation are approximately symmetrically disposed on eastern and western flanks. The sequence runs from lowland wet tropical forest, through tropical montane forest, cooler mountain cloud-forest (variously termed tierra fria, selva nublada, Nebelwald, and moss forest) consisting of dense stands of ferns, epiphytes,

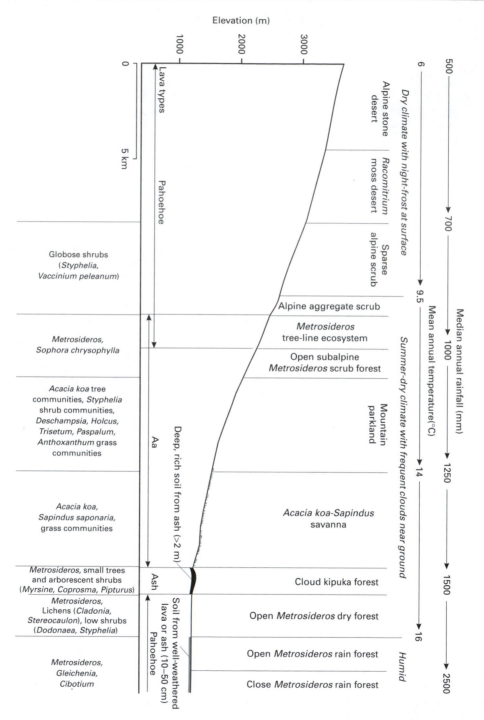

Figure 6.5 Profile diagram of the Mauna Loa transect showing vegetation, climate, and substrate. (Adapted from Mueller-Dombois and Bridges, 1981.)

and bamboo, to microthermal tropical grassland (páramo) above the tree-line. Beyond the snow-line lies the nival zone. The central Andes is a dry zone and the aridity influences the vegetation. On the coastal plain, the aridity is greatest in the Atacama desert, one of the driest places on Earth. The arid zone supports desert vegetation: thorn bush, succulents, and loma vegetation supported by the garua, a dense mist that drifts northwards along the Peruvian coast and is associated with the cold Humboldt Current. At higher elevations, the great high plateau, or altiplano, is covered by puna vegetation. Three types of puna occur along the humidity gradient running north to south. The northerly moist or humid puna crosses the Andes as a swath running from north-west to south-east. It consists of an uninterrupted mat of grassy plants. Bunch grasses with resinous evergreen shrubs and cushion plants dominate the dry puna. Such xerophytic plants as cactus and thorny shrubs characterise desert puna. The eastern flank of the central Andes has a similar set of vegetation belts to those in the humid tropical zone, though the cloud forest in this central region is called *ceja de la montaña* or 'eyebrow of the mountain'. The southern Andes comprise a seasonally dry zone and a temperate zone. Moving south, eastern slopes become drier and western slopes become wetter. Along the coast immediately south of the Atacama Desert is an area of thorn scrub. This gives way to a sclerophyllous woodland and scrubland growing under a Mediterranean climate, similar to analogous formations growing elsewhere in the world but dotted with tall-growing trees. At higher elevations an extensive belt of desert puna gives way southwards to zones of mountain grassland and mixed forests. Comparable latitudes on the eastern flanks become progressively drier than on the west in moving south. Consequently, the vegetation becomes scrubbier and more xerophytic. The southern tip of the Andes, south of Valvivia, carries southern beech (*Nothofagus*) forest and other types of cool temperate forest on the western slopes.

Altitudinal faunal zones

Animal communities are zoned according to altitude, but not so neatly as are plant communities. Animals tend to associate themselves with particular types of plant community, so it is not surprising that their communities show some accordance with altitudinal vegetation zones. The example of vegetation and animal communities on Mauna Loa, Hawaii, will illustrate the complexities of altitudinal fauna zoning.

Altitudinal bird and rat ranges on Mauna Loa

The flora and fauna along an elevational transect on Mauna Loa, running 35 km from 3660 m near the summit to 1190 m near the summit of Kilauea, have been studied in great detail as part of the International Biological Programme (Mueller-Dombois *et al.* 1981). The broad ecological conditions along the transect are summarised in Figure 6.5. For vegetational analysis, samples (relevés) were laid out nearly continuously along 25 km of the transect, from the slope base at 1190 m to the upper alpine scrub limit at 3050 m. Statistical analysis of the data revealed seven zones of woody plants. Zones I and II, the alpine and subalpine transect zones, extend through areas of lava-rock outcrop that contain only small amounts of fine soil material. They are occupied primarily by shrub species, though the lehua (*Metrosideros collina* subspecies *polymorpha*), a tree species, appears in Zone II. Zones III and IV, mountain parkland and savannah, are underlain by lava covered by a blanket of volcanic ash that thickens and becomes more continuous downslope. In these zones, shrub-species diversity decreases, tree-species diversity increases, and herbaceous life forms, especially grasses, become more abundant. In Zone V, lehua dry forest, most of the shrubs, except those confined to high altitudes – 'ohelo (*Vaccinium peleanum*) and pukiawe (*Styphelia douglasii*) – reappear while the other tree species disappear. In Zones VI and VII (open and closed lehua rainforest), ferns become prevalent, shrubs associated with rock outcrops decrease in Zone VII, and lehua is dominant only on deeper soil in the rainforest.

Twenty-two bird species were recorded on the Mauna Loa transect (Mueller-Dombois *et al.* 1981). These were grouped, using the Sørensen community coefficient based on presence–absence data, into six well-defined avian communities. These are indicated in Figure 6.6 as transect zones. The distribution along the transect of one indigenous bird (*Pluvialis dominica*) and eight endemic birds is portrayed in Figure 6.6. Four species groups were recognised. Group 1 has one member – the Hawaiian thrush or 'oma'o (*Phaeornis obscurus obscurus*). This is the only bird living beyond the tree-line in the sparse alpine scrub. It also lives in the closed rainforest. Its distribution is therefore rather odd, occupying the ends of the transect but not the intervening areas. Its absence from the middle part of the transect appears to result from competition with another frugivore, the exotic leiothrix (*Leiothrix lutea*). Group 2 comprises a spatially heterogeneous set of open-area birds.

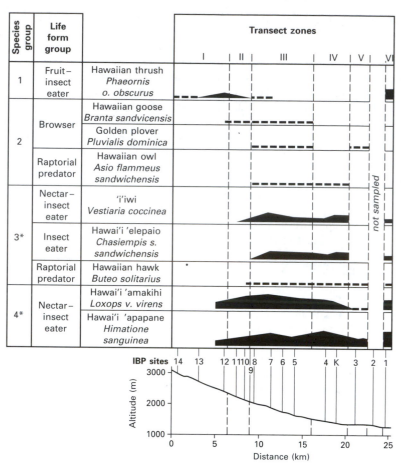

Figure 6.6 Distribution of native bird species along the Mauna Loa altitudinal transect. The six transect zones were established using the Sørensen index. The *black area* for each species shows abundance (birds per 40 ha), the maximum abundance being 7 units (for example, *Himatione sanguinea* at site 4): 7 units = 160–300 birds; 6 units = 120–160 birds, 5 units = 70–120 birds, 4 units = 30–70 birds, 3 units = 15–30 birds, 2 units = 5–15 birds, and 1 unit = 2–5 birds. The *dashed line* represents one bird per 40 ha. Numbers with *asterisks* are computer-generated species groups. K is the closed kipuka forest in savannah zone IV. (Adapted from Mueller-Dombois *et al.*, 1981.)

The Hawaiian goose or nene (*Branta sandvicensis*) occurs from the tree-line throughout the subalpine forest and mountain parkland zones. The golden plover or kolea (*Pluvialis dominica*), an indigenous bird, occurs mainly in the mountain parkland and open lehua dry-forest. It is absent from the savanna zone, probably because it dislikes the tall grass growing there. The Hawaiian owl or pueo (*Asio flammeus sandwichensis*) is confined to the mountain parkland and savannah zones. Group 3 comprises three endemic species ranging from the lower subalpine scrub to

the rain forest. The Hawaiian hawk or 'io (*Buteo solitarius*) is found throughout the range but the distribution of both the Hawai'i 'elepaio (*Chasiempis sandwichensis sandwichensis*) and 'i'iwi (*Vestiaria coccinea*) is interrupted by the lehua dry-forest zone. The differences in distribution are attributable to differences in general behaviour and feeding habits: the 'elepaio and 'i'iwi favour colonies of koa (*Acacia koa*) trees or other forest groves with closed canopies, neither of which grow in the lehua dry-forest zone. Group 4 consists of the two most abundant native honeycreepers – the Hawai'i 'amakihi (*Loxops virens virens*) and the 'apapane (*Himatione sanguinea*). The wide distribution of these species indicates broad tolerance of temperature and rainfall regimes. The slight fall in 'apapane density in the lehua dry-forest suggests that it prefers closed tree canopies.

Several conclusions may be drawn from the elevational distribution of these birds on Mauna Loa. First, distinct patterns of bird distribution are related to the distribution of major vegetation types along the transect. Second, the pattern of bird distribution is correlated with environmental factors, including vegetation: witness the upper limits of the 'apapane and Hawai'i 'amakihi that end suddenly at the tree-line. Above the tree-line, no individual 'apapane or Hawai'i 'amakihi were seen, but just below the tree-line they live at high densities. The 'i'iwi and Hawai'i 'elepaio exhibit distinct density changes where savannah vegetation changes to open lehua dry forest. Third, vegetation structure exerts such a profound influence upon bird species' distributions that the distributional effect of competition between species is of little importance. This contrasts with a similar altitudinal gradient in the Peruvian Andes where competition plays a significant role (Terborgh and Weske 1975). The difference between the two transects may reflect the contrasting settings – Mauna Loa is an island setting with low avian diversity; the Peruvian Andes is a continental setting with high avian diversity.

Altitude and animal diversity

Altitude undoubtedly has an influence on animal species' diversity. This is seen in a study of amphibian ranges and distributional patterns on the Western Ghats, part of the Malabar biogeographical region in south India (Daniels 1992). The Western Ghats run unbroken for 1600 km between latitudes 8°N and 21°N, save for the 30-km-wide Palghat Gap at around 11°N. They are very rich in amphibian species (frogs, toads, and caecilians), containing 117 species, 89 of which are endemic. Climatic effects on amphibian species' diversity are expressed through altitudinal and latitudinal gradients. Species diversity is greater in the southern half of the Western Ghats, south of 13°N, and in hills of low-to-medium elevation, than it is in the northern half of the area and in high hills. This suggests that the presence of high hills does not add to amphibian species diversity in the Western Ghats, a finding contradicting other work (Inger *et al.* 1987). The differences in diversity appear to relate to more widespread rainfall and less climatic variability

in the southern area. Unlike birds and flowering plants in the area, where endemics are found chiefly on the higher hills, endemic amphibians are confined largely to the lower altitudinal range, with the majority in the range 800 to 1000 m.

Tree-lines

Tree-lines are steep ecotones that mark an eye-catching change of dominant life-form that accompanies a switch from moderate to harsh environmental conditions. They are seen in the circumpolar boundary between taiga and tundra vegetation in the Northern Hemisphere (latitudinal tree-lines) and in the boundary between subalpine vegetation and low-growing alpine vegetation on mountains (upper altitudinal tree-lines). In some situations, as where a forested mountain stands in a desert, inverted or lower altitudinal tree-lines occur. Some tree-lines are abrupt, with no more than a few metres separating forested and treeless areas; some are gradual, taking hundreds of kilometres to change from continuous forest to treeless terrain. They are all equally difficult to explain, though there is no shortage of plausible mechanisms.

Upper tree-lines

Altitudinal and latitudinal tree-lines may comprise three separate 'lines', to which a variety of terms are applied (see Wardle 1974, 1993; Hustich 1979; Heikkinen *et al.* 1995). The Waldgrenze is the limit of tall, erect trees growing at normal forest densities. It is also called the forest line, forest limit, and, in the United States, the timberline. The Baumgrenze lies beyond the Waldgrenze and is the limit at which individuals recognisable as trees (sometimes defined as trees more than 2 m high) grow. It is also known as the tree-line. Beyond the Baumgrenze lies the tree-species line, which marks the altitude or latitude at which tree species will grow, though in a deformed state. In many cases, the tree-line and tree-species line coincide, as in the southern beech forests of New Zealand. In other cases, all three lines may coincide, and where this occurs, the discontinuity between forested terrain and treeless terrain is sharp.

The Kampfzone is a band lying between the timberline and the tree-species line. In English, Kampfzone means 'battle zone', or perhaps a better rendition would be 'stress zone'. Trees live in the Kampfzone, but, owing to the severity of the environmental conditions, they have a deformed, dwarf, or

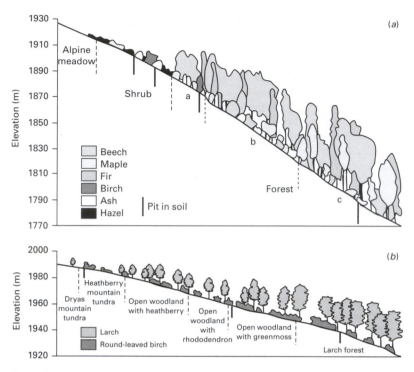

Figure 6.7 Gradual and abrupt tree-lines. (a) Upper boundary of the fir–beech forest on the southern slope of Assara Mountain in the west Caucasus Range. The letters **a**, **b**, and **c** denote forest subzones. (b) Upper boundary of the larch forest on the north-eastern slope of the Balakhtyn-Shele Range, East Sayans. (Adapted from Armand, 1992.)

prostrate appearance, all of which features signify stress. Environmental factors inducing the stress include breakage by snow and ice, downslope movement of snow, deep and persistent snowpacks (which are a weighty burden for trees to bear), snow fungi, strong prevailing winds impacting on the forest edge, abrasion by ice or sand, frost, winter dessication, and damage by grazers and browsers (Wardle 1993). Trees in the Kampfzone are described as Krummholz, meaning 'crooked wood'. Flagging is a common form of deformity: the crown of a 'flagged' tree grows mainly on the lee side of the trunk so that the tree resembles a flag flying in the wind. In Austria, in the high core of the central Alps, such as around Mount Patscherkofel (2247 m), the Waldgrenze runs at about 1950 m (Pears 1985, 135). Norway spruce (*Picea abies*) and European larch (*Larix decidua*) grow robustly up to this height. The Baumgrenze lies at about 2150 m and is marked by the last appearance of the stately arolla or stone pine (*Pinus cembra*). Dwarf specimens of Norway spruce, European larch,

and arolla pine are found in the Kampfzone. A word of caution is necessary here – some trees grow as dwarf forms whatever the environmental conditions. In the central Alps, the dwarf mountain pine (*Pinus mugo*) thrives in the Kampfzone but retains its dwarf habit even when grown at lower altitudes. Its Krummholz character is therefore genetically determined and not the outcome of environmental stresses.

Tree-lines are not always sharp. In the Caucasus Mountains, for instance, they may be sharp or diffuse (Armand 1992). On the southern slope of the Assara Mountain, the transition from forest to alpine meadow has two steps: the shrub belt, in which hazel (*Corylus avellana*) is the dominant species, lies above the forest line but within 50 m it is replaced by alpine meadow (Figure 6.7(a)). On a north-eastern slope of the Balakhtyn-Shele Range, in the East Sayans, the forest line is diffuse (Figure 6.7(b)). Siberian larch (*Larix sibirica*) forest, with occasional Siberian pine trees (*Pinus sibirica*) and a subcanopy of round-leaved birch (*Betula rotundifolia*), gives way through about

100 m of elevation to mountain tundra with dryad (*Dryas oxiodonta*), blueberry (*Vaccinium uliginosum*), crowberry (*Empetrum nigrum*), and others. The results of the study show that, in these two mountains, a gradual altitudinal change of climate causes in one case a sharp replacement of a tree formation with a herbaceous formation, but in the other case a gradual change. The sharp transition occurs as several steps, but each step is clearly visible in the field.

Lower tree-lines

Inverted or lower tree-lines occur on high mountains in arid zones (e.g. Gerrard 1990, 51). In dry environments, high mountains ordinarily stand out as 'oases' – they are often better watered than the surrounding desert. Upper and lower tree-lines are found on such mountains, the first being produced by a lack of heat and associated factors, the second by a lack of water. The outcome is a belt of forest that encircles arid-zone mountains. High mountains in arid environments tend to sit in a transition zone between humid areas and an arid core. Lower forest limits, therefore, drop towards the humid zone on one mountainside and rise towards the arid core on the opposite side. Moreover, the lower tree-line normally rises more rapidly than the upper tree-line so that the forest belt progressively thins, and may even disappear, away from the humid zone. The lower and upper tree-lines on a transect from the Punjab, across the north-west Himalayas, to the Karakorams exhibit this feature (Troll 1973a). On the Pir Punjal Range, between the Punjab and Kashmir Basins, on the Nanga Parbat and Rakaposhi Massifs, the lower tree-line sits at 2700 m and the upper tree-line at 3800 m. In the drier main range of the Hunza Karakoram, the equivalent tree-line elevations are 3400 m and 3900 m. In the even drier Luphar Group, the forest belt vanishes at around 4000 m.

In major Himalayan valleys, the wind circulation typically involves air subsiding over valley bottoms. Accordingly, valley-side slopes and ridges are wet, while valley bottoms are dry. This, the 'Troll effect', produces a lower tree-line – the valley bottoms are too dry for trees to grow and instead support grassland. Forests grow on the wet valley sides up to the upper tree-line. So the forest belt is confined to valley-side slopes and is prevented from extending upwards by cold and from extending downwards by dryness.

Inverted tree-lines also occur in topographic lows where cold air commonly drains into, and fills, mountain valleys to create frost hollows. This phenomenon is not very common but has been reported from Australia and the United States. On Mount Buffalo, Victoria, Australia, frost hollows occurring at about 1520 m prevent the growth of snow gums (*Eucalyptus pauciflora*) and allow herbs, grasses, or dwarf-shrub vegetation to flourish. In Nevada, temperature inversions in the zone of single-leaf piñon pine (*Pinus monophylla*) and bigberry juniper (*Juniperus osteosperma*) lead to a lower tree-line at about 1470 m and treeless valley bottoms (Billings 1954). The upper tree-line stands at about 2067 m, which means that ridges are treeless. The belt covering the valley-side slopes between the lower and upper tree-lines, known as the thermal zone, favours the growth of piñon pine and juniper.

What causes tree-lines?

Much debate surrounds the causes of tree-lines (e.g. Stevens and Fox 1991). Present tree-lines result from a diverse range of processes. These processes operate at different scales and vary through time. Climate is the prime control, setting the broad limits to tree growth, but local factors may raise or lower a tree-line from its climatically determined level. For example, in the Chilean Andes, massive slope disturbances, chiefly the result of volcanic activity, impose severe constraints on the altitude of forest growth, and forest composition and structure (Veblen *et al.* 1977). In the European Alps, the climatic tree-line sits above the actual tree-line, largely owing to the felling of trees and the grazing of domestic animals. In the central Swiss Alps, the highest Holocene tree-line was at 2400–2450 m, or 50–100 m higher than the uppermost specimen of stone pine (*Pinus cembra*) today, between 9000 and 4700 years ago (Tinner *et al.* 1996). After 4700 years ago, human activity began to reduce the forest limit. In northernmost Fennoscandia, the interaction of climate and grazing shapes tree-line woodlands (Oksanen *et al.* 1995). Grazing by reindeer and sheep is the main factor determining the transition from woodlands and forests to heaths. Extensive birch (*Betula*) brushwoods form the tree-line in areas where summer grazing is light. The birch trees in these woods form 'tables', with most of the leaves growing a metre or two above the ground. Such table-formed birch trees are rarely found in grazed areas where the uppermost birch stands are savannah-like, with relatively large trees, the leaves of which nearly all grow more than two metres from the ground. In tree-line regions where grazing pressure is reduced, a rise of the tree-line might be expected.

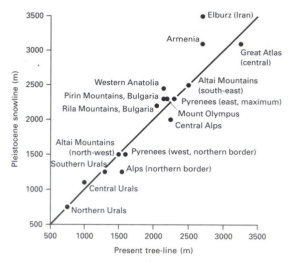

Figure 6.8 Pleistocene snowline versus present tree-line. (Adapted from data in Troll, 1973b.)

This process is happening in the Northern Corries area of the Cairngorms, Scotland, where Scots pine (*Pinus sylvestris*) has been colonising the subalpine zone following a reduction in grazing and browsing, possibly by deer, around 1960 (French *et al.* 1997). The result is the formation of a natural subalpine scrub zone. Scots pine colonisation is most prominent near to the forest, at altitudes around 600–700 m, in heather (*Calluna vulgaris*) moor or lichen-rich dwarf *Calluna* heath, on well-drained mineral soils with at most a shallow organic horizon. If grazing and browsing pressure should remain low, a natural tree-line may develop, similar to the only other widespread natural tree-line presently known in Scotland, at Creag Fhiaclach, also in the Cairngorms.

Present upper tree-lines bear a fairly consistent relationship with Pleistocene snowlines, which are usually defined as the level at which cirques form (Troll 1973b). In temperate zones, the present tree-line and the Pleistocene snowline coincide (Figure 6.8). In arid zones, such as Armenia and Iran, the Pleistocene snowline was well above the present tree-line. In oceanic climates, such as the Alps and Atlas Mountains, the Pleistocene snowline was below the present tree-line.

As a rule of thumb, the height at which most tree-lines occur is marked by the 10°C isotherm for the warmest month (Daubenmire 1954). But it is difficult to unravel the causes of climatic tree-lines as there are so many climatic factors involved, some of which

act directly and others indirectly. To identify the decisive factors hindering tree growth in a particular situation is therefore a formidable task. Indeed, trying to single out one decisive factor may be an unavailing exercise, as a tree-line is more likely to reflect a combination of adverse climatic factors. The length and warmth of the growing season at tree-canopy level are commonly critical factors, largely because they determine the ability of shoots to grow, mature, and harden against the winter cold. But several other factors are consequential (Table 6.1). Nonetheless, the sensitivity of tree-lines to climate is beyond question. This sensitivity is revealed in tree-line and alpine vegetation on oceanic islands in the tropical and warm-temperate climatic zones (Leuschner 1996). On these islands, three vegetation types dominate the tree-line forests: conifers, heath woodland, and broad-leaved trees. Conifers dominate on the Canary Islands and Yakushima (Japan). Heath woodland dominates on the Azores, Madeira, Réunion, Njazidja (formerly Grande Comore), and Bioko (formerly Fernando Póo). Broad-leaved trees dominate on the Hawaiian islands, Juan Fernandez islands, and Tristan da Cunha. Three vegetation types also dominate the subalpine and alpine belts. First, dry sclerophyllous scrub prevails on island mountains exposed to the trade winds: the Canaries, Cape Verde islands, Hawaiian islands, Réunion, and Njazidja. The aridity on these mountain tops above the tree-line results from a temperature inversion restricting the ascent of humid air-masses farther upslope. The peaks of other island mountains are located either in the wet equatorial and monsoonal regions, or in the temperate westerly zones without an effective inversion layer. These climates favour two kinds of vegetation. Mesic to wet alpine grassland and mire communities dominate on Juan Fernandez, Tristan da Cunha, Yakushima, and Bioko. Mesic to wet alpine heathland communities dominate on Pico (in the Azores), and partly on Madeira and Tristan da Cunha.

Interestingly, the tree-line on tropical and warm-temperate islands sits some 1000 to 2000 m lower than it does on continental mountains at similar latitudes. This 'tree-line depression' may result from several factors, two of which – the immaturity of volcanic soils and the absence on remote islands of tree species adapted to life at high altitudes – are not climatic. The other two factors causing the tree-line depression are climatic. First, island peaks exposed to trade winds with stable temperature inversions tend to experience dry conditions. Second, on islands,

Table 6.1 Factors that influence the upper limits of trees and forests in the temperate zone

Process	Contributory factors	
	Climatic	Genetic, physiological, and other non-climatic
Maintenance of forest limit	Carbon gain balancing carbon loss[a] Shoot growth and leaf production balancing shoot and leaf losses[a]	Adequate seed production at the timberline Adequate dispersal to the timberline Vegetative reproduction Tree death balanced by tree regeneration[a]
Rising timberline	Rising global temperatures Increasing atmospheric carbon dioxide levels	Introduction of, or invasion by, hardier tree species
Woody plant establishment beyond forest limit	Winter protection under snow cover Sheltered microsites Daytime warming near the ground	Rosette-tree form in tropics maximises warmth and affords protection from frost Inherently shrubby growth forms Ability of trees to survive dieback Spread by layering shoots Light-demanding seedlings Seed dispersal and establishment
Falling timberline	Falling global temperatures	Over-cutting of timber Destruction of seedlings by grazers and browsers Catastrophic events destroying forest microclimates
Depression of 'climatic' timberline	Cold climate at high latitude Cool oceanic summers Nocturnal temperature inversion Cold-air drainage in gullies and cirques Shady aspects Persistent snowbanks	Insufficient cold tolerance Intolerance of strong insolation Cold and wet soils

[a] Carbon balance, tree production, and tree mortality are largely determined by climatic factors.
Source: Adapted from a diagram in Wardle (1993).

the vertical temperature gradient is modified by only a small mountain mass, while on continents it is modified by large mountain masses. This factor results from the Massenerhebung (mass–elevation) effect: large mountain masses tend to create their own climates that promote higher tree-lines than smaller or isolated mountains. Thus, tree-lines in the massive mountains occupying the siliceous core of the central Alps are found at higher elevations than on the less massive, calcareous mountains (the South Tyrol and the Northern Alps) at roughly the same latitude. On tropical islands, the tree-line depression varies from around 400 m on Réunion to 2000 m on Gran Canaria and La Palma (Leuschner 1996).

Plant species and altitude

The effects of altitude on plants are complex because many environmental factors change together as elevation increases. Nonetheless, a good correlation usually exists between altitude and the abundance and range of plant species. A major thrust of research concerns

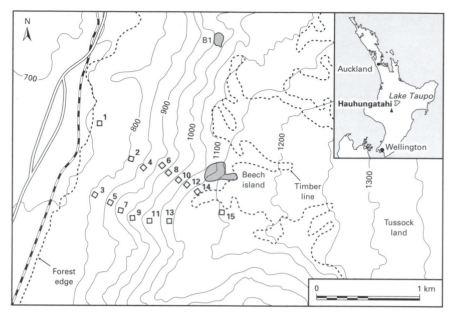

Figure 6.9 The location of Mount Hauhungatahi (inset), New Zealand, and the location of sample stands on its flanks. Even-numbered stands are on the gully transect. Odd-numbered stands are on the ridge transect. (Adapted from Druitt *et al.*, 1990.)

the arrangement of vegetation along altitudinal gradients. A key issue is whether the ranges of plant species have randomly overlapping distributions that form a continuum, or whether they form discrete communities with sharp ecotones. Plants in the mountain forests on Mount Hauhungatahi will illustrate the problem.

Mount Hauhungatahi: a case study

Mount Hauhungatahi lies in North Island, New Zealand, near the western edge of Tongariro National Park (Colour Plate 3). It is an extinct andesitic volcano that rises from a plateau at 700 m to a summit at 1519 m. A study conducted in February 1985 gathered data on all woody plants and ferns along two west-slope transects – one in a gully and one along a ridge – running from 750 m to the tree-line at about 1160 m (Figure 6.9) (Druitt *et al.* 1990). Figure 6.10 shows the altitudinal ranges of all woody species. The number of woody species is fairly stable up to about 1000 m, after which altitude species numbers drop (Figure 6.11). Individual species ranges overlap, but two marked changes of species composition reveal two ecotones, one at around 1050 m and another

at the tree-line (Figure 6.11). The tree-line occupies the zone from 1150 to 1200 m. Some 33 species change in this zone. All forest species, save the alpine celery top pine or mountain toatoa (*Phyllocladus alpinus*), attain their upper limits in this zone and are replaced by alpine shrubs. The lower ecotone forms a band at 1000 to 1050 m, in which 10 species are gained and seven species lost, yielding a turnover of 17 species. This is the highest turnover anywhere below the tree-line and it marks a manifest change in species composition as montane forest gives way to subalpine forest. Many lowland and montane canopy species reach their upper altitudinal limits in this ecotone: examples are rimu or red pine (*Dacrydium cupressinum*), matai (*Podocarpus spicatus*), and black maire (*Nestegis cunninghamii*). On the other hand, many subalpine canopy species, such as mountain toatoa (*Phyllocladus alpinus*) and pink pine (*Dacrydium biforme*), make their first appearance in this ecotone. The mean altitudinal range of the 28 woody species that had upper and lower limits on the section of the mountain studied was 171 ± 79 m.

The abundance of the species along the transects also suggested the two ecotones. Species abundances were measured as importance values. Smoothed

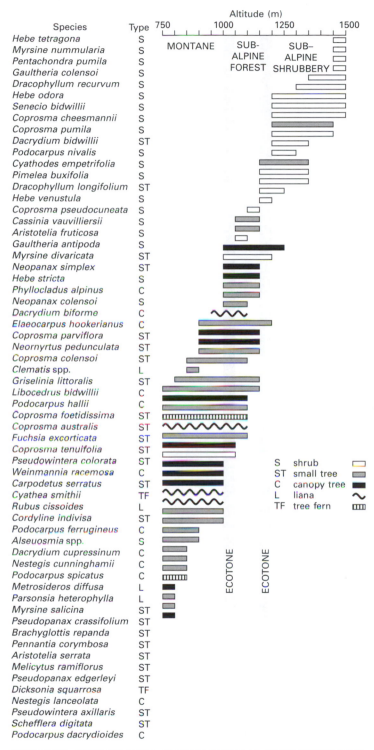

Figure 6.10 Altitudinal ranges of woody species on Mount Hauhungatahi, New Zealand. (Adapted from Druitt *et al.*, 1990.)

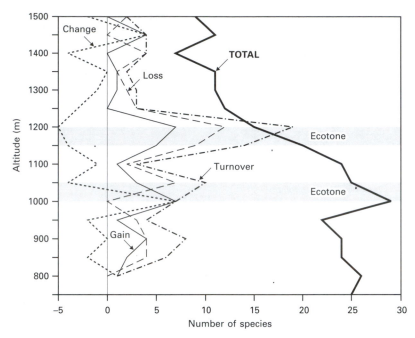

Figure 6.11 Total species and species losses, gains, change, and turnover on Mount Hauhungatahi. Notice the turnover peaks corresponding to two ecotones. (Adapted from Druitt *et al.*, 1990.)

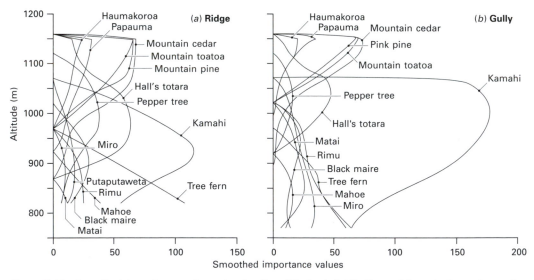

Figure 6.12 Smoothed importance values showing the altitudinal distributions of important woody species along the Mount Hauhungatahi transects: (a) ridge transect; (b) gully transect. (Adapted from Druitt *et al.*, 1990.)

importance values are given in Figure 6.12. In the gully transect, kamahi (*Weinmannia racemosa*) dominates lower and middle altitude stands, Hall's totara (*Podocarpus hallii*) has a sharp peak at 970 to 1050 m, and three gymnosperms – mountain toatoa, pink pine, and mountain cedar (*Libocedrus bidwillii*) – dominate upper altitude forests (Figure 6.12(b)). Most of the species' curves overlap, but some species have a common upper and lower limit, as in the case of rimu, matai, and black maire, which end at about 1000 m, and pink pine and mountain toatoa, which start at 1000 m. This zone accords with the data depicted in Figure 6.11 and suggests a change from podocarp–hardwood forest to subalpine forest at 1000–1050 m. Also seen in Figure 6.12(b) is the kamahi–Hall's totara forest at 950–1050 m. Species importance values along the ridge transect show a more equal share of importance than along the gully transect (Figure 6.12(a)). The tree fern *Cyathea smithii* is common at lower altitudes, along with kamahi and, to a lesser degree, rimu and black maire. Hall's totara occupies a wider altitudinal range than it does in the gully, and co-dominates middle-altitude stands with kamahi and pepper tree (*Pseudowintera colorata*) and upper-altitude stands with the subalpine tree species, mainly mountain pine (*Dacrydium bidwillii*), mountain cedar, and mountain toatoa. The Hall's totara zone and the podocarp–hardwood–subalpine forest boundary are less evident in the ridge transect, but several species have ranges ending at around 970–1020 m and subalpine species, including papauma (*Griselinia littoralis*) and mountain toatoa, increase noticeably in importance between 1020 and 1070 m.

Detrended correspondence analysis (DCA) ordinations of forest stands and subplots, using both density and basal area data, reinforced three findings. First, it confirmed the complete change of species at either end of the transects. Second, it supported the separation of subalpine forest stands from the rest. Third, it reinforced the altitudinal order of species shown in Figure 6.10. In addition, DCA revealed a highly significant correlation between stand position on the first DCA axis and mean altitudes: correlation coefficients for stand density, stand basal area, subplot density, and subplot basal area were all significant at $p < 0.001$. This fact, taken together with the dominance of the first eigenvalues, points to a close match between vegetation changes and altitude changes.

Two caveats should be added at this point. First, later work on the mountain revealed that the Taupo tephra eruption, which occurred about 1718 years ago, disturbed some of the communities and caused a shift in species abundance (Horrocks and Ogden 1998; see also Horrocks and Ogden 2000). In particular, mountain cedar, which had already been increasing at all sites owing to a more variable climate over the previous 1000 years or so, filled small canopy gaps in montane and subalpine forest created by the eruption and came to dominate the subalpine forest. It also spread upwards and downwards from the montane forest–subalpine forest ecotone after the Taupo eruption. The present large trees in the upper montane zone (Colour Plate 4) are probably the second or at most third generation of trees since the Taupo event, and the current lack of cedar reproduction at these lower limits suggests that readjustments are still taking place (Odgen, personal communication 2001). The disturbance may also have prompted the invasion of the montane forest by kamahi. Second, the montane forest–subalpine forest ecotone also approximately coincides with a change of substrate from sedimentary rocks to volcanic rocks, although the change is not as abrupt as might be supposed owing to the slumping of the higher-situated volcanic rocks over the lower-situated sedimentary rocks (Odgen, personal communication 2001). These findings do not wholly undermine the climatic explanation of vegetation zones on Mount Hauhungatahi, but they do show that climate acts conjointly with other environmental factors and that the vegetation change is not solely dictated by climatic influences.

Latitudinal complications

Altitudinal vegetation boundaries based upon species limits vary with latitude. The height and degree of isolation of a mountain may also affect them (Grubb 1971), as may aspect and slope angle. A study of elevational effects on two oak species – Gambel oak (*Quercus gambelii*) and shrub live oak (*Quercus turbinella*) – in the American South-west was particularly penetrating (Neilson and Wullstein 1983, 1986). Cold temperatures, and thus decrease in elevation northward along the 'polar front' air-mass gradient, tend to control the upper elevational boundaries. Growing season (summer) rainfall, which increases to the south along the 'Arizona monsoon' air-mass gradient, tends to fix the lower bounds. Further south, the lower elevational boundary rises again in response to higher evapotranspiration rates outweighing the increased availability of summer rainfall. Transplant experiments of over 700 oak seedlings through the region showed that, where the regional climate is not stressful, the seedlings survived equally well on all aspects and elevations and under different levels of cover. As temperature and rainfall constraints converged at the northern boundary of the species, seedling survival was progressively more constrained to specific slopes, elevations, and cover characteristics.

Animal species and altitude

The effects of altitude on animals, like those on plants, are complex. Factors influencing altitudinal ranges of animal species include climate, vegetation, substrate, and history. Vegetation is itself strongly influenced by climate, so some elevational zoning of animals might arise through the medium of vegetation. Climate also has a direct impact on the altitudinal range of many animal species: the tolerance range of many species extends over a limited altitudinal band, and only a few species are broad enough generalists to thrive in all mountain environments. So, although made complicated by vegetational effects, altitudinal climatic influences upon animals are often powerful and manifest in the range and abundance of particular species. To illustrate this point, the examples of altitudinal rat distributions on Mauna Loa in the Hawaiian Islands, and animals in old-growth forest in Oregon, will be considered.

Animals and elevation in Oregon

The effect of altitude on the distribution of animals is evident in a study of old-growth sites within the H. J. Andrews Experimental Ecological Reserve, which lies in the Willamette National Forest, Oregon, USA (Harris *et al.* 1982). Fifteen sites were studied. Records of species occurrence, habitat (primary and secondary – primary being sites where a species displays all its life history functions), wetness (very wet, wet, dry, very dry), elevation, and successional stage (1 regeneration, 2 seedling–sapling, 3 pole timber, 4 saw-timber, 5 large saw-timber, 6 old growth) were taken for each site. Elevation was the foremost variable for predicting the species richness on a site. Some 108 species of amphibians, reptiles, and mammals live in the western Oregon forests. Of these, 95 live at 1500 feet while just 32 live at 7000 feet. The relationship between species richness and elevation is curvilinear and highly significant (Figure 6.13). The number of amphibian and reptile species falls sharply, and the number of mammal species more gradually, with increasing elevation. This means that amphibians and reptiles (and birds) become less abundant relative to mammals at higher altitudes. The decline in vertebrates may result partly from a reduction of hardwood forests and partly from a decrease in land area

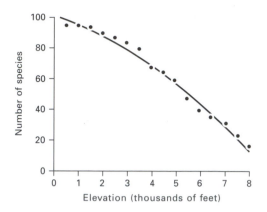

Figure 6.13 Species richness versus elevation in old-growth sites within the H. J. Andrews Experimental Ecological Reserve, Willamette National Forest, Oregon, United States. (After Harris, 1984, 58.)

(especially on cone-shaped mountains). The altitudinal species ranges overlap. Nearly all the amphibians and reptiles occur at sea level, the only major exception being the Cascades frog (*Rana cascadae*), which is not found below 3000 feet (Figure 6.14). Some of them, such as the painted turtle (*Chrysemys picta*) and the western aquatic garter snake (*Thamnophis couchi*), are confined to relatively low altitudes; others, such as the Pacific tree-frog (*Hyla regilla*), live across the full altitudinal range. A similar pattern is found in the mammals, though altitudinal zonation is more noticeable (Figure 6.15). The Pacific shrew (*Sorex pacificus*) is confined to sites below 1000 feet, while six species, including the wandering shrew (*S. vagrans*) and coyote (*Canis latrans*), are found at all elevations. Eight species, including the wolverine (*Gulo luscus*) and northern water shrew (*S. palustris*), are not found below 4000–4500 feet.

Rats on Mauna Loa

Four species of commensal murid rodents, all introduced by humans, were recorded along the Mauna Loa transect (p. 169). The Polynesian rat (*Rattus exulans*) arrived with Polynesian settlers. The roof rat (*R. rattus*), Norwegian rat (*R. norvegicus*), and house mouse (*Mus musculus*) are all derived from Eurasian and American stock that were carried on ships to the Hawaiian Islands some time after their discovery in 1778. The Norwegian rat tends to live in close association with humans and was not recorded on the Mauna Loa transect. The other three species overlap broadly in their altitudinal ranges and do not

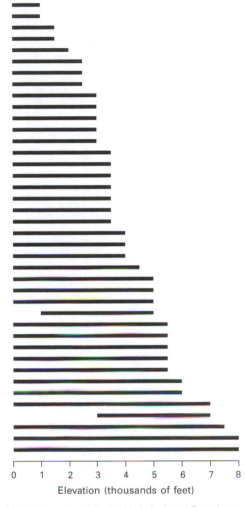

Western pond turtle	*Clemmys marmorata*
Painted turtle	*Chrysemys picta*
Del Norte salamander	*Plethodon elongatus*
California slender salamander	*Batrachoseps attenuatus*
Western aquatic garter snake	*Thamnophis couchi*
Western fence lizard	*Sceloporus occidentalis*
Southern alligator lizard	*Gerrhonotus multicarinatus*
Ringneck snake	*Diadophis punctatus*
Olympic salamander	*Rhyacotriton olympicus*
Dunn's salamander	*Plethodon dunni*
Black salamander	*Aneides flavipunctatus*
Common kingsnake	*Lampropeltis getulus*
California mountain kingsnake	*Lampropeltis zonata*
Western red-backed salamander	*Plethodon vehiculum*
Siskiyou mountain salamander	*Plethodon stormi*
Red-legged frog	*Rana aurora*
Sharp-tailed snake	*Contia tenuis*
Western yellow-bellied racer	*Coluber constrictor mormon*
Gopher snake	*Pituophis melanoleucus*
Western rattlesnake	*Crotalus viridis*
Larch mountain salamander	*Plethodon larselli*
Foothill yellow-legged frog	*Rana boylei*
Western terrestrial garter snake	*Thamnophis elegans*
Rubber boa	*Charina bottae*
Clouded salamander	*Aneides ferreus*
Oregon salamander	*Ensatina eschscholtzi*
Western skink	*Eumeces skiltonianus*
Oregon slender salamander	*Batrachoseps wrighti*
Long-toed salamander	*Ambystoma macrodactylum*
Pacific giant salamander	*Dicamptodon ensatus*
Rough-skinned newt	*Taricha granulosa*
Spotted frog	*Rana pretiosa*
Northwestern garter snake	*Thamnophis ordinoides*
Northwestern salamander	*Ambystoma gracile*
Northern alligator lizard	*Gerrhonotus coeruleus*
Common garter snake	*Thamnophis sirtalis*
Cascades frog	*Rana cascadae*
Tailed frog	*Ascaphus truei*
Western toad	*Bufo boreas*
Pacific treefrog	*Hyla regilla*

Elevation (thousands of feet)

Figure 6.14 Altitudinal ranges of amphibians and reptiles in old-growth sites within the H. J. Andrews Experimental Ecological Reserve, Willamette National Forest, Oregon, United States. (Adapted from Harris, 1984, 56.)

form a distinct spatial group (Figure 6.16). Most abundant is the common house mouse which ranges from the alpine zone to the closed rainforest (and on down to sea level). It is most abundant in the middle of the transect which carries closed herbaceous or grass vegetation with scattered trees (savannah zone) or tree groups (mountain parkland). Upslope and downslope of these zones the abundance of the house mouse declines. Downslope this decline occurs at the lower end of the savannah zone and is associated with the transition to closed forest. Upslope it is associated with a change from mountain parkland with closed grass cover patches, to open subalpine scrub forest with almost no herbaceous undergrowth. It is the only

rodent to occupy the sparse alpine scrub zone where it builds secure nests in the pahoehoe lavas and survives on a meagre supply of 'ohelo and pukiawe fruits (p. 169) and insects. Clearly, the ecological optimum of the species is in vegetation with sunny grassland habitats. The roof rat was not trapped beyond the tree-line. Its peak abundance lies in closed forest vegetation and in open forest with dense fern undergrowth, and it was not observed in the lehua dry-forest zone. Thus, although the house mouse and roof rat have overlapping distributions along the altitudinal transect, their ecological optima do not coincide. The Polynesian rat shows a distribution limited by altitude, being rarely found above the rainforest.

Figure 6.15 Altitudinal ranges of mammals in old-growth sites within the H. J. Andrews Experimental Ecological Reserve, Willamette National Forest, Oregon, United States. (Adapted from Harris, 1984, 57.)

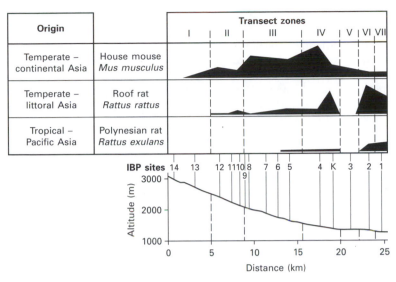

Figure 6.16 Distributions of three rodent species along the Mauna Loa transect. For each species, the black areas show abundance in equal units of five animals trapped per year, with a maximum of 9 units = 41–45 animals (e.g. *Mus musculus* at site 4); 8 units = 36–40 animals, 7 units = 31–35 animals, 6 units = 26–30 animals, and so on. The horizontal line represents one animal. K is the closed kipuka forest in the savannah zone IV. (Adapted from Mueller-Dombois *et al.*, 1981.)

Life and landform

Aspect and plants

The north-facing–south-facing divide

Slope exposure affects local climate and microclimate, light, windiness, and soil conditions. These environmental factors in turn influence vegetation. North-facing and south-facing slopes in extratropical regions commonly display the starkest contrasts. The slopes concerned range in scale from mountainsides, through sand dunes, to the sides of anthills.

European botanists looked at plant distribution in relation to slope exposure and topography in the mid-nineteenth century (see Braun-Blanquet 1932). In the United States, botanists working mainly in the western part of the country during the first quarter of the twentieth century attributed tree distributions to variations in altitude and aspect (e.g. Turesson 1914). Later studies in the eastern United States uncovered the importance of aspect, working through soil moisture and temperature, to plant communities (e.g. Potzger 1939). Microclimatic differences were clearly the ultimate cause of aspect-controlled plant distribu-

tions, but debate surrounded the prime microclimatic agents involved. Suggestions included air and soil temperature, insolation, winter snowfall, soil moisture, and the sand-to-clay ratio. On north-facing and south-facing slopes of Cushetunk Mountain, New Jersey, seasonal variations in temperature and moisture gradients near the soil surface were the main determinative factor (Cantlon 1953). The sand-to-clay ratio appears to reflect past and present microclimatic conditions and its value in the B horizon soils in southern Blue Ridge Gorge, North Carolina, out of a range of environmental factors recorded on north-facing and south-facing slopes, was the best predictor of six tree-species distributions (Mowbray and Oosting 1968). In Colorado, USA, slope aspect, acting through the surface radiation balance and wind speed, influenced evapotranspiration and soil desiccation in the alpine tundra (Isard 1986). South-facing and east-facing sites upon a fellfield (blockfield) knoll received, respectively, 18 and 14 per cent more net radiation than the north-facing slope on clear summer days. West-facing slopes experienced moderate radiation loads but the highest wind speeds. During two dry periods, water loss by evapotranspiration for east-facing, west-facing, and north-facing slopes was 80, 80, and 60 per cent, respectively, of the evapotranspiration from the south-facing slope. In consequence, soil at the south-facing site dried faster than soil on the other slopes of the knoll. The differing, topoclimatically determined soil-moisture levels affected vegetation distribution. For instance, mountain avens (*Dryas octopetala*) favoured the moister northern slopes on Niwot Ridge.

Whatever the primary causes of exposure-related vegetation differences, their effects can be spectacular. The Lägern is an 800-m-high, east–west running mountain range made of limestone, near Baden, in the Swiss Jura. The north-facing and south-facing slopes are separated by a narrow mountain ridge, only some 50 cm wide. The microclimate on the north-facing side is cold and supports beech (*Fagus sylvatica*) forest with subalpine elements such as alpine penny-cress (*Thlaspi alpestre*) and green spleenwort (*Asplenium viride*). The warm south-facing slope supports warmth-loving oak (*Quercus pubescens*) woodland with such southern elements as pale-flowered orchid (*Orchis palleus*), wonder violet (*Viola mirabilis*), and bastard balm (*Melittis melissophylum*). This dramatic botanical contrast on either side of the ridge is equivalent to an altitudinal difference of 1 km and a latitudinal difference of 1000 km. On coastal sand dunes in the Netherlands, north-facing slopes support dense stands of boreal crowberry (*Empetrum nigrum*), while south-facing slopes support open cover of grey hair-grass (*Corynephorus canescens*) and lichens. More subtle differences in aspect-related topoclimates are commonly recorded. In south-eastern Ohio, microclimatic differences between north-east-facing and south-west-facing valley sides lead to mixed-oak association on south-west-facing slopes and a mixed mesophytic plant association on the moister north-east-facing slopes (Finney *et al.* 1962). On Cushetunk Mountain, New Jersey, life-form spectra on north-facing slopes, south-facing slopes, and the entire ridge differ significantly (Cantlon 1953).

Slope exposure tends to produce distinctive patterns in plant species distributions. In the Northern Hemisphere, plants with a southern distribution tend to grow on south-facing slopes at the northern edge of their range, while plants with a northern distribution tend to grow on north-facing slopes towards the southern end of their range. As a rule of thumb, a plant close to the edge of its distribution range tends to find itself on a slope that looks towards the centre of the range. For example, the wild strawberry (*Fragaria vesca*) is restricted to south-facing slopes in northern Norway, prefers level areas and gentle slopes in temperate Europe, and occurs mainly on north-facing slopes in the Mediterranean lowlands (Stoutjesdijk and Barkman 1992, 77).

As well as affecting vegetation distribution, aspect affects flowering phenology and seed development of plants. The mean height of three-month-old messmate stringybark (*Eucalyptus obliqua*) seedlings varies with aspect on Mount Oberon, Wilson's Promontory, Victoria, Australia (Ashton and Kelliher 1996). Seedling height on warm, north-west aspects is about 3 cm, while on cool, south-east aspects it is about 26 cm. Near Innsbruck, Austria, two alpine sedges growing at 2200–2300 m showed phenological and developmental differences according to topoclimate (Wagner and Reichegger 1997). The two sedges are the curved sedge (*Carex curvula*) and the cushion sedge (*C. firma*), which are key species of grassland communities in the upper alpine belt of the European Alps. Curved sedge was found growing on Mount Patscherkofel (2247 m), an isolated mountain 6 km south of the Inn valley, while cushion sedge was found on Mount Hafelekar (2334 m) in the northern mountain range 4 km north of the Inn valley. Curved sedge was studied at three sites and cushion sedge at two sites (Table 6.2). Both sedge species exhibited striking differences in their flowering phenology between the sites. The timing of their reproductive development was largely influenced by the time of snowmelt. On early-thawing sites, the prefloration period (the time between snowmelt to the onset of flowering) was about 20 days for curved sedge (south-facing and west-facing sites) and 28 days for cushion sedge (west-facing site). On late-thawing sites, the prefloration period was 14 days for curved sedge (north-facing site) and 8 days for cushion sedge (north-facing site). These differences in anthesis (the date of flowering) influenced the length of the postfloration period (the time between flowering and the formation of mature fruits). Curved sedge on the west-facing Mount Patscherkofel site bloomed at the earliest date and took 60 days to produce mature fruits, while the later-flowering plants on the north-facing site needed 50 days to complete seed development. The equivalent figures for cushion sedge on Mount Hafelekar were 70 days (west-facing site) and 55 days (north-facing site). This compensation for a shorter growing season by shortening the time of prefloration is common in many arctic and alpine plants (e.g. Bliss 1971).

Some plants can control the amount of radiation received by their bodies by turning to the Sun. Some

Table 6.2   Study sites of sedges in the Austrian Alps

Aspect and description	Altitude (m)	Slope (°)	Distribution of sedges
Mount Patscherkofel South-facing slope on a windswept ridge with shallow snow in winter	2240–2245	20–25	Tussocks of curved sedge sparsely distributed among rocky outcrops and scree
North-facing slope in a wind-protected hollow, 50 m (horizontally) from the ridge site. In spring, snow lasts 2–3 weeks longer than on the ridge	2225–2200	28–30	Tussocks of curved sedge scattered among the vegetation
West-facing slope with moderate wind exposure and snow cover during the winter	2185–2200	20	Curved sedge tussocks near their lower limit compete with alpine grasses and ericaceous dwarf shrubs
Mount Hafelekar West-facing slope exposed to sun and wind with little or no snow cover in winter and ground frost lasting far into spring	2310–2320	18–22	Cushions of cushion sedge
North-facing slope with snowdrifts of 2–3 m in winter and spring	2310–2320	30–32	Cushions of cushion sedge distributed along a gradient from the margins of the snowpack down to where the snow remains longest, which is their survival limit

Source: Adapted from information in Wagner and Reichegger (1997).

species of lupin (*Lupinus*), such as the yellow bush lupin (*L. arboreus*), track the Sun during the day, at least as seedlings. The desert or Arizona lupin (*L. arizonicus*) folds its leaves when it is dry and exposes the narrow side to the Sun to lower the radiation received (Forseth and Ehrelinger 1982). The prickly lettuce (*Lactuca serriola*) arranges its leaves in a north–south direction to avoid maximum exposure to heat and light. Subtropical trees, such as the Peruvian peppertree (*Schinus molle*) and Australian gum tree (*Eucalyptus*) species, have their hanging leaves orientated north–south.

Leeward and windward slopes

Aspect determines exposure to prevailing winds. Leeward slopes, especially on large hills and mountains, normally lie within a rain shadow. Rain-shadow effects on vegetation are pronounced in the Basin and Range Province of the United States: the climates of the Great Basin and mountains are influenced by the Sierra Nevada, and the climates of the prairies and plains are semi-arid owing to the presence of the Rocky Mountains. In the Cascades, the eastern, leeward slopes are drier than the western, windward slopes. Consequently, the vegetation changes from western and mountain hemlock (*Tsuga heterophylla* and *T. mertensiana*) and Pacific silver fir and subalpine fir (*Abies amabilis* and *A. lasiocarpa*) to western larch (*Larix occidentalis*) and ponderosa pine (*Pinus ponderosa*), and finally to sagebrush (*Artemisia* spp.) desert (Billings 1990).

Aspect, slope inclination, and altitude

Aspect commonly works in partnership with other topographic factors, and particularly with altitude and slope inclination, to influence plant distributions (e.g. Holland and Steyn 1975). Slope exposure modifies the altitudinal range of juniper woodland (*Juniperus excelsa polycarpos*) in the Hajar mountains, north Oman (Gardner and Fisher 1996). Juniper is confined to the highest areas (the central massif of Jebel Akhdar and the outlying mountains of Jebel Qubal and Jebel Kawr), where it mostly forms open woodlands. It grows on exposed slopes from 2100 m to the highest peak at 3009 m, and there is no upper tree-line. On shady, north-facing slopes, juniper trees grow down to 1375 m. Evergreen broadleaved forest on Mount Lopei, Taiwan, displays a single dominant altitudinal gradient between 540 m and 1320 m. However, wind exposure associated with topography accounted for additional species variation within altitudinal bands, with the effects of exposure being more pronounced at high elevations (Hsieh *et al.* 1998). In the Front Range, Colorado, USA, altitudinal vegetation zones are dominated by ponderosa pine (*Pinus ponderosa*) in the lower-montane zone, mixed Douglas fir (*Pseudotsuga taxifolia*) and ponderosa pine in the upper-montane zone, Engelmann spruce (*Picea engelmannii*) and subalpine fir (*Abies lasiocarpa*) the subalpine zone, and sedge (*Kobresia*) meadow-tundra in the alpine zone (Billings 1990). Aspect influences the distribution of the trees in all zones. The ponderosa pine stands are open and park-like on south-facing slopes, but are dense on north-facing slopes and are often admixed with Douglas fir. In the zone where ponderosa pine and Douglas fir occur together, the pines are dominant on the south-facing slopes and Douglas fir on the north-facing slopes, but Douglas fir becomes more common on south-facing slopes as elevation increases. At around 2600 m, lodgepole pine (*Picea contorta*) and quaking aspen (*Populus tremuloides*) occur in fairly pure stands on north-facing slopes where fire has occurred in the past. From about 3050 m, Engelmann spruce and subalpine fir are dominant.

Under some circumstances, relief and slope gradient accentuate the effects of aspect. The Cascades span over 800 km in Washington and Oregon, USA. They are highly dissected by stream channels and slopes are steep. The main range is aligned north–south and the principal streams and secondary ridges run east–west. This topographic pattern augments the importance of high latitude and slope exposure (Harris 1984, 17). North-facing slopes are mainly wetter and cooler, with snowpack forming earlier in the autumn; south-facing slopes are much drier and hotter, and have a longer growing season. These environmental factors influence regeneration, forest succession, and stand structure. Slope exposure affects the harshness of the site, which in turn affects the speed with which the site is invaded and thus the age-structure of a stand. On north-facing slopes, stand regeneration occurs rapidly with high stocking; these stands tend to become dense and even-aged, and have little understorey. East-facing, west-facing, and finally south-facing slopes have progressively longer establishment periods and increasingly depart from even-aged conditions. The result is a highly variable canopy structure and composition with a denser and more varied understorey. Aspect also affects snowpack and thaw rates, which affect runoff, which governs stream and riparian strip characteristics.

Aspect and animals

Geographical ranges of some animal species are influenced by slope exposure. In the mountainous regions of the western USA, valleys tend to lie on an east–west axis. Accordingly, south-facing slopes are drier and warmer than adjacent north-facing slopes. These microclimatic differences strongly influence the distribution of animals and plants. For instance, in the steep-sided mountains of southern California, where the 'climax' vegetation is chaparral, the biotas on adjacent north-facing and south-facing slopes are altogether different. Some species of small mammal, such as the San Diego pocket mouse (*Chaetodipus fallax*), are confined to south-facing slopes, and others, such as the dusky-footed woodrat (*Neotoma fuscipes*), are restricted to north-facing slopes (Vaughan 1954).

Animals adjust to small-scale differences of aspect. In North America, golden eagles (*Aquila chrysaetos*) prefer to nest on cliffs with a southern exposure, especially in cooler climates (Mosher and White 1976; McGahan 1968). Many animals can control the amount of radiation received by their bodies by turning to the Sun. Animals simply turn the body axis to expose the broadest side to gain more radiation. Ants and termites use the microclimatic differences of exposure to good advantage. Termites are tropical animals. Their nests are vertical with a broad side orientated in a north–south direction. The advantage of this arrangement is that the nest receives relatively most radiation in the early morning and late afternoon, when the Sun is not so intense, and least radiation during the hottest part of the day. Ants have a temperate and boreal distribution. Their nests are dome-shaped. They build nests with steeper sides at higher latitudes, so adjusting to the lower elevation of the Sun.

Detailed studies of alpine marmots (*Marmota marmota*) in the French Alps revealed a significant role for slope exposure in habitat preference and growth rates (Allainé *et al.* 1994, 1998). Two high-altitude sites in the Vanoise National Park were investigated – the Réserve de la Grande Sassière (La Sassière) and the Vallon de la Lenta (La Lenta). The altitudinal range at La Sassière is 2300–2800 m, and for La Lenta it is 2100–2400 m. Slopes are gentler and plant cover greater at La Lenta, and aspect is east–west, as opposed to north–south in La Sassière.

Disturbance at La Sassière is chiefly occasioned by hikers harassing the marmots, but at La Lenta it is mainly the result of agricultural practices, especially hay harvesting. One study (Allainé *et al.* 1994) found that marmot settlement, which occurred in 59 of the 88 quadrats measured at La Sassière and in 26 out of the 35 quadrats measured at La Lenta, was significantly affected by slope (only at La Sassière), exposure to the Sun (at both sites), and plant cover (at both sites). At La Sassière, the following percentage of quadrats had marmot settlements: 48 per cent of north-facing quadrats, 71 per cent of valley quadrats, and 79 per cent of south-facing quadrats. The equivalent figures for La Lenta were 63.6 per cent of west-facing quadrats, 38 per cent of valley quadrats, and 100 per cent of east-facing quadrats. Taking all environmental factors into account, it was concluded that alpine marmots prefer sites with a southern or eastern aspect (where snow melts relatively early), intermediate slopes (15–45°), moderate plant cover (25–75 per cent), and a low level of human disturbance.

Later work at La Sassière looked at the effects of several environmental factors on the post-weaning growing pattern of wild juvenile alpine marmots (Allainé *et al.* 1998). Factors considered were the year of birth, exposure to the Sun (aspect) in the home range (classified as south-facing, valley, or north-facing), litter size, and sex of young. Measured components of the growing pattern were juvenile mass at emergence from the natal burrow (which results from pre-weaning growth), post-weaning growth rate, and the length of the active season during which growth occurs. Mass at emergence and post-weaning growth rate varied according to year of birth, were higher in small litters, and were higher in males; they were also higher in south-facing home ranges than in north-facing home ranges. For a given year, the mass of juveniles at emergence fell by about 60 g from southern to northern exposure, while the post-weaning growth rate dropped from about 23 g/day to about 11 g/day from southern to northern exposure. Exposure also influenced the date of emergence, which occurred on average 6 days earlier on southern exposures than in the valley, and 12 days earlier in southern exposures than in northern exposures. The dates of emergence of juveniles were similar to the dates of emergence of adults from hibernation. Exposure also affected the length of the active season, which was longest on southern exposures. In summary, juveniles born on southern exposures emerged heavier, grew faster, and benefited from a longer time to gain weight than juveniles born on northern exposures.

Slopes

Slopes of all inclinations affect microclimates to greater or lesser degrees. When combined with slope exposure, slope gradient often explains certain features of animal and plant distributions, as some of the examples in the previous section revealed. Two aspects of slope merit further attention – steep slopes and linked sequences of slopes (toposequences or catenas).

Steep slopes

Steep slopes affect animals directly by offering a distinct habitat and indirectly through local climatic effects. Under the right conditions, steep slopes are associated with the growth of thermals, uprising columns of air that can assist flying animals. Thermals support soaring birds such as vultures, gulls, and eagles (e.g. Pennycuick 1973). In India, vertical walls in towns absorb early-morning solar radiation, warm up, and produce heated air that starts to rise. Vultures begin soaring from towns an hour earlier than they do in the open country (Cone 1962).

Steep slopes influence animals by their physical presence and by their contribution to topographic heterogeneity in an area. Several mammals are adapted to living on steep slopes in mountainous terrain. In Europe, the chamois (*Rupicapra rupicapra*) and the alpine ibex (*Capra ibex ibex*) are well adapted to life in precipitous and rocky terrain. The chamois is found in the Alps, Pyrenees, and Carpathians. The alpine ibex, an adroit jumper adapted to cliffs, lives in several European and North African mountains. Some large birds of prey nest on high ledges in mountainous terrain. In Europe, the golden eagle (*Aquila chrysaetos*) nests high on rocky crags or a cliff face, though it will sometimes nest in a tall, exposed tree.

Cliffs increase topographic heterogeneity of landscapes and affect physical and biological processes. In grasslands, cliffs create an array of vegetative characteristics that are only observed on cliff sites (Ward and Anderson 1988). Although heterogeneity may occur in hilly terrain without cliffs, cliffs provide benefits for some wildlife species that hilly terrain alone cannot provide. For instance, they provide secure nesting sites, and afford protection against predation and extreme environmental conditions. Cliff sites in the Hanna Basin, which occupies 610 km^2 of south-central Wyoming, USA, had a higher diversity of vegetation and were more heterogeneous (with higher topographic roughness factors and structurally more complex and 'patchier' vegetation) than sites without cliffs (control sites). Owing partly to the greater land-scape heterogeneity, the abundance and diversity of small mammals and male birds were greater on cliff sites, too. The six most common small mammals living on the cliffs were, in order, deer mouse (*Peromyscus maniculatus*), least chipmunk (*Eutamias minimum*), Wyoming ground squirrel (*Spermophilus elegans elegans*), bushy-tailed woodrat (*Neotoma cinerea*), white-tailed prairie dog (*Cynomys gunnisoni*), and northern grasshopper mouse (*Onychomys leucogaster*). On control sites, the most common species were, in order, silky pocket mouse (*Perognathus flavus*), white-tailed prairie dog, north-ern grasshopper mouse, Wyoming ground squirrel, and least chipmunk. The six most common male birds found on cliff sites were, in order, Brewer's sparrow (*Spizella breweri*), rock wren (*Salpinctes obsoletus*), green-tailed towhee (*Pipilo chlorurus*), vesper sparrow (*Pooecetes gramineus*), sage thrasher (*Oreoscoptes montanus*), and horned lark (*Eremophila alpestris*). The same six species were the commonest species in control sites but in a different rank order – horned lark, vesper sparrow, Brewer's sparrow, sage thrasher, rock wren, and green-tailed towhee.

Also, in the Hanna Basin, 'artificial' cliffs are created by mining operations that leave a 'highwall'. Mined areas are restored by levelling ground to pro-duce a rather monotonous landscape that tends to leave little cover, shelter, or food sources for wildlife. An alternative practice is to permanently retain a final highwall (Ward and Anderson 1989). An investiga-tion on the effect of a south-facing sandstone highwall, some 10 m tall at the highest point and 425 m long, revealed an influence on the richness (the number of species), diversity (as measured by the Shannon–Wiener function), and abundance of vegetation, small mammals, and birds. Vegetation species richness and diversity were higher near the highwall compared with two adjacent sites 150 m in front of and behind the highwall. Vegetation abundance, measured as cover, was higher at the adjacent sites, largely due to the high abundance of Russian thistle (*Salsola* spp.). Small mammal abundance, richness, and diversity were greater on the highwall. Deer mouse (*Peromyscus maniculatus*) was twice as abundant near the highwall than at adjacent sites, and bushy-tailed woodrat (*Neotoma cinerea*) and least chipmunk (*Eutamias minimus*) were trapped only near the highwall. On three occasions during the study, four pronghorn antelope (*Antilocapra americana*) were seen at the foot of the highwall, feeding or drinking from an intermittent pond. Numerous striped skunk (*Mephitis*

mephitis) were discovered under three of the rockpiles, suggesting that they use the rocks for cover. Male bird abundance, richness, and diversity were gener-ally greatest in front of the highwall. During the 1985 breeding season, one pair of the following species nested in the highwall: mountain bluebird (*Sialia currucoides*), violet-green swallow (*Tachycineta thalassina*), rock wren (*Salpinctes obsoletus*), and mourning dove (*Zenaida macroura*). In addition, a pair of American kestrels (*Falco sparverius*) tried to nest on the highwall in 1984, but one of the adults died. In conclusion, the evidence suggests that the highwall afforded secure nesting sites for at least five bird species, and den sites for three mammal species. The three mammal species, rock wren, and mourning dove made dens and nests on other substrates, but the violet-green swallow, mountain bluebird, and Amer-ican kestrel only used the highwall and would have been missing had the highwall been completely backfilled.

Catenas

Milne's pioneering work on catenas, and the studies that followed in its wake, included changes of vegeta-tion, as well as soils, over undulating topography. Vegetation catenas have made something of a come-back in ecology, with many examples recently appearing in the literature (e.g. Walker *et al.* 1989; Walker and Everett 1991; Giblin *et al.* 1991; Degórski 1994; Shaver *et al.* 1996; Swanson 1996; Oliveira-Filho *et al.* 1997).

A good example of a vegetation catena comes from arctic Alaska, where toposequences are found on old floodplains of the Sagavanirktok River, in the north-ern foothills of the Brooks Range (Giblin *et al.* 1991; Shaver *et al.* 1996). Figure 6.17 depicts a topose-quence running some 200 m across a sequence of old floodplain terraces. The total change in elevation is roughly 20 m. The entire sequence is underlain by permafrost. In summer, soil water moves as through-flow above the top of the permafrost. The vegetation found along such toposequences varies enormously, depending on the age of the surface, slope inclina-tion, and position along the slope of the river terrace. The toposequence shown in Figure 6.17 is typical of those found in the foothills region of the North Slope of Alaska. Six vegetation types occur. Tussock tundra occupies the uplands. The sequence then runs hilltop heath, hillslope shrub and lupin community, footslope *Equisetum* community, wet sedge tundra community, and riverside willow community. The

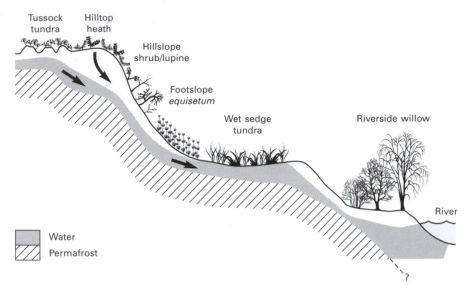

Figure 6.17 The Sagavanirktok River toposequence, Alaska. The six communities, starting with tussock-tundra in the uplands, occur on a staircase of old river floodplains. Almost the entire sequence sits upon permafrost. The relief of the toposequence is about 20 m and the length 100–200 m. (Adapted from Giblin *et al.*, 1991.)

chief plant species in the communities are summarised in Table 6.3.

The interior Alaskan soil catena mentioned in the previous chapter (p. 131) supports a vegetation catena (Swanson 1996). The catena lies in the boreal forest zone, not far from the forest–tundra ecotone, with lowland tundra lying 120 km to the west in the Selawik Valley. Different vegetation types along the catena correspond with the chief soil groups (dry and rocky soils, moist soils with permafrost, and wet soils with permafrost). The dry and rocky soils occupy convex to plane slopes, shoulders, and the crests of bedrock hills. They support closed or open forests of paper birch (*Betula papyrifera*), white spruce (*Picea glauca*), or black spruce (*P. mariana*). The shrub layer includes lingonberry (or Northern mountain cranberry) (*Vaccinium vitis-idaea*), prickly rose (*Rosa acicularis*), swamp redcurrant (*Ribes triste*), and marsh Labrador-tea (*Ledum palustre*). Splendid feather moss (*Hylocomium splendens*) or reindeer lichens (*Cladina* spp.) are also present. Moist soils with permafrost occupy slightly convex to plane slopes and crests of bedrock-cored hills. They are covered with open black spruce forest with low shrubs of Labrador-tea, lingonberry, and bilberry (or alpine blueberry) (*Vaccinium uliginosum*), and with feather moss and reindeer lichens. Wet soil with permafrost are found

on concave lower slopes of bedrock hills. Their vegetation cover is widely scattered, stunted black spruce trees with Labrador-tea, blueberry, lingonberry, and crowberry (*Empetrum nigrum*) as low shrubs, and with various mosses, including peat moss (*Sphagnum* spp.) and feather moss, and reindeer lichens.

Tropical vegetation catenas have also been studied. A good example comes from the margins of the Rio Grande, in the municipality of Conquista, Minas Gerais, south-eastern Brazil (Oliveira-Filho *et al.* 1997). Here, researchers explored the relationships between topography, soil taxonomic categories, and tree-species distribution in a riverside semi-deciduous forest, which was part of a 160-ha forest fragment known as the Mata dos Dourados. The forest was sampled along two transects comprising 50 contiguous 15 m × 15 m quadrats, giving a total area of 1.125 ha (Figure 6.18). All individual non-climbing plants within the quadrats with a circumference at the stem of 15.7 cm or more were recorded. A quick but more extensive survey was carried out in the whole forest fragment covering the same soil habitats in order to assess the extension of soil–species relationships found in the transects. This survey used the categories abundant, common, frequent, occasional, and rare. Each transect started at the river margin and extended upslope to include two soil catenas

Table 6.3   Vegetation communities along a north Alaskan toposequence

Type of plant	Tussock tundra	Hilltop heath	Hillslope shrub and lupin	Footslope *Equisetum*	Wet sedge tundra	Riverside willow
Deciduous erect	Swamp birch (*Betula nana*) Alpine blueberry (*Vaccinium uliginosum*)	Alpine blueberry In depressions: Swamp birch Gray-leaf willow (*Salix glauca*) Alpine blueberry	Alpine blueberry Woolly willow (*Salix lanata*)	Alpine blueberry Swamp birch Woolly willow		Gray-leaf willow Woolly willow Alpine blueberry
Deciduous prostrate	Black bearberry (*Arctostaphylos alpina*)	Black bearberry	White mountain-avens (*Dryas integrifolia*)	White mountain-avens		Red fruit bearberry (*Arctostaphylos rubra*)
Evergreen	Marsh labrador-tea (*Ledum palustre*) White arctic mountain heather (*Cassiope tetragona*)	Northern mountain cranberry (*Vaccinium vitis-idaea*) Black crowberry (*Empetrum nigrum*) In depressions: Marsh labrador-tea Northern mountain cranberry	White arctic mountain heather Black crowberry	White arctic mountain heather		
Graminoid	Tussock cotton-grass (*Eriophorum vaginatum*) Bigelow's sedge (*Carex bigelowii*)			Sedge (*Carex*, 3 species)	Tufted clubrush (*Trichophorum caespitosum*) Sedge (*Carex*, 9 species)	
Forb			Arctic lupin (*Lupinus arcticus*) Richardson's brookfoam (*Boykinia richardsonii*)	Horsetails (*Equisetum*, 3 species)		Arctic lupin
Moss	*Sphagnum* spp. *Dicranum* spp. *Aulacomnium turgidum*	*Dicranum* spp. In depressions: *Hylacomium splendens* *Aulacomnium turgidum* *Dicranum* spp. *Rhytidium rugosum*	*Hylacomium splendens*	*Dicranum* spp. *Tomenthypnum nitens* *Hylacomium splendens* *Aulacomnium turgidum*	*Dicranum* spp. *Bryum* spp. *Tomenthypnum nitens*	*Hylacomium splendens* *Drepanocladus* sp. *Tomenthypnum nitens* *Dicranum* spp.

Source: Adapted from Giblin *et al.* (1991).

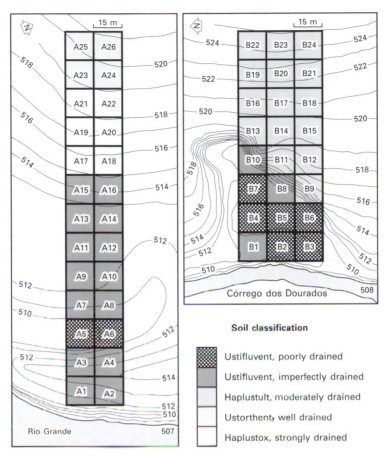

Figure 6.18 Topographic map showing the location and layout of two vegetation and soil transects on the margins of the Rio Grande, Minas Gerais, Brazil. Contours are in metres. Sample quadrats are labelled A1 to A26 and B1 to B24. The soil classification of the quadrats is given at the bottom right of the diagram. (Adapted from Oliveira-Filho *et al.*, 1997.)

(Figure 6.18). Both catenas contained a hydrosequence of soils that were better drained in upslope sites than in downslope sites. Soil catena 1 included poorly drained and imperfectly drained Ustifluvents, strongly drained Haplustoxes, and well-drained Ustorthents. Soil catena 2 included poorly drained and imperfectly drained Ustifluvents and moderately drained Haplustults.

Eigenvalues for the first four axes of a canonical correspondence analysis (CCA) were 0.68, 0.50, 0.36, and 0.23, suggesting that the first two axes were sufficient to separate the species (Figure 6.19). The first axis was best correlated with an Ustifluvent dummy variable, followed by drainage and vertical distance to the watercourse. Soil texture and the sum of bases (a measure of soil fertility) also scored significantly,

and are highly correlated with the differentiation between alluvial soils and upland soils. CCA axis 2 was best correlated with the three upland-soil dummy variables. Tree species' distribution within the ordination axes indicated a correlation with the soil habitats, but with soil drainage class as the leading factor, and not soil chemical properties. Thus, the distribution of tree species was related to the hydrosequence running from poorly drained soils to strongly drained soils. To examine this relationship further, the 55 most abundant species were classified according to their growth strategy – pioneer species, shade-tolerant climax species, and light-demanding climax species. Chi-square analysis showed that light-demanding and shade-tolerant species favour different soil habitats. Light-demanding

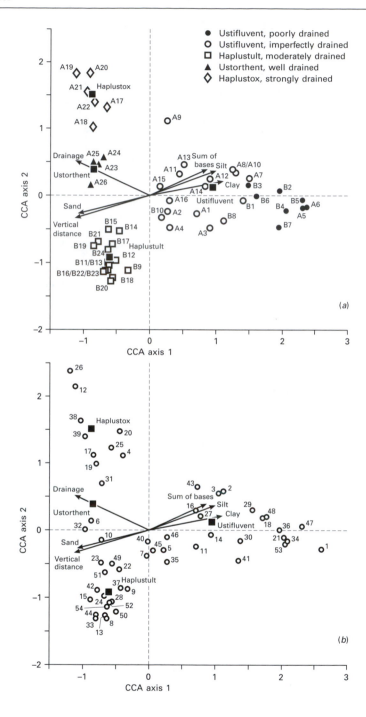

plants preferred poorly drained, imperfectly drained, and strongly drained sites. Shade-tolerant species preferred moderately drained to well-drained sites.

The results of this research suggest that, in the catenas investigated, vegetation responded far more to differences in soil-water and associated groundwater regimes than it did to differences in soil texture or soil fertility. According to the orthodox explanation, tropical vegetation catenas result from hillslope topography determining soil texture, soil nutritional status, and groundwater regimes (e.g. Richards 1996). In the Brazilian catenas, upland soils were indeed coarser-textured and poorer in nutrients than the alluvial soils. However, these differences were not due to the direct influence of downslope catenary processes. They probably resulted from the alluvial soils having formed by rapid accretion of finer sediments and nutrients lost from upstream areas by drainage and flood-waters.

Microtopographic influences of animals and plants

Some animals and plants are responsive to small-scale variations in topographic factors. Such species are able to exploit small-scale variations in environ-

Figure 6.19 Ordination biplots from CCA for south-east Brazilian vegetation toposequences. (a) Sample quadrat and environmental variables in two-dimensional space of the first two CCA axes. Quadrats are identified by their letters in Figure 6.18. (b) Species and environmental variables in two-dimensional space of the first two CCA axes. The species, identified by numbers 1 to 54, are as follows. 1 *Ficus obtusiuscula*. 2 *F. tomentella*. 3 *Calophyllum brasiliense*. 4 *Cariniana estrellensis*. 5 *Duguetia lanceolata*. 6 *Qualea multiflora*. 7 *Eugenia florida*. 8 *Copaifera langsdorffii*. 9 *Cryptocarya aschersoniana*. 10 *Galipea multiflora*. 11 *Miconia latecrenata*. 12 *Ceiba speciosa*. 13 *Virola sebifera*. 14 *Guarea guidonia*. 15 *Tabebuia serratifolia*. 16 *Unonopsis lindmanii*. 17 *Astronium graveolens*. 18 *Alchornea glandulosa*. 19 *Casearia gossypiosperma*. 20 *Plinia grandifolia*. 21 *Cecropia pachystachya*. 22 *Pouteria gardneri*. 23 *Guapira tomentosa*. 24 *Pterogyne nitens*. 25 *Trichilia catigua*. 26 *Ficus enormis*. 27 *Calyptranthes lucida*. 28 *Terminalia glabrescens*. 29 *Deguelia hatsbacchii*. 30 *Nectandra nitidula*. 31 *Ixora warmingii*. 32 *Schefflera morototoni*. 33 *Vochysia magnifica*. 34 *Genipa americana*. 35 *Protium heptaphyllum*. 36 *Picramnia sellowii*. 37 *Chrysophyllum gonocarpum*. 38 *Siphoneugena densiflora*. 39 *Platycyamus regnellii*. 40 *Apuleia leiocarpa*. 41 *Acacia glomerosa*. 42 *Cheiloclinum cognatum*. 43 *Licania apetala*. 44 *Nectandra oppositifolia*. 45 *Mollinedia widgrenii*. 46 *Croton priscus*. 47 *C. urucurana*. 48 *Stylogyne ambigua*. 49 *Myrsine umbellata*. 50 *Siparuna cujabana*. 51 *Psychotria vauthierii*. 52 *P. sessilis*. 53 *P. carthagenensis*. 54 *Stryax pohlii*. (Adapted from Oliveira-Filho *et al.*, 1997.)

mental conditions associated with the microtopographic variation. For plants, soil moisture and pH levels, which tend to reflect other soil properties, are often key factors. For animals, small-scale differences in relief may be significant – tiny ridges, for example, may serve as a refuge from water flowing in tiny valleys during a storm. Small-scale differences in aspect, with concomitant differences of microclimates, are used by some animal species. This section will explore the subtle adaptations of some plants and animals to microtopographic variations.

Plants and microtopography

In Norfolk, England, sandy anthills built by the meadow ant (*Lasius flavus*) are found in saltmarshes. The sea heath (*Frankenia laevis*), a Mediterranean plant, prefers the south-facing side of the anthills (Woodell 1974). Still in England, the grassland terrain of calcareous mires often consists of hummocks and hollows. Some plant species are associated with different parts of the microtopography (Kershaw 1964, 80). The long-stalked yellow sedge (*Carex lepidocarpa*) and glaucous sedge (*C. flacca*) favour the hollows, while tormentil (*Potentilla erecta*) prefers the hummocks. These preferences are difficult to detect by eye, but are plainly brought out by a Student's *t*-test of mean frequencies of the three species in hummocks and in hollows. All three mean frequencies are significantly different at $p = 0.001$. The distribution of the species appears to relate to differences of pH, which is slightly acidic (5.6–5.8) on the hummocks and neutral to slightly alkaline in the hollows where base-rich water collects after a period of heavy rain. In some English permanent grassland, the microtopography consists of a series of furrows and ridges. Soil drainage conditions differ on the ridges and furrows and these differences affect the distribution of three buttercup species (Harper and Sagar 1953). Creeping buttercup (*Ranunculus repens*) is most abundant in furrows, bulbous buttercup (*R. bulbosus*) is most abundant on ridges, and meadow buttercup (*R. acris*), although found on ridges, is most abundant on the sides of ridges.

On talus slopes in the high Andes, Venezuela, the distribution of caulescent (stemmed) rosette plants, commonly referred to as frailejónes owing to their supposed resemblance to friars, is partly related to microtopographic differences in soil moisture. A transverse topographic profile shows how vegetation and topography interact (Figure 6.20). Tall frailejón

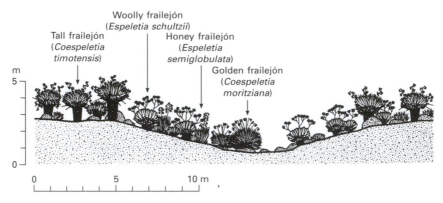

Figure 6.20 A vegetation 'mini-toposequence' across a swale within a talus slope, Piedras Blancas, north-west Venezuelan Andes (above 3700 m). Larger granitic blocks on the ground surface are shown. (After Pérez, 1995.)

(*Coespeletia timotensis*) and some October or woolly frailejón (*Espeletia schultzii*) grow on the drier soils of the high ground. Woolly frailejón, although found on the high ground, prefers the sloping swale margins where it grows alongside a few specimens of honey frailejón (*E. semiglobulata*). Lone rosettes of golden frailejón (*C. moritziana*) grow within crevices on boulders along the moist swale, which showed signs of incipient solifluction.

Pit-and-mound topography is produced by the catastrophic uprooting of forest canopy trees. Persistent tree uprooting can create large areas of land covered by pit-and-mound topography. A single wind event can disturb 4.3 per cent of the ground within small areas of the southern Appalachians, and continued wind disturbance creates single-tree gaps and larger forest openings that cumulatively affect a sizeable portion of the southern Appalachian mountain landscape (Greenberg and McNab 1998). Small-scale variations in forest-floor topography created by tree-throw produce microsites that help to maintain species diversity in upland and wetland forest communities. The effects on species diversity act through small variations in soil chemistry, texture, moisture, and temperature. Microtopography may also influence the differentiation of plant species associations, the differential occurrence of species within wetland sites, the survival and growth of plant seedlings, and competitive interactions between and among species.

In wetlands, microsites include dead branches, trunks, moss-covered or lichen-covered sediments, and various types of peat and muck substrates. The elevation of microtopographic features is likely to be a 'master variable' in determining the relative abundance and spatial distribution of chemical and physical conditions in wetlands (Ehrenfeld 1995a). Microtopography affects shrub and tree seedling distributions in undisturbed stands of Atlantic white cedar (*Chamaecyparis thyoides*), and in stands disturbed by blowdowns and by fire, within swamps of the New Jersey Pinelands, USA (Ehrenfeld 1995a; see also Ehrenfeld 1995b). In all types of cedar swamp, the microtopographic features are created out of woody debris that acts as sites of sediment and organic-matter accumulation and places where dense mats of fine woody roots develop. These woody structures become covered with mosses and are colonised by new shrub and tree seedlings (Colour Plate 5). The relative elevation of the forest floor varies over 1 m. Burned sites have a lower microrelief, while blowdown sites, in which several to many adjacent trees topple at the same time, have a higher microrelief than undisturbed swamp. The main species in all sites was Atlantic white cedar, but low densities of red maple (*Acer rubrum*) and sweetbay (*Magnolia virginiana*) were present. Highbush blueberry (*Vaccinium corymbosum*) and sheep-laurel (*Kalmia angustifolia*) were more abundant on burn sites. Shrubs and tree seedlings were usually excluded from low-elevation microsites, but their densities were roughly the same at intermediate and higher elevations. Despite the diversity of microhabitats in the swamp, the shrub and tree seedling species appeared not to evince any preference for particular microhabitats, nor to any particular range of elevation. However, in other kinds of wetland, the distribution of bryophyte species is related to microtopographic gradients (e.g. Vitt and Slack 1984).

The effects of microsite conditions upon tree regeneration are evident in Nelson Swamp, a 607-ha, forested minerotrophic peatland in Madison County,

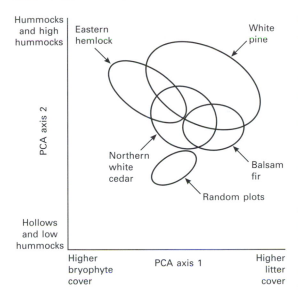

Figure 6.21 Principal components analysis (PCA) of bryophyte, litter, and muck cover among seedling and random plots in Nelson Swamp, Madison County, central New York, United States. The ellipses define 95 per cent confidence zones for the tree species. (Adapted from St Hilaire and Leopold, 1995.)

central New York, United States (St Hilaire and Leopold 1995). The bryophyte layer, which forms the 'seedbed' for conifer seedlings, is related to hummock-and-hollow microtopography and the occurrence of partly decomposed woody material. PCA revealed that balsam fir (*Abies balsamea*), northern white cedar (*Thuja occidentalis*), eastern hemlock (*Tsuga canadensis*), and white pine (*Pinus strobus*) seedlings have overlapping habitats, but are separated from random sites by a second axis that is related to microtopography (Figure 6.21). PCA axis 1 defined a gradient from high bryophyte cover to high litter cover. PCA axis 2 defined a gradient from high cover of muck and delicate thuidium moss (*Thuidium delicatulum*) to high cover of the leafy liverwort, three-lobed bazzania (*Bazzania trilobata*), which represented a microtopographic gradient from hollows and low hummocks to hummocks and high hummocks. Experiments showed that bryophyte substrate affects the germination of different tree species, but has no effect on their first-year survival rate. For example, in the field, balsam fir germinated better on hypnum moss (*Hypnum imponens*) than on the splendid feather moss (*Hylocomium splendens*) and Girgensohn's peatmoss (*Sphagnum girgensohnii*).

Plant distribution patterns in polar desert environments commonly accord with microsites produced by geomorphic processes. In Devon Island, Canada, the high arctic landscape is characterised by patterned ground (Anderson and Bliss 1998). Sorted nets and stripes and unsorted mud-boils with minimal development of cryptogamic crusts prevail. In this landscape, the plant habitat is divided by geomorphic processes into discrete and repeated spatial units. Patterned-ground microsites were classed as centre, transition, and stony border of sorted nets and stripes. They were further subdivided by the presence or absence of a mat of bryophytes, lichens, and algae, often referred to as a cryptogamic crust as it is composed of cryptogams. Plants species growing on stone-net and stone-stripe microsites without crusts were non-randomly distributed. For example, icegrass (*Phippsia algida*) preferred centre and transition sites, tufted alpine saxifrage (*Saxifraga caespitosa*) grew mainly on transition sites, and longstalk starwort (*Stellaria longipes*) was the main occupant of border sites, where it favoured moss polsters among stones with their small pockets of soil. Mosses, though potential competitors for vascular plants, hold water and nutrients in stony border sites and release acids that promote weathering. The stones themselves tend to have a supply of moisture beneath them and they ameliorate the cold temperatures. Even isolated stones in the mineral soil centres were more likely to have plants growing adjacent to them than was the surrounding mineral soil. Nevertheless, centre sites were the least hospitable owing to frost heaving of the soil. On microsites with crusts, the only species to have a non-random distribution were Bering chickweed (*Cerastium alpinum*), which greatly preferred large-stone borders, purple mountain saxifrage (*Saxifraga oppositifolia*), which occupied mainly large-stone and small-stone borders, icegrass, which occurred mainly on crusted centres, and two-flowered rush (*Juncus biglumis*), which occurred primarily on crusted centres and was the only species living on mud-boils. Extensive cryptogamic crusts form in limited areas that receive surface water snowflush for part of the summer. They alter their surrounding environment and offer a microclimate and soil conditions suitable for the growth of vascular plants.

Animals and microtopography

Microtopography wields an influence upon some animals. The grass-cutting attine ant *Acromyrmex landolti* excavates nests in the lowland tropical savannah of South America. A study in eastern Colombia conducted over a period of 27 months (February 1991 to April 1993) showed that microtopography plays a main role in the siting of nests (Lapointe *et al.* 1998). Maps of colony location and savannah microtopography displayed a clustered distribution associated with microrelief. Colonies were clustered on slightly raised ridges that were most likely caused by surface-water runoff.

The fungus-cultivating termite *Macrotermes bellicosus* builds nest mounds in the African savannah belt and fashions them to achieve optimum thermoregulation for the habitat. The mounds are impressive structures, up to 8 m tall and occurring at densities of up to 83 mounds per hectare. In the Comoé National Park, Ivory Coast, the mound architecture in the cooler, but thermally more stable, gallery forest is different from the mound architecture in the shrub savannah (Korb and Linsenmair 1998). Gallery-forest mounds are dome-shaped with thick walls and hardly any protruding structures. Scrub-savannah mounds have thin walls and are highly structured with many turrets and ridges. To see whether these differences in mound architecture were due to thermoregulation, eight shrub-savannah mounds and seven gallery forest mounds were heated artificially and their dimensions and relevant physical properties measured (Korb and Linsenmair 1998). The results showed that thermal properties were different in each mound and depend upon mound dimensions, and especially upon the external structure. They also showed that mounds in the gallery forest had higher heat capacities than mounds in the shrub savannah. So considered as a whole (outside and inside), mounds had a higher thermal inertia in the cooler, but thermally more stable, forest habitat than in the warmer, but thermally more variable, shrub savannah. In the gallery forest, for much of the day and year ambient temperatures fall below the optimum 30°C level for fungus cultivation and the growth and development of termites. Consequently, the termites need to insulate their nests to reduce heat loss. The insulation is achieved by building dome-shaped mounds with thick walls and a topologically simple surface. When shading trees in the gallery forest were removed, the termites began to make the mound surfaces topologically complex. So, the different architectures used by termites in shrub savannah and gallery forest appear to be an adaptation to differing thermal conditions. However, the structural adaptations to cooler conditions do not fully compensate for lower ambient temperatures. Nest temperatures for gallery-forest colonies are about 28°C, 2°C below the optimum, which reduces lifetime reproductive success compared with the shrub-savannah colonies.

Surfaces

The distribution of vegetation may be affected by contour curvature, which varies according to the vergency of slope flowlines. Flowlines are convergent in coves (hollows), divergent over noses (spurs), and straight or parallel over sideslopes. These different patterns of vergency affect soil moisture independently of slope aspect. Coves tend to be wet, noses dry, and sideslopes intermediate. Vegetation distribution registers these differences in soil moisture. The influence of contour curvature on vegetation distribution is not well researched, but the classic study by Hack and Goodlett (1960) and the later resurvey of their site (Osterkamp *et al.* 1995) demonstrate the potential importance of this topographic factor. The growing number of projects using digital elevation models to help understand vegetation distribution normally include a contour curvature element.

Hack and Goodlett (1960) explored the forest of Little River Basin, north-western Virginia. They found a close association between forest type and topographic position, while acknowledging that the relationships they discovered were modified by aspect. Coves favoured northern hardwood forests, sideslopes oak forests, and noses pine forests. Northern hardwood forests in coves were composed of basswood (*Tilia heterophylla*), sugar maple (*Acer saccharum*), yellow birch (*Betula alleghaniensis*), black birch (*B. lenta*), pignut hickory (*Carya glabra*), and red oak (*Quercus rubra*). Eastern hemlock (*Tsuga canadensis*) and eastern white pine (*Pinus strobus*) were major components in cool and wet coves. Oak forests on sideslopes consisted of red oak, chestnut oak (*Q. prinus*), and black oak (*Q. velutina*). These forests also occupied noses where pines of the yellow pine forest type were absent. Yellow pine forests on noses consisted mostly of pitch pine (*P. rigida*) or table mountain pine (*P. pungens*), or both. On rocky nose sites, these pines were commonly accompanied by chestnut oak. These vegetation associations were drawn up on the basis of presence–absence data of selected tree species – they did not take into account species abundance, as indicated by species density and basal area.

To re-examine Hack and Goodlett's observations, a survey considered vegetation, topographic position, soil moisture, and aspect in ten 6 × 30 m blocks along a transect at constant elevation (Osterkamp *et al.* 1995). Dominant tree species and mean soil moisture levels in each block are listed in Table 6.4. The dominant species in coves, on valley sides, and on

Table 6.4 Dominant tree species and soil-moisture content along an equal-elevation transect, Little River Basin, Virginia

Block	Contour curvature[a]	Aspect	Dominant tree species[b]	Mean soil moisture[c] (%)
1	Straight (sideslope)	North	Red maple, dogwood	25.8
2	Convergent (cove)	North	Black oak, chestnut oak, yellow birch	23.6
3	Convergent (cove)	Variable	Witch-hazel, yellow birch	18.8
4	Convergent (cove)	South	Black oak, witch-hazel	26.8
5	Straight (sideslope)	South	Chestnut oak, basswood, witch-hazel, red maple	22.6
6	Divergent (nose)	South	Chestnut oak, witch-hazel	14.3
7	Divergent (nose)	South	Chestnut oak, hemlock, white pine	20.7
8	Divergent (nose)	South-east	Hemlock, black oak, red oak	19.5
9	Straight (sideslope)	East	Hemlock, chestnut oak	15.3
10	Straight (sideslope)	East	Chestnut oak, red maple, black oak	16.4

[a] The transect defined a reverse-S plan form with blocks representing nose, sideslope, and cove environments.
[b] Listed in decreasing order of importance value for each block. For scientific names, see text.
[c] Three to four soil samples were collected at intervals within each block at average depths of 30 mm and analysed gravimetrically for moisture content. The values are expressed as mean percentages of sample weights within blocks.
Source: Adapted from Osterkamp *et al.* (1995).

noses were similar to Hack and Goodlett's findings. Dogwood (*Cornus florida*) and witch-hazel (*Hamamelis virginiana*) were locally abundant on the transect. Hack and Goodlett (1960) may have included these with unmeasured understorey shrubs. Importance values for sugar maple, pitch pine, and black birch did not exceed 25 on the transect. The transect investigation suggests that local conditions may also influence species abundance. Chestnut oak favoured fairly dry, rocky sites and is common on noses and sideslopes. However, it dominated some cove sites where concentrated runoff has deposited coarse, well-drained sandstone rubble devoid of a fine-grained matrix. Similarly, red maple (*Acer rubrum*) tolerated a wide spectrum of environmental conditions, though it was commonest on sideslopes.

Another study in the same area adopted methods similar to those used by Hack and Goodlett. A 410 m, along-contour transect was measured as a chain of blocks, each block being 6 m wide and either 15 m or 30 m long. Nine blocks occurred in coves, six on sideslopes, and one on a nose. The species and diameter of all trees with a diameter of at least 5 cm and a height of at least 1.5 m were recorded in the block. Basal areas, stem density, and importance values were computed from these data. The block data, together with block data from Hack and Goodlett's study, were subjected to ordination by DCA (Figure 6.22). The ordination brings out the three forest types identified by Hack and Goodlett along DCA axis 1. This axis captures a moisture gradient ranging from wet condi-

tions in coves to dry conditions on noses. DCA axis 2 picks out two classes of cove blocks – a group dominated by basswood and sugar maple, and a group dominated by yellow birch and eastern hemlock. The environmental interpretation of this axis is unclear. Possible correlates are (1) disturbance by humans or fire and (2) temperature, hemlock requiring cooler temperatures than the other cove species. Also, basswood and sugar maple may be a little more tolerant of dry conditions than yellow birch and hemlock, and so establish themselves during periods of moisture stress. Once established, they thrive under normal moisture conditions but perhaps not so well as yellow birch and hemlock.

Integrated topographic effects

Altitude, latitude, aspect, slope, and other landform attributes exert individual influences upon animals and plants. However, they act in concert and it is instructive to investigate multivariate effects in the field. A study of myrtle beech (*Nothofagus cunninghamii*) in the cool temperate rainforest of south-eastern Australia demonstrated the multivariate subtlety of plant–environment interactions (Lindenmayer *et al.* 2000). Topographic, climatic, disturbance, and community factors all influenced the distribution of myrtle beech. Five variables were discovered to be significant in explaining the presence of myrtle beech: the age of the overstorey eucalypt stands, the dominant species of overstorey eucalypt tree, topographic position, slope

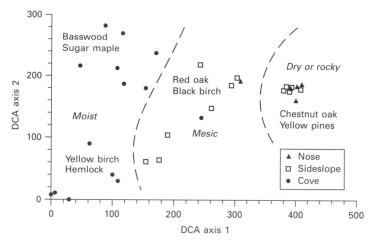

Figure 6.22 A DCA ordination of transect blocks in the same area as Hack and Goodlett's classic study, from which the block data are included in the analysis. DCA axis 1 defines a moisture gradient from moist coves, through mesic sideslopes, to dry or rocky noses. Key tree species are shown. (Adapted from Osterkamp *et al.*, 1995.)

gradient, and the estimated amount of rainfall in the warmest quarter of the year. Myrtle beech was more likely to occur in old-growth eucalypt stands, in gullies rather than on mountainsides, and in places with high rainfall in the warmest quarter of the year. It was present on flatter areas dominated by alpine ash (*Eucalyptus delegatensis*) and shining gum (*E. nitens*) but on steeper slopes dominated by Australian mountain ash (*E. regnans*).

A study of high-Andean vegetation along environmental gradients in north-western Patagonia also shows the value of a multivariate approach (Ferreyra *et al.* 1998a, 1998b). The work was carried out within Nahuel Huapi National Park, which lies in the western Neuquén and Río Negro provinces, on the border with Chile, at about latitude 41°S. The relief of the area is highly dissected. The highest peaks stand at around 3500 m, and the lowest valleys, many of which contain glacial lakes, sit at 600–700 m (Colour Plate 6). Westerly winds from the Pacific Ocean produce a west-to-east moisture gradient, which at lower elevations produces a change of vegetation from temperate humid and mesic forests, through xeric transitional woodland, to Patagonian steppe with tussock grasses and shrubs. At high elevations, deformed lenga (*Nothofagus pumilio*) trees grow to around 1700 m and give way to the alpine, high-Andean zone. Three mountain tops were chosen – Mount Tronador (3554 m), Mount López (2076 m), and Mount Meta (2120 m) – all belonging to massifs of similar size

and lying at roughly 30-km intervals along the west–east precipitation gradient. Precipitation, measured at nearby valley weather stations and corrected for altitude, was 2007 mm on Mount Tronador, 1442 mm on Mount López, and 1034 mm on Mount Meta.

Vegetation was surveyed in 166 relevés of 10 × 10 m, using stratified sampling that combined successive 100-m belts with different slopes and aspects. Limited accessibility on parts of Mounts López and Tronador meant that only leeward slopes, which had east, north-east, and south-east aspects, were sampled. Altitude, aspect, slope, and substrate were recorded for each sample area. Substrate was categorised as massive rock, blocks (diameter larger than 30 cm), large fragments (diameter 10–30 cm), medium fragments (diameter 1–10 cm), and small fragments (diameter less than 1 cm). In total, 166 vegetation samples and 117 plant species (discounting single appearances) were classified using TWINSPAN, and were subjected to ordination using DCA. CCA then revealed relationships between the floristic variation and environmental factors, namely longitude, altitude, relative east aspect (measured as sine of aspect), relative north aspect (measured as cosine of aspect), and percentage of the substrate classes. Analyses were run for all relevés, for relevés with easterly aspects on all three mountains, and for the relevés on Mount Meta, which cover all aspects, not just easterly ones. TWINSPAN gave two community groups with different community types in each (Table 6.5).

Table 6.5 Community identified by TWINSPAN of data from Mounts Tronador, López, and Meta, north-western Patagonia, Argentina

Community	Main characteristic species		Other frequent species	
Group I.1 *Senecio portalesianus*	Suffructifose forb	*Senecio portalesianus*	Forbs	*Nassauvia dentata* *N. revoluta*
			Other forbs	*Lucilia alpina* *Stenodraba pusilla* *Valeriana philippiana* *Adesmia longipes* *Epilobium australe* *Leceria papillosa*
			Graminoids	*Poa tristigmatica* *Deyeuxia erythrostachya* *Luzula chilensis*
			Shrub	*Gaultheria pumila*
Group I. 2 *Gaultheria pumila* and *Baccharis magellanica*	Dwarf shrubs	*Gaultheria pumila* *Baccharis magellanica*	Dwarf shrub	*Empetrum rubrum*
			Forb	*Quinchamalium chilense*
			Graminoids	*Luzula chilense* *Agrostis meyenii* *Poa tristigmatica*
			Suffructifose forbs	*Senecio portalesianus* *S. julietti* *S. poeppigii*
Group II.3 *Nassauvia pygmaea* and *N. revoluta*	Forbs	*Nassauvia pygmaea* *N. revoluta*	Forbs	*Nassauvia pygmaea* *N. revoluta*
			Graminoids	*Luzula chilense* *Poa tristigmatica*
			Suffructifose forbs	*Senecio boelckei* *S. julietti*
Group II.4[a] *Gaultheria pumila* and *Senecio argyreus*	Shrubs	*Gaultheria pumila* *Senecio argyreus*	Shrubs	*Berberis empetrifolia* *Baccharis magellanica* *Chiliotrichum rosmarinifolium*
	Forb	*Perezia bellidifolia*	Grass	*Poa tristigmatica*
			Suffructifose forb	*Senecio baccharidifolius* *Tristagma nivale*
			Other forbs	*Hypochoeris tenuifolia* *Erigeron andicola* *Sisyrinchium patagonicum* *Oreopolus glacialis* *Viola cotyledon*

[a] On pronounced slopes, this community has other frequent forbs: *Ranunculus semiverticillatus* and *Oxalis adenophylla*. Near the summit, it has two frequent suffructifose cushion forbs: *Azorella madreporica* and *Oxalis erythrorhiza*; and several frequent forbs: *Armeria chilensis*, *Valeriana moyanoi*, *Loasa nana*, *Moschopsis subandina*, *Viola sacculus*, *Onuris graminifolia*, *Nassauvia lagascae*, *Senecio boelckei*, and *Colbanthus lycopodioides*.
Source: Adapted from Ferreyra *et al.* (1998a).

Ordination patterns produced by DCA and CCA were very close. Eigenvalues for the first two CCA axes were 0.48 and 0.39. Axis 1 scored strongly on geographical longitude and easterly aspect, suggesting an environmental gradient from the western and rainiest east-facing relevés to the eastern and driest west-facing relevés. Substrate particle-size was also associated with this axis, but to a lesser extent than the other two variables. CCA axis 2 scored strongly on altitude, relative north aspect, small particle-size substrate, and medium particle-size substrate. This gradient is a combination of topography and substrate. The gradient runs from the highest and south-facing relevés on substrate of medium size to the lowest and north-facing relevés on sandy substrate.

The ordination results suggest that two environmental gradients exert a dominating control on high-mountain vegetation. The first gradient is moisture availability. This is a composite gradient consisting of a longitudinal precipitation gradient and a wind-exposure gradient, as defined by relative east aspect. The interplay of the longitudinal precipitation gradient and the wind-exposure gradient produces some interesting effects on vegetation. Mount Meta lies at the arid end of the longitudinal precipitation gradient. Nonetheless, its leeward slopes are moist enough to support vegetation that is more akin to the vegetation on the westernmost mountains with higher precipitation than it is to the vegetation on Mount Meta's east and west slopes. The second environmental gradient is topoclimatic. It combines altitude with relative north (sunny) aspect to create a continuous thermal gradient ranging from cold sites (high altitude, or southerly aspect, or both) to warm sites (lower altitude, or northerly aspect, or both).

The four communities obtained with TWINSPAN corresponded to four distinctive habitat types: humid and cold habitat, humid and warm habitat, arid and cold habitat, and arid and warm habitat. Humid and cold habitat, found at low altitudes (1600–1800 m) on south-east slopes of Mount Tronador and on the highest altitudinal belts (about 1800–2100 m) on Mounts Tronador and Lopéz, favoured Community 1 with *Senecio portalesianus* (Colour Plate 7). This community has unique elements that are typical of the southern Andes. Humid and cold habitat, located in the lower belts (1600–1900 m) of Mounts Tronador and Lopéz and in the lower belt (1700–1900 m) of Mount Meta, supported Community 2 with *Gaultheria pumila* and *Baccharis magellanica*. Arid and cold habitat, which lies in the highest belt of Mount Meta

(1900–2100 m) and on southern aspects in its lowest belt (1700–1900 m), is associated with Community 3, with *Nassauvia pygmaea* and *N. revoluta*. Arid and warm habitats, found in the lower two belts on northern and western slopes of Mount Meta (1700–1900 m), favours Community 4 with *Gaultheria pumila*, *Senecio argyreus*, and *Perezia bellidifolia* (Colour Plate 8). This community is richest and has elements in common with dry-steppe communities found at lower elevations to the east.

This study nicely shows why multivariate analysis captures the basic dimensions of environmental relationships so much better than univariate techniques. However, care must be exercised in reading too much into the findings. As was found on Mount Hauhungatahi in New Zealand (p. 179), although current climatic factors appear to be the major force governing vegetation gradients, historical events during the Pleistocene epoch almost certainly had a role in moulding the present vegetation patterns in the high Andes. Indeed, variation in southern Andean mountain vegetation is better viewed as the result of the interaction between modern climate and historical events (Ferreyra *et al.* 1998a).

Landscape patches

The foregoing material in this chapter has dealt with life in particular places, along toposequences, and over ground surfaces. The landscape is a three-dimensional body and a wealth of studies probe the intricacies of geographical space and how it affects animals and plants. The concluding sections of this chapter examine the connections between life and landscapes, starting with patches and progressing through corridors, networks, and mosaics.

Patch size

General considerations

The well-known species–area curve anticipates that species diversity would be related to patch size. In simple terms, large patches house more species than small patches, the species–area curve having a steep slope at a small area but levelling out fairly sharply at a 'minimum area point', beyond which very few species are added as area increases. Other factors – isolation, patch age, patch diversity, and many more

– all contribute to species diversity, but patch size is of overarching importance. Such exceptions as exist appear to stem from unusual circumstances, as when another variable covaries with patch size (Helliwell 1976), or where there are no specialist interior species present (Matthiae and Stearns 1981). Importantly, species–patch-size relationships are affected by latitude and longitude. Bird species in small woods within agricultural landscapes of the Netherlands, the United Kingdom, Denmark, and Norway displayed species–area relationships (Hinsley *et al.* 1998). Species richness across all woods declined with increasing latitude, as did the proportion of resident species; the proportion of migrant species rose. In addition, the slopes and intercepts of the species–area curves declined with increasing latitude. So, not only were there fewer species available to colonise woods at higher latitudes, but the gain of species richness for a unit increase in wood area was smaller than at lower latitudes.

For birds, insectivores are more patch-size sensitive than seed-eaters or omnivores, and long-distance migrants are more patch-size sensitive than residents or short-distance migrants. In a study of bird species in 16 small, isolated woodlands in North Humberside, England, four species out of 49 recorded had minimum area requirements (McCollin 1993). These species were the great tit (*Parus major*) and spotted flycatcher (*Muscicapa striata*), with minimum area requirements of about 1 ha, and the jay (*Garrulus glandarius*) and turtle dove (*Streptopelia turtur*), with minimum area requirements of about 4 ha. Chaffinch (*Fringilla coelebs*), blue tit (*Parus caeruleus*), robin (*Erithacus rubecula*), wren (*Troglodytes troglodytes*), and blackbird (*Turdus merula*) occurred in all 16 woodlands, which suggested that their minimum area requirements were less than 0.73 ha, the area of the smallest wood.

On the whole, species restricted to large patches are far commoner than species restricted to small patches. Nonetheless, several small-patch-restricted species are known. Grey squirrels (*Sciurus carolinensis*) in New Jersey, USA, are common in small woodland patches but scarce or absent in large patches (see Forman 1995, 60). The reason for this pattern is probably that small patches do not contain the owl predators that large patches do. In New Zealand, six species of leafy mistletoes (family Loranthaceae) occur in small patches because large patches support introduced herbivorous brushtailed possums (*Trichosurus vulpecula*) (Ogle and Wilson 1985); and large (up to 8-cm diameter) carnivorous land snails (*Paryphanta b. busbyi*) are more likely to survive in small patches because in large patches they are prey to wild pigs (Ogle 1987). In New Guinea, several island bird species are restricted to small islands because on larger islands they come into competition with other birds (Diamond 1973).

Patch size wields a strong influence on the occurrence of animal and plant species. It also affects such ecological processes as breeding success, nest parasitism, and nest predation. The volume of research into species diversity and patch size is huge. Several 'case studies' will be examined that are chosen to illustrate various aspects of landscape 'patchiness' and its influence on animals and plants.

Oak woodlots in New Jersey

An early study of species–patch-area relationships looked at different groups of species (trees, mosses, mushrooms, seed-eating birds, and insect-eating birds) within old mixed-oak woodlots surrounded by open fields on the central New Jersey Piedmont, USA (Forman *et al.* 1976; Galli *et al.* 1976; Forman 1995, 60). The woods are dominated by white oak (*Quercus alba*), black oak (*Q. velutina*), and grey oak (*Q. borealis*), with less-abundant species being American beech (*Fagus grandifolia*), yellow poplar (*Liriodendron tulipifera*), red maple (*Acer rubrum*), sweet pignut (*Carya ovalis*), and white ash (*Fraxinus americana*). Black cherry (*Prunus serotina*) and sweet cherry (*P. avium*) are associated with the woodland edge. Thirty woods were sampled, ranging in size from 0.01 ha (a large, old tree) to 24 ha (the largest woodland remnant in the area). Expectedly, all species groups had more representatives in large woodlots than in small woodlots. As the foliage-height diversity was unrelated to the size of wood, it seemed reasonable to suggest that the increase in bird species richness was due to area itself, and not to habitat diversity. Interestingly, the shape of the species–patch-size curves differed considerably and had a range of minimum-area points. For trees and seed-eating birds it was 2 ha, for mosses 4 ha, for mushrooms 6 ha, and for insect-eating birds 40 ha. These data suggested that the bird community in the woodlots was the most sensitive to woodlot size, and would be the first to suffer from further shrinkage.

Another finding in the New Jersey woodlot study was that some species were patch-size independent, while others were patch-size restricted. Patch-size-independent species (those that lived in small and large patches) accounted for half the 35 bird species. They included the starling (*Sturnus vulgaris*), the common grackle (*Quiscalus quiscula*), and the indigo bunting (*Passerina cyanea*). The other bird species lived only in large, or in medium and large, patches. They included the blue jay (*Cyanocitta cristata*), the red-eyed vireo (*Vireo olivaceus*), and the eastern wood pewee (*Contopus virens*). Patch-size-restricted birds had individual patch-size requirements, so that as woodlots became progressively smaller, the species' populations diminished, and one by one the species disappeared. The small woodlots housed only patch-size-independent species, which were confined to edges in large woodlots.

Forest patches and forest herbs, Blue Ridge Province, North Carolina

Scott M. Pearson and his colleagues studied the effects of forest fragmentation on cover-forest herbs in the French Broad River Basin, Buncombe and Madison Counties, North Carolina, which are in the Southern Blue Ridge Province, USA (Pearson *et al.* 1998). Fourteen patches of closed-canopy, mesic, cove-forest in the altitudinal range 600 to 920 m were sampled with 4-ha study plots, each of which was divided into 1-ha subplots. The plots were set down in patches ranging from 5 ha to more than 10 000 ha. There were eight small (<25 ha) plots, three isolated (>200 ha) plots, and three large plots within continuous forest. A control on the environmental characteristics was achieved by limiting the investigation to closed-canopy, mesic, deciduous forest. Mesic forests were defined as communities on north-facing or sheltered slopes with one of the following trees dominant in the canopy: yellow poplar or tulip-tree (*Liriodendron tulipifera*), sugar maple (*Acer saccharum*), American basswood (*Tilia americana*), yellow buckeye (*Aesculus octandra*), bitternut hickory (*Carya cordiformis*), and sweet birch (*Betula lenta*). American beech (*Fagus grandifolia*), northern red oak (*Quercus rubra*), and red maple (*Acer rubrum*) were commonly, but not always, present. Small stands of rosebay rhododendron (*Rhododendron maximum*) and eastern hemlock (*Tsuga canadensis*) were avoided. Seventeen herb species, selected as good representatives of cove-forest communities, were recorded as cover and density estimates. Soil samples were also taken and analysed.

The survey revealed that the coverage and density of herb species were greater in large patches (>200 ha) than in small patches (<10 ha). Eight of the 17 species displayed a greater abundance in large patches. The eight species were wild ginger (*Asarum canadense*), false goatsbeard (*Astilbe biternata*), toothwort (*Cardamine diphylla*), yellow mandarin (*Disporum lanuginosum*), spotted mandarin (*D. maculatum*), bloodroot (*Sanguinaria canadensis*), foam flower (*Tiarella cordifolia*), and bellwort (*Uvularia grandiflora*). Spotted mandarin and bellwort, which are dispersed by ants, were more likely to be absent from small patches than from large patches. Wind-dispersed species, such as ferns and composites, were not affected by patch size and isolation. Hairy chickweed (*Stellaria pubera*) was more abundant in small patches.

Several factors may be invoked to explain the species–area relationships of herbs in the cove forests. First, habitat fragmentation is apt to disrupt the dynamics of herb populations in smaller patches, possibly leading to local extinctions. Six species that were more likely to occur in larger patches had limited seed-dispersal abilities that may prevent them from colonising isolated patches. Of these, wild ginger, spotted mandarin, bellwort, foam flower, and Canada violet (*Viola canadensis*) are dispersed by ants; while false goatsbeard and foam flower produce many small seeds that fall close to the parent plant. Good dispersers, such as the maidenhair fern (*Adiantum pedatum*), rattlesnake fern (*Botrychium virginianum*), and broad-leaved golden rod (*Solidago flexicaulis*), were unaffected by patch size. A second environmental factor that might help to explain the species–area relationships is habitat differences between small and large patches. The only soil factor that varied significantly with patch size was related to organic matter content. To be sure, humic matter affected the coverage and density of many species. This is because humic soils have higher moisture retention, improved aeration and tilth, and more nutrients. Plainly, if small patches have less humic matter than large patches, this will have knock-on effects in the herb community. Disturbance may affect the species–area relationships. Although there were no signs that the patches had been cultivated over the last 75 years, some small patches were adjacent to land currently or formerly in agricultural use, which renders them more prone to being used for grazing and as a source for forest products than is the case for the larger plots. Such disturbance might have affected herb populations by increasing mortality rates and by degrading the habitat. So, although forest patch-size does go some way towards explaining the cover and density of herb species, habitat fragmentation, habitat quality, and

Table 6.6 Type, area, and distance from continuous forest of forest patches, southern Campos Gerais, Brazil

Site	Type	Area (ha)	Distance from continuous forest (m)
A	Continuous forest	840	0
B	Linked by forest corridor	40	100
C	Linked by forest corridor	20	200
D	Linked by forest corridor	12	600
E	Linked by forest corridor	10.5	400
F	Isolated	10	3000
G	Isolated	9	2000
H	Isolated	8.5	800
I	Isolated	6.5	2500
J	Isolated	4	1500
L	Isolated	1.5	500
M	Isolated	0.5	1000

Source: Adapted from Dos Anjos and Boçon (1999).

human disturbance may have had an impact on the species–area pattern.

Birds in temperate rainforest patches, southern Brazil

Patch–area relationships are evident in a study of the number of bird species living in patches of mixed temperate rainforest set within grassland of the Campos Gerais region, Paraná State, southern Brazil (Dos Anjos and Boçon 1999). In this region, forest patches survive as part of a once continuous forest cover that has been fragmented by logging. An 840-ha patch and 11 smaller patches, ranging from 0.5 ha to 40 ha, were censused from September to December 1995 in the Fazenda Santa Rita and Vila Velha State Park (Table 6.6). The dominant species in rainforest patches are Paraná pine (*Araucaria angustifolia*), brave pine (*Podocarpus lambertii*), branquilho (*Sebastiana commersoniana*), imbúia (*Ocotea porosa*), and yellow canela (*Nectandra grandiflora*). In sum, there were 189 bird species – 13 open-area species, 51 edge species, and 125 forest species. The most abundant species were white-browed warbler (*Basileuterus leucoblepharus*), rufous-bellied thrush (*Turdus rufiventris*), scaled woodcreeper (*Lepidocolaptes squamatus*), golden-crowned warbler (*B. culcivorus*), rufous-browed peppershrike (*Cyclarhis gunjanensis*), and olive spinetail (*Cranioleuca obsoleta*). The open-area species used the forest for roosting or for nesting (or for both), but not for foraging.

Bird species diversity was strongly correlated with forest patch-size ($r^2 = 0.85$). Forest (interior) species diversity decreased as forest patch-size decreased, but this loss in biodiversity was partly offset by an increase in the number of edge species with decreasing patch-size. These changes were reflected in an increase in the ratio of edge-species to interior-species with decreasing patch size. Many of the forest species lost were understorey species, probably because canopy species are better able to fly between patches if need be. Of the understorey species that were lost, many were leaf insectivores. Large trunk insectivores, such as the lineated woodpecker (*Dryocopus lineatus*), were recorded only in the continuous forest, while only small or medium-sized twig and trunk insectivorous species, such as the olive spinetail, were recorded in the smallest patches. Frugivores were found in patches of all sizes, although their relative abundance was low in small patches. Omnivores can switch diet from fruit to insects and vice versa, and this adaptability perhaps helps to explain their increasing contribution to the avifauna in smaller patches. Carnivores and nectarivores were represented by few species in all sites. Forest-patch isolation modified the overall species–area relationships. The smaller forest patches that were linked to the 840-ha continuous-forest patch were more like the continuous-forest patch in species composition and diversity than they were to the small isolated patches.

Nest-patches in Missouri

A variant of the species–patch-size relationship is the 'nest-patch'. A nest-patch is the habitat patch directly surrounding a bird's nest. Take the case of the nest-patches of the yellow-breasted chat (*Icteria virens*) in dense shrub thickets at Thomas Baskett Wildlife Research and Education Center, near Ashland, Boone County, Missouri, USA (Burhans and Thompson 1999). The features of nesting-habitat patches influence nest predation rates, parasitism by brown-headed cowbirds (*Molothrus ater*), and nest-site selection. A female cowbird is like a female cuckoo in that she finds the nest of another species, here yellow-breasted chats, to incubate her eggs and feed her offspring. She waits for the host to leave its nest and then quickly lays one or more eggs, which the host commonly accepts as one of its own clutch. The cowbird offspring take less time to incubate, so gaining a size advantage over the host's offspring at feeding time. Research questions posed in the yellow-breasted chat study were as follows. (1) Are chats' nests in large thickets (patches) more likely to fledge young than in small patches or single shrubs and trees? (2) Are chats' nests in patch interiors more likely to fledge young than in edge sites? In small patches, which have an average diameter of 5.5 m, nests parasitised by cowbirds were more likely to suffer predation than nests that were not parasitised in large patches. However, nests in large patches were more likely to become parasitised by cowbirds, as were nests with more large stems nearby. Chats tended to nest in patches with more small stems than those in unused sites. They seemed to prefer large patches, where they bore lower nest-predation rates. Although nests in larger patches might expect to sustain higher brood parasitism rates, higher rates of nesting success counterbalanced the reduction of fitness because the mean number of chat young that fledged was the same in small and large patches alike.

Patch size and breeding success in scarlet tanagers, New York

Forest patch-size and habitat characteristics affect the breeding success of scarlet tanagers (*Piranga olivacea*) in western New York, USA (Roberts and Norment 1999). During their breeding season, scarlet tanagers are foliage-gleaning insectivores and nest in the tree canopy, between 2 and 20 m above the ground. A study was conducted in 1995 and 1996 to assess how forest area and forest isolation affected scarlet tanager density, pairing success, and fledging success. Twenty forest stands were studied. Six were of less than 10 ha, seven were in the range 10 to 50 ha, five were in the range 50 to 150 ha, and two were greater than 1000 ha. All sites were habitat patches located in fragmented landscapes, save the two largest sites, which were in continuous forests. Male tanager densities were similar at all sites, but no tanagers lived in forest patches smaller than 10 ha. In the sites occupied, territory size did not differ among males. Pairing success and fledging success both varied with forest area, being generally higher in larger forest patches. Pairing success exceeded 75 per cent in all forest size-classes, and 100 per cent of the observed males were paired in continuous forest sites. Fledging success increased significantly with area and was highest (64 per cent) in continuous forest sites. Principal components analysis (PCA) and stepwise multiple regression were used to examine the relationships between tanager breeding variables and habitat characteristics. The habitat was characterised using 13 variables – total forest area, per cent core area, forest

cover within 1 km, tree density, shrub density, canopy cover, canopy height, stand age, tree species richness, tree species diversity, oak (*Quercus* spp.) density, total basal area, and maple (*Acer* spp.) density. As many of the variables were highly correlated with one another, PCA was used to reduce them to a smaller set that captured the main patterns of variation in the data. Four principal components had eigenvalues greater than 1.0 ($\lambda > 1.0$) (Table 6.7). Factor loadings on the axes were rather low, but interpretation was possible. PCA axis 1 represented forest patch area and surrounding forest cover within 1 km of a patch; PCA axis 2 stood for tree density; PCA axis 3 involved canopy height and tree species diversity and was taken to represent forest development; and PCA axis 4 was canopy height and oak (*Quercus* spp.) density. The new variables were then used in a stepwise multiple regression to analyse tanager–habitat relationships (Table 6.8). The results showed that male tanagers breeding in forest patches with higher canopy cover and lower density of oaks had higher pairing success than males in patches with lower canopy cover and higher density of oaks; and that males breeding in larger forest patches with more surrounding forest cover had higher fledging success than males in small patches with less surrounding forest cover. They indicated that breeding density is not the best indicator of habitat quality for forest-interior Neotropical migrants, and that large tracts of continuous forest are significant for sustaining populations of scarlet tanagers.

Patch shape

In some landscapes, patch shape is superior to patch size as a predictor of species diversity. In 19 suburban woodlots of four adjacent cities in southern Ontario, Canada, woodlot edge-length was a better predictor of bird species richness and abundance than woodlot area, perhaps because most of the species that colonise suburban woodlots, including the grey catbird (*Dumetella carolinensis*), are edge species (Godfryd and Hansell 1986). Pikas (*Ochotona princeps*) living on nine small rock-slides on St Joe Baldy, Benewah County, Idaho, USA, are also sensitive to patch shape (Bunnell and Johnson 1974). For each rock-slide, area, perimeter, elevation, and mean rock-size were measured, as well as the pika population size. The rock-slides ranged in area by a factor of 12 (292 m² to 3740 m²) and in perimeter by a factor of four (73 m to 325 m); elevation and mean rock size showed relatively little variation. Stepwise

Table 6.7 Vegetation and landscape factors generated by PCA for forest stands near Brockport, western New York, USA

	Principal Component 1	Principal Component 2	Principal Component 3	Principal Component 4
Interpretation of Principal Components	Forest patch area Forest cover within 1 km	Tree density	Forest development	Canopy cover Density of oak species
Factor				
Eigenvalue (λ)[a]	4.453	3.125	1.949	1.300
Proportion of total variance (%)	34.3	24.0	15.0	10.0
Cumulative variance (%)	34.3	58.3	73.3	83.3
Variable				
Total forest area	**0.443**	−0.063	−0.113	0.116
Tree density	0.193	**−0.458**	0.179	−0.010
Shrub density	−0.183	0.363	0.252	0.159
Canopy cover	0.044	0.210	−0.454	**0.539**
Canopy height	0.063	0.212	**0.570**	0.052
Stand age	0.085	0.378	0.296	0.312
Tree species richness	**0.376**	0.212	−0.119	−0.288
Tree species diversity	0.219	0.230	**−0.407**	−0.174
Oak (*Quercus* spp.) density	0.201	0.228	0.148	**−0.575**
Total basal area	**0.395**	0.117	0.168	0.281
Maple (*Acer* spp.) density	0.054	**−0.485**	0.177	0.152
Per cent core area	**0.369**	−0.153	0.001	0.153
Forest cover within 1 km	**0.440**	−0.026	0.109	0.027

[a] Only factors with eigenvalues >1.0 are shown.
Source: Adapted from Roberts and Norment (1999).

Table 6.8 Stepwise multiple-regression analyses of principal component scores against 1995–1996 mean scarlet tanager breeding variable within each forest patch of the western New York study

Variable	Pairing success	Fledging success
Principal Component 1 (Forest patch area; forest cover within 1 km)	0.12	*0.44*[a]
Principal Component 2 (Tree density)	0.06	0.10
Principal Component 3 (Forest development)	0.04	0.07
Principal Component 4 (Canopy cover; density of oak species)	*0.31*	0.03

[a] No other variables entered the model.
Notes: Bold italics = significant at $p < 0.01$;
italics = significant at $p < 0.05$.
Source: Adapted from Roberts and Norment (1999).

linear regression showed that perimeter length accounted for 81 per cent of the variation in pika population ($r^2 = 0.901$), the regression coefficients of the other variables not differing significantly from zero. The population size, y, may be predicted from perimeter length, x_1 (m), in the following way:

$$y = -0.11 + 0.016x_1 \qquad (6.1)$$

The coefficient 0.016 shows that each 100 m of perimeter supports an average of 16 pikas, and each pika territory then occupies about 62 m of perimeter. Perimeter length is probably more important a factor in determining pika population size than area because the animals living on long-perimeter slides are optimally placed for harvesting food in the adjacent matrix and for taking cover from larger predators in the rock-slide areas.

Toshihide Hamazaki (1996) conducted an interesting field experiment to see the effect of patch shape on the density of the garden, greenhouse, or hothouse millipede (*Oxidus gracilis*). This cosmopolitan species lives under leaves, logs, and rocks. It prefers humidity levels over 70 per cent and temperatures in the range 13 to 18°C. In moving, it takes more care in avoiding undesirable environments than it does in seeking optimum conditions (Cloudsley-Thompson 1951). The field experiment was run on a farm in Madison County, Georgia, USA. It used 1-cm-thick plywood boards fashioned to represent habitats of different shapes. Five shapes were used, each with an area of 900 cm²: 30 × 30 cm, 15 × 60 cm, 10 × 90 cm, 5 × 180 cm, and 2.5 × 360 cm. The plywood boards were placed in a hardwood forest and an old field. The forest floor was covered with leaves, twigs, and branches and supported no live herbs or grasses. The old field was about seven years old. Chinese lespedza (*Lespedza sericea*) and camphorweed (*Heterotheca subaxillaris*) were the dominant species. To lessen microenvironmental variability at each board, standing dead vegetation from the previous year in the old field and leaf and stem litter on the forest floor were removed before the study began. Three blocks, or locations, were placed in the two environments, and five boards were laid in each block in mid-April 1993. Millipedes were censused under boards around noon in each sampling period. Three sampling periods were defined. Period 1 comprised 16 sampling days from 2 May to 16 June, when most of the millipedes were larvae. Period 2 was the 40 sampling days from 16 June to 31 July, when adult millipedes were present. Period 32 was the 18 sampling days from 9 August to 26 August. Humidity and temperature under the 30 × 30 cm board and in the ambient air were measured at around the same time as the millipede counts.

The results showed that the mean number of millipedes was higher in total and more variable in the old field (Figure 6.23). However, the difference in millipede abundance between the two environments was not statistically significant. The pooled mean numbers of millipedes under boards of different shapes were significantly different in the old field and in the forest. In the old field, the pooled mean number of millipedes was positively correlated with board length and perimeter length, but the correlation was not significant in the forest. In the old field, the mean number of millipedes appearing under boards at the three locations was significantly different, but this was not the case in the forest. The locations with the greatest millipede population varied with patch shape (Figure 6.23). In the old field, the square patch had roughly the same number of millipedes at each location, more elongated patches supported more millipedes at location 2, and the most elongated patch had the maximum millipede population at location 3.

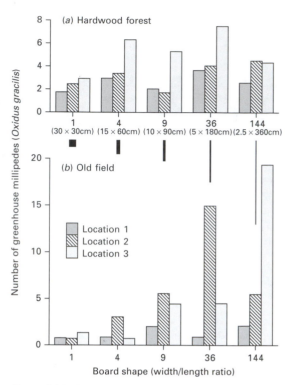

Figure 6.23 Pooled mean numbers of greenhouse millipedes (*Oxidus gracilis*) under five boards of different shapes at three locations at two sites. (Adapted from Hamazaki, 1996.)

The temporal variation in the pooled mean number of millipedes present in each sampling period was variable. Taken togther with the spatial variability in population size, the findings confirm that patch shape does affect population size. However, the relationships are more complicated than theory might suggest. The populations under each board are not resident populations – the millipedes simply find sites with suitable humidity and temperature during the day and then move elsewhere at night. Some millipedes were tracked over tens of metres during a night, and a mark and recapture study found that no marked animals were recovered under the boards on the next day of sampling – the millipede population is mobile and has a high turnover of individuals at a given location. The differences observed between old field and forest partly relate to the more stable temperature and humidity conditions in the forest where millipedes can also find alternative refuges under logs, fallen branches, leaves, and rocks. The stable forest environment with its plentiful supply of alternative refuges clouded the relationship between millepede abundance and patch shape. In the more variable environment of the old field, where few alternative refuges were available, the connection between patch shape and millipede abundance was much plainer to see.

Patch edges

A variety of ideas and views are covered by the term 'edge effects' (Lidicker 1999). This section will consider three topics: edge and interior species, habitat selection in edges, and edges as ecological traps.

Edge species and interior species

A distinction is commonly drawn between edge species and interior species. Interior species live in the core of a habitat. They actively avoid the habitat edges if they are able to meet their resource needs within their territories or home ranges. English examples include the great spotted woodpecker (*Dendrocopos major*) and nuthatch (*Sitta europaea*), bark gleaners that live in the interior of British woods (Fuller 1988). Breeding ovenbirds (*Seiurus aurocapillus*) commonly live at lower densities adjacent to forest edges than in forest interiors. In an extensively forested region of Vermont, USA, territory densities on seven study plots were 40 per cent lower within edge areas, which were from 0 to 150 m from unpaved roads, than within interior areas, which were 150 to 300 m from roads (Ortega and Capen 1999). Territory size decreased away from roads, suggesting that the reduced ovenbird densities in forest edges might result from lower habitat quality. Not all small mammals are willing to use edge habitats or the surrounding matrix. A Brazilian example comes from a forest–farmland edge in two fragments of Brazilian Atlantic forest in Sergipe (Stevens and Husband 1998). Species diversity increases significantly into the forest, and no small mammals were captured outside the forest in the surrounding farmland matrix. Of 671 captures made, only 43 were made along the forest–farmland edge, 39 of which were captures of just two species.

Edge species use a habitat edge. Two types of edge species are recognised: the first are intrinsically edge species, and the second are ecotonal species (McCollin 1998). Ecotonal species occur near the edge because the edge habitat suits them. They are not dependent on adjacent habitats for food, shelter, or anything else. Edge species *per se* live near edges because the adjacent habitat provides resources. For instance, in highly fragmented agricultural landscapes, bird species living in woodland edges next to open country depend upon food resources offered by farmland. Examples are the rook (*Corvus frugilegus*) and carrion crow (*C.*

corone corone), which feed mainly on grain, earthworms and their eggs, and grassland insects, with the crow also taking small mammals and carrion; and the starling (*Sturnus vulgaris*) which feeds on leatherjackets and earthworms in the upper soil layers of pasture (McCollin 1998).

Studies commonly reveal the prevalence of edge species in small patches and interior species in large patches. Plainly, large patches contain more 'interior' than 'edge'. Indeed, there is a minimum patch size below which there is no 'interior', the whole patch being an 'edge' (Levenson 1981). If edge widths are 10 to 50 m, then woodlands larger than 100 ha are mainly interior habitat. In England and Wales, a mere 1.9 per cent (by number) of 100-ha-plus woodlands are considered of interest for conservation but account for 26.6 per cent of the area (Spencer and Kirby 1992). Many surviving woodland fragments lie in the range 1 to 5 ha (44.0 per cent by number, 9.6 per cent by area) and are 'edge' habitat.

Habitat selection in edges

Much research effort is expended on trying to elucidate how species respond to edge habitats. Generalisations are apt to mislead because species respond individualistically to edge effects (Murcia 1995). This is borne out by a study of ground-layer vegetation of a forest–old-field-edge gradient at the Hutcheson Memorial Forest Center, near Millstone, New Jersey, USA (Meiners and Pickett 1999). The site was once two agricultural fields separated by a hedgerow of large red oaks (*Quercus rubra*) running north-east to south-west. The north-western side of the hedge at the time of the study was a young forest dominated by red maple (*Acer rubrum*) and pin oak (*Q. palustris*), which developed after the field was abandoned in the 1950s. About 2 m of the field on the south-eastern side was also abandoned at that time, but the rest was abandoned in 1986 and at the time of the study was dominated by herbaceous perennial species and a few scattered trees and shrubs. In all, 104 vascular plant species were observed within the sampling plots, 35 of which were exotic. Species showed individualistic responses to the forest edge, with peak abundances at different spatial positions relative to the edge and no overall 'edge' response. A principal components analysis resulted in three axes that together explained 63.2 per cent of the variation within the data set. The first two PCA axes were significantly correlated with distance from the forest edge. PCA axis 1 was interpreted as multiflora rose (*Rosa multiflora*) cover and PCA axis 2 as distance from the forest edge.

PCA axis 3 separated plots into those that were dominated by Canada golden rod (*Solidago canadensis*) and those that were dominated by early golden rod (*S. juncea*) and located farther from the forest edge. All population-level and community-level attributes – species richness, species diversity, total cover, between-plot heterogeneity, and exotic cover – varied along the edge gradient.

Four main explanations try to account for habitat selection by birds in forest edges: an individualistic use of resources and patches; biotic interactions; microclimatic habitat modification; and a change in vegetation structure (McCollin 1998). These explanations apply equally to mammals. Resource use by individual species is reflected in adaptive strategies. In woodland, habitat generalists are geared to exploiting woods and the external habitat matrix, which they are able to cross, and are likely to benefit from the extra edge habitats created by forest fragmentation. These species are intrinsic edge species and tend to be most abundant in edge environments (Murcia 1995). A case in point is the wood mouse (*Apodemus sylvaticus*), a typical inhabitant of ecotones at woodland edges. A study of the winter distribution of wood mice in Spain revealed the high abundance of this species in fragmented forest edges (García *et al.* 1998). The fragmented forests, which ranged in area from 0.02 to 400 ha, were located near Santa Elena de Jamuz–La Bañeza, in north-western Léon Province, and near Villatobas-Corral de Almaguer, in southern Toledo Province. Wood mice were trapped in 27 woodlots in the northern area, and 28 woodlots in the southern area. The pitfall traps were located at a range of distances from forest edges in large forests, in small woodlots, and in the agricultural matrix surrounding the woodlots and the forests at sites both close to forest edges and far from them. Wood mouse abundances in both study areas were greatest at forest edges, followed by woodlots, and forest interior. Lower abundances were found in croplands, and these decreased away from woodlot edges and forest edges. The chief conclusions of the research were that wood mouse distribution is affected by edge effects, and that the species behaves as an intrinsic, soft-edge species. The wood mouse is able to exploit two contiguous habitats, forest and cropland, that provide complementary food and shelter: the cropland offers plentiful food, while the forest provides good shelter.

Habitat selection is sometimes modulated by species interactions. In forest–old-field edges at or near the Institute of Ecosystem Studies, south-eastern New York, USA, meadow voles (*Microtus pennsylvanicus*) were commonest in old fields distant from the forest edge, where grasses and herbs dominated, shrub cover was low, and the plant community was structurally simple (Manson *et al.* 1999). White-footed mice (*Peromyscus leucopus*) were commonest near the forest edge where shrubs dominated the microhabitats and the plant community's structural complexity was higher. But white-footed mice used the old-field interiors when meadow vole density was low, which points to competitive displacement of mice by voles, the mice being forced into edge habitats. In birds, some species choose to nest close by habitat edges where food supplies are greater, despite running the risk of higher nest-predation rates. These findings suggest that competitive interactions, which are known to determine the preferences of organisms within habitats, may also operate along habitat interfaces, too, where they may influence edge permeability or the landscape-boundary resistance to animal movements along edges.

Microclimates change in moving from open fields to closed-canopy woodland. As a rule, wind speed lessens, air temperatures fall, and humidity rises. In very approximate terms, the woodland interior microclimate is attained at a distance in from the woodland edge equivalent to three times the canopy height. The edge microclimate undoubtedly affects habitat selection but the topic is little studied (see McCollin 1998). Vegetation structure influences habitat selection. For birds, it affects the provision of nest sites, food, and song-posts (McCollin 1998). Patent evidence for this comes from the changing composition of bird species in woodland recovering from coppicing (e.g. Fuller 1992). The tree pipit (*Anthus trivialis*) and whitethroat (*Sylvia communis*) arrive in the early stages after coppicing. Such migrants as willow warbler (*Phylloscopus trochilus*), chiffchaff (*P. collybita*), blackcap (*S. atricapilla*), garden warbler (*S. borin*), and nightingale (*Luscinia megarhynchos*) are common at the thicket stage. Residents such as the robin (*Erithacus rubecula*) become more abundant after a new closed canopy forms. These would be joined by other species – dunnock (*Prunella modularis*) in the establishment phase, and great tit (*Parus major*), for example – depending upon the geographical location of the coppice.

Edges as 'ecological traps'

Species diversity in edge habitats tends to be high, and frequently includes a number of exotic species. Forest edges running along old-field boundaries in upstate New York, USA, were significantly more abundant in exotic species and locally rare species than in the forest interior, but were poorer in tree seedlings (Goldblum and Beatty 1999). In south-central Sweden, some ancient oak–hazel (*Quercus robur–Corylus avellana*) woodlands around Uppsala lie in a conifer forest matrix. Structural differences between the centres and edges of these ancient wood-lands lead to considerable differences in the occur-rence and abundance of vascular plants and birds (Hansson 2000). The abundances of dead trees, fallen trunks and branches, and stumps were not higher in the edges, but the abundance of young deciduous seed-lings and saplings was. In the woodland edge, bird species diversity in 1998 varied in direct proportion to numbers of large, old trees. Plant species numbers did not relate to the number of large trees, but where the proportion of large, old trees was high, invasion of such plants as wavy hair-grass (*Deschampsia flexuosa*), hairy wood-rush (*Luzula pilosa*), bilberry (*Vaccinium myrtillus*), and cowberry (*V. vitis-idaea*) plants from the surrounding coniferous forest was curtailed.

Edge habitats typically support a higher diversity of herbivores and predators than adjacent habitats. The herbivores are attracted by the 'rich pickings' of food supplies, and the predators are lured by the abundance of herbivores. This generalisation is some-times called the 'ecological trap hypothesis', which purports to explain the richness of species in edge habitats (Gates and Gysel 1978). The ecological trap hypothesis has several interesting aspects. A notably interesting aspect of the ecological trap hypothesis is the effects of higher herbivore abundance on plants. Elevated rates of herbivory in edge habitats put pres-sure on edge plants. Some edge plants survive the herbivore onslaught by being less palatable or less sensitive to trampling.

Many gamebirds commonly live at higher dens-ities in edges than in patch interiors and provide a banquet for predators. Hawks, cats, canines, and other predators often centre their foraging in edge habitats (e.g. Andrén and Angelstam 1988, 1993). The reasons why prey and predators concentrate in edges are not clear (e.g. Andrén 1995). Increased predation rates in edges might arise from predators using edges as conduits, from active hunting in response to the higher prey density, from higher predator abundances in edges compared with forest interiors, or from a more diverse predator community in edges (see Dijak and Thompson 2000). Landscape patterns complicate local nest-predation effects within edges. An invest-igation in the American Midwest showed that in highly fragmented landscapes, nest-predation rates were

high in forest interiors and agricultural edges; in moderately fragmented landscapes, nest-predation rates were low in forest interiors but high in agricultural edges; and in severely fragmented landscapes, nest-predation rates were low in both forest interiors and agricultural edges (Donovan *et al.* 1997).

Predation rates vary with edge type. Three types of edges in a bottomland forest along the Roanoke River in North Carolina, USA, had different levels of pre-dation during the 1996 breeding season (Saracco and Collazo 1999). The three edge types were: (1) a forest–farm edge, which has a sharp exterior edge; (2) a forest–river edge, which has an abrupt interior edge; and (3) a levee–swamp edge, which has a gradual interior edge at the boundary of the two dominant communities in the floodplain – cypress-gum swamps and coastal plain levee forests. Predation rates of northern bobwhite (*Colinus virginianus*) eggs and clay eggs were significantly higher along forest–farm edges than along the other two edges, where the predation rates were roughly the same. Taken with higher avian predator abundance on forest–farm edges, the pat-tern of egg predation indicated that avian predators exerted more predation pressure along these edges, a finding consonant with other studies where agricul-tural encroachment into forested landscapes may have a deleterious effect upon breeding birds.

All generalisations have their exceptions, and not all edges display elevated predation rates. In bushland sites, ranging in size from 3.8 ha to 14 717 ha, within the urban area of Sydney, Australia, nest predation of the eastern yellow robin (*Eopsaltria australis*) was unaffected by edge effects (Matthews *et al.* 1999). Twenty artificial nests were constructed from halved tennis balls covered with coconut fibre and lined with forest oak (*Casuarina* spp.) needles to mimic eastern yellow robins' nests. Each nest contained two dummy eggs made of non-toxic, blue-green modelling clay. The eggs were distributed around the edges and within the centre of 24 bushland sites. After 15 days of exposure, the nests and contents were collected. Nest were classed as 'depredated' if one or both eggs were damaged or removed. The impression left in the eggs enabled the predators to be identified. The chief four predators were birds, rats, most likely black rats (*Rattus rattus*), brown antechinus (*Antechinus stuartii*), and ringtailed possums (*Pseudocheirus peregrinus*). Nest predation rates were high, averaging 70.6 per cent, but were unaffected by patch size or distance from fragment edges. This finding goes against the general view that nest-predation rates in patch edges are higher than in the patch centres, and may relate to the ubiquity of generalist predators and the degree of habitat modification throughout remnants of bushland in an urban environment.

Corridors

Corridors serve several ecological roles. They are at once habitats, conduits, and filters. As habitats, corridors of all kinds tend to have a high species diversity with edge and generalist species dominating. As conduits, they provide routes for wildlife and for humans, the rate of movement depending on such factors as the species moving, the type of corridor, the corridor width and length, the number of 'narrows' and gaps, the number of entrances and exits, the curviness, the patchiness, the degree of criss-crossing with other corridors, and the strength of any environmental gradients present. As filters, corridors permit some species to cross but bar others. The filtering process may create separate patches on either side of the corridor, each with a different species composition.

Roads and trails

Roads and roadsides as refuges and conduits

Roadsides, and even roads themselves, provide habitats that are different from surrounding habitats. They act as wildlife refuges for some species. In particular, they often support 'edge' species from surrounding forest (e.g. Lynch and Saunders 1991). They are also potential avenues of movement for various groups of animals and plants through what otherwise might be uninviting terrain. Grassy roadsides provide routes for grassland species across forested and intensive agricultural regions. The length of the routes can be enormous: the Interstate Highway System in the United States has created about 70 000 km of potential movement corridors.

Movement may take place on the road surface and on the vehicles that move along it, in the open space above the road, and in the roadside habitat. Vehicles move a surprisingly large number of plant seeds along roads (e.g. Clifford 1959). In a survey of a car-wash flora from the Shell Canberra Superwash, Australia, 259 plant species were identified, of which 19 were not native to, naturalised in, or cultivated in Canberra or the nearby New South Wales tablelands (Wace 1977). Fungal spores are also transported in mud and soil clinging to vehicles. The cinnamon fungus

(*Phytopthora cinnamoni*) spread through the forests of southern Australia in this manner (Weste 1977). Seed transport by vehicles leads to a large variety of pioneer and adventive plants in roadsides, and in particular the immediate road edges. The open road and roadside, both of which harbour predators and other hidden dangers (such as vehicles), are generally little used by animals. Some frogs and snakes are dispersed by 'hitching a ride', though this is not a common process, a documented case being the establishment of the spotted grass frog (*Limnodynastes tasmaniensis*) at Kununurra, in north-western Australia, some 1800 km from its south-eastern Australian homeland (Martin and Tyler 1978). Tracks and lightly used roads are often employed, especially at night, by such predators as the red fox (*Vulpes vulpes*), wolf (*Canis latrans*), dingo (*C. familiaris dingo*), cheetah (*Acinonyx jubatus*), and lion (*Panthera leo*). They use them as clear pathways, uncluttered with vegetation, along which to move and hunt. Metalled road surfaces are generally avoided by animals as conduits, though moose (*Alces canadensis*) sometimes amble along them at night (and charge at cars). Bats use open spaces above roads as flight paths and foraging spaces. In Queensland, Australia, they use forestry roads through rainforests as conduits (Crome and Richards 1988).

Four types of movement are associated with roadside habitats: local foraging movements; dispersal between separated populations; long-distance migration; and local or geographical range expansion (Bennett 1991). In the south-western part of Western Australia, Carnaby's cockatoo (*Calyptorhynchus funereus latirostris*) sometimes uses roadside vegetation as a route for regular foraging movements within its large home range, while broad strips of roadside vegetation enable it to move between woodland remnants (Saunders 1990). The dispersal of the arboreal sugar glider (*Petaurus breviceps*) from its home range is aided by roadside natural strips in Gippsland, Victoria, Australia (Suckling 1984). The sugar glider lives in forest remnants within farmland, but also resides in roadside vegetation that links forest patches. Recorded dispersal movements of young sugar gliders, which involve distances of up to 1.9 km, all took place along roadside strips. Birds use wooded strips alongside roads during long-distance migration, pausing to shelter or forage. Nine bird species use roadside vegetation as a corridor for nomadic and dispersal movements near Ongerup, Western Australia (Newbey and Newbey 1987). Range expansion along roadsides appears to be common in small mammals and plants.

Purple loosestrife (*Lythium salicaria*), a perennial, herbaceous wetland plant from Eurasia, invaded the eastern United States through several ports in the middle to late 1800s. The early road systems, railroads, and the first state and federal highway networks were not used as migration routes. But the modern superhighway is suspected of offering new disturbed habitats and two means of dispersal – aquatic dispersal (by surface sheetflow and flow along roadside ditches) and airborne dispersal (by wind currents stirred up by the passage of high-speed trucks and normal winds blowing across exposed areas). The distribution of purple loosestrife colonies along a 402-km section of the New York State Thruway (Interstate 90) between Albany and Buffalo, New York, confirms this suspicion (Wilcox 1989). Range expansion in small mammal species is exemplified by the spread of vole species along dense grassy highways in central Illinois, USA (Getz *et al.* 1978). Two species were looked at – the prairie vole (*Microtus ochrogaster*) and the meadow vole (*M. pennsylvanicus*), also known as the eastern meadow mouse. The prairie vole has long lived in all of Illinois, while the meadow vole is currently expanding its range in the central part of the State. Both are grassland species, but the meadow vole needs more cover, hence it is common in mowed country roadsides while the prairie vole is not. Three interstate systems were built near Champaign–Urbana during the 1960s and 1970s. Extensive trapping in the area in 1976, in conjunction with earlier records, enable dispersal patterns to be detected. Most records of meadow vole occurrence show a close association with the interstate roads, as well as dense strips of grassy vegetation extending away from the roads, drainage ditches, and railroads. The prairie vole remained the predominant microtine rodent in dense grassy sites not connected to the interstate highway and along mowed country roads. The meadow vole expanded its range by 90–100 km across Illinois farmland in about six years, using the continuous and dense grassy roadside along the interstate highway.

Roads as barriers and filters

Roads act as barriers or filters to animals that would cross them. Almost every animal, from spiders and beetles to kangaroos and deer, is a potential road-crosser. In wetlands, roads pose barriers to the free movement of aquatic animals, and may divide and isolate populations. The chief road components deter would-be crossers: the bare road surface; the altered roadside habitat; and the noise, movements, emission, and lights that are part of road traffic (Bennett 1991). On wide roads with much traffic, all three elements conspire to create a daunting and formidable barrier to wildlife. An overgrown and narrow track will present hardly any barrier.

In Germany, it has been found that grassy field-tracks have no significant effect on the movement of carabid beetles and lycosid spiders, but paved and gravel field-tracks (and railway) stimulate movements along corridors and reduce the corridor-crossing rate (Mader 1984; Mader *et al.* 1990). Given that the density of paved agricultural roads is increasing in modern agricultural landscapes, it seems that the dispersal of ground-foraging arthropods, and their immigration to isolated patches of natural or semi-natural habitats, will be hampered. Small mammals will cross roads, but only if the roads are not too wide. In Germany, none of 121 marked specimens of the yellow-necked mouse (*Clethrionomys glareolus*) would cross a 6-m, two-lane paved highway, though many movements paralleling the highway were recorded (Mader 1984). In south-eastern Ontario, Canada, the clearance of habitats on either side of a road was a key factor in determining road-crossing by small-sized (less than 700 g) and medium-sized (700–14 000 g) mammals (Oxley *et al.* 1974). Small forest-adapted species – the eastern chipmunk (*Tamias striatus*), red squirrel (*Tamiasciurus hudsonius*), and white-footed mouse (*Peromyscus leucopus*) – crossed roads with up to a 35-m clearance, but were reluctant to cross wider roads with clearances up to 137 m. In contract, medium-sized mammals – snowshoe hare (*Lepus americana*), woodchuck (*Marmota monax*), raccoon (*Procyon lotor*), eastern grey squirrel (*Sciurus carolinensis*), striped skunk (*Mephites mephites*), common muskrat (*Ondatra zibethica*), and porcupine (*Erethizon dorsatum*) – frequently crossed roads with a clearance of less than 15 m, but were never observed crossing highways with a wide clearance. The eastern chipmunk was the only species to suffer high mortality rates on narrow roads. In a different study, large and mobile mammals, such as the mountain goat (*Oreamnos americanus*), caribou (*Rangifer tarandus*), and elk (*Cervus elaphus*) were observed to cross most roads, but were normally loath to do so and some displayed signs of fear (e.g. Singer and Doherty 1984). However, generalisations about mammal size and width of road crossed are very much a rule of thumb, since variation between species is considerable. Some small mammals occupy home ranges that compass both sides of a road (Bennett 1990).

In cases where a species will not cross a road, the population becomes divided, the interbreeding rates on either side of the corridor change, and two subpopulations evolve. Genetic differentiation in the two subpopulations may ensue. In Britain, roads separating common frog (*Rana temporaria*) populations have led to genetically differentiated subpopulations (Reh 1989). In the Saar–Palatinate lowlands, Germany, land use and topographic distance influence the genetic structure of common frog populations (Reh and Seitz 1990). In particular, strong deviations from the Hardy–Weinberg equilibrium were found, and the degree of homozygosity was higher than expected. The populations were affected within 3–4 km of motorways and railways, which acted as substantial barriers to frog movement. In the mid-western USA, the Mississippi River separates two red fox (*Vulpes vulpes*) subpopulations (Storm *et al.* 1976). Differentiation of neighbourhoods by a green belt, or a main road, or a railway within cities is a human analogue of this process.

A danger of crossing roads is being struck by vehicles. Road kill takes an enormous annual toll on wildlife. A million vertebrates are killed every day on roads in the USA. There is even a book called *Flattened Fauna: A Field Guide to Common Animals of Roads, Streets and Highways* (Knutson 1987). In north-central Idaho, the mortality of elk (*Cervus elaphus*) increased with road density (Unsworth *et al.* 1993). Techniques to lower the toll include the use of reflectors, mirrors, repellents, bait, fencing of different kinds, one-way gates (to escape from fenced roadsides), lighting, wildlife crossing signs (for motorists, not the wildlife), and animated warning signs for motorists. All are of moderate or little success. Far more successful are overpasses (bridges), underpasses (tunnels), and the seasonal closure of a road. Underpasses are generally successful and are used, for instance, for badgers (*Meles meles*) in Great Britain and for mountain goats (*Oreamnos americanus*) in Montana, USA. The highest road-kill rates occur on two-lane main roads with high speeds, but multi-lane highways have a greater ecological impact, removing more native habitat and creating barriers that many animals are disinclined to cross or even go near (Forman 1995, 168).

Trails

Trails are used by some plants as conduits. The invasion of cheatgrass (*Bromus tectorum*) over huge, dry areas of north-western North America took place largely along cattle trails and railroad corridors (Mack 1981). Many terrestrial mammals make trails within their home range, which they use for foraging and so forth. Humans are inveterate trail-makers. They use their trails for movement, either on foot, on horseback, or on a motorcycle or other kind of vehicle. The environmental impacts of trails will be considered in the next chapter.

Powerlines

Electricity transmission lines, gas lines, oil lines, and dykes tend to be fairly straight with sharp boundaries, and to have a fairly constant width over which disturbance or maintenance is evenly distributed. They favour edge and generalist species. In a forested Tennessee landscape, in the USA, almost all the birds in powerline corridors were edge species (Anderson *et al.* 1977). Just two forest interior species – the scarlet tanager (*Piranga olivacea*) – and the red-eyed vireo (*Vireo olivaceus*) – ranged into narrow powerline corridors. The abundance of most edge species decreased with increasing corridor width, whereas the abundance of two open-field species – field sparrow (*Spizella pusilla*) and yellow-breasted chat (*Icteria virens*) – showed a marked increase in abundance as corridor width increased.

In the Mojave Desert, California, a study explored the relationships between linear rights-of-way (paved highways and transmission powerlines) and raven (*Corvus corax*) and red-tailed hawk (*Buteo jamaicensis*) populations (Knight and Kawashima 1993). Control transects were established, being defined as areas with neither powerlines nor highways and within 3.2 km of a transect centre, the 3.2 km being derived from the home ranges of the birds. Using a Bell Jet Ranger helicopter, 97 transects were surveyed, between 06:20 and 19:20 hours from 14 May to 1 June 1989. Of the 1684.8 km surveyed, 462.5 km were transmission powerlines, 404 km were paved highways, and 818.3 km were control transects. Available perch and nest sites – trees, cliffs, transmission towers, telephone poles, signs, highway overpasses, and others – were noted. Ravens were as common along highway transects as they were along powerline transects, and were more abundant along both than they were along control transects. Raven nests were more abundant along powerline transects than along highway and control transects. Both red-tailed hawks and their nests were more abundant along powerline transects than along highway and control transects. Ravens used power poles as nest sites more than expected based on availability, but did not use them as perch sites more than expected. Red-tailed hawks used power poles for both nesting and perching more than expected based on availability. Ravens appear to be more abundant along highways due to the carrion produced by vehicles. Ravens are facultative scavengers and a road kill is a ready meal. Red-tailed hawks hunt live prey and would not be attracted by the road carrion. Both ravens and red-tailed hawks appear to be more common along powerlines, owing to the superior perch and nest sites that powerlines afford. Powerline towers are tall and may give predators a wide field of vision. Nests built in tall towers may benefit from greater cooling, owing to the increased openness around the nest, while the beams and latticework offer sturdy nest-site anchors to guard against dislodgement by extreme winds.

Powerlines act as strong filters, largely because they make a noise (a loud 'buzzing hum'), especially in wet weather, that deters would-be crossers (Gates 1991). To some crossers they are a hazard. In south-central Nebraska, USA, spring-migrating sandhill cranes (*Grus canadensis*) may crash into powerlines. A study marked alternating spans of powerline with 30-cm-diameter yellow aviation balls (Morkill and Anderson 1991). The number of cranes flying over marked and unmarked powerline spans did not differ, but cranes reacted more often to marked than unmarked spans, mainly by gaining altitude or by changing the direction of flight. More dead cranes were found under unmarked than marked powerline spans.

Hedgerows and other wooded strips

Hedgerows and windbreaks are line corridors dominated by edge species living at high densities. Forest-interior species are normally present, albeit in low numbers. Width is a key factor in determining the species richness and abundance of many species living in hedgerows and windbreaks, though vertical structure also exerts a major influence on bird species diversity. Hedgerows support a large variety of gamebirds. If a hedgerow is associated with a wall, fence, ditch, or soil bank, then an even greater species richness is encouraged. A ditch attracts amphibians and reptiles, while the sunny side of a soil bank encourages drought-tolerant species. Some of the species harboured by a hedgerow or windbreak may frequent adjacent fields where they eat crops. Conversely, many birds that live in fields use hedgerows for perching or for foraging. The high density of animals in hedgerows draws in predators from the surroundings.

Hedgerows and other wooded strips are used as conduits by a variety of wildlife. The high number of road-kill victims found where a woodland strip is broken by a road attests to this fact. Not all species will cross gaps in hedgerows, however. The dormouse (*Muscardinus avellanarius*) is an arboreal habitat specialist that is averse to crossing even narrow gaps in hedgerows, though dormice are prepared to move across grass fields (Bright 1998). Plants may also move along wooded strips, though the evidence is patchy (e.g. Forman 1991). A recent study of remnant and regenerated hedgerows in Tompkins County, in the Finger Lakes region of central New York State, USA, indicated that forest herbs colonise hedgerows from source areas (Corbit *et al.* 1999). Hedgerows were categorised as remnant or regenerated, and attached (to a forest) or isolated, after examining aerial photographs from 1938 and 1980, along with available land-use history maps. A large portion (40 per cent) of forest herbs was observed in the hedgerows. The regenerated hedgerows had grown up spontaneously between open fields over the last 50 years, but they did not differ significantly in the richness or abundance of forest herbs from the remnant hedgerows. This lack of differences presumably means that the forest herbs have moved along the regenerated hedgerows from attached forests, and that the hedgerows act as corridors. The fact that the species composition of forest herbs in hedgerows attached to forest stands bore a strong affinity with the species composition of the adjacent stand, for both remnant and regenerated hedgerows, supported this idea. The idea was also supported by a decline in richness of forest herbs along the hedgerows away from the source areas (forest stands).

Stream and river corridors

Stream or riparian corridors play a starring role in many ecosystems, exerting a weighty influence over animals and plants. Many species rely on stream corridors for food and water, for shelter, for travel and rest, and for reproduction. The varied habitats and excellent food base (water plants, herbs and shrubs bearing berries and seeds, leafy foliage) foster high species diversity. A reason for this richness of wildlife and wildlife habitats is the fact that riparian ecosystems receive water, nutrients, and energy from upstream ecosystems. These upstream additions allow greater species richness and help to maintain a fairly constant supply of resources (Harris 1984, 142). Permanent lakes and rivers support many amphibians and aquatic birds and mammals. They also contain fish and other aquatic organisms that form the seat of several food webs.

The chief habitats in stream corridors are riverbank and floodplain. Riverbank habitats are characterised by changing water levels, changing soil moisture levels, and erosion and deposition of sediment. Floodplains are often rich in wetland habitats, different types being dictated by the frequency and degree of inundation – marsh or bog, shrub swamp or thicket, forested swamp, vegetation of rarely flooded levees, ridges, and hummocks. All these floodplain habitats share periodic flooding, poor soil drainage, occasional surface deposition of sediment, and nutrient-rich soils. In dry regions, such as Arizona, riparian corridors sometimes form 'linear oases' (thin green lines), which are supported primarily by groundwater and contain rare species. In tropical grasslands, biodiverse 'gallery forest' meanders across the plains, providing water, food, and shade for many species in a grassland matrix (e.g. Redford and Fonseca 1986).

Riverbank plants tend to be either tolerant of, or resistant to, disturbance, or else opportunistic. The opportunists often have seeds dispersed by water (hydrochores) and germinate rapidly. In England, alder (*Alnus glutinosa*) and purple willow (*Salix purpurea*) are resistant species, while Indian balsam (*Impatiens glandulifera*) is an opportunist. Some common trees, including black poplar (*Populus nigra*), possess characteristics of resistance and opportunism. Riverbank animals also cope with the changing conditions. Some, such as river otters, weasels, duck-billed platypuses, dippers, and kingfishers, live in riverbank burrows and feed on the nearby store of aquatic invertebrates and fish. Beavers alter the nature of the corridor by building dams (e.g. Remillard *et al.* 1987). Some plants are favoured by beaver ponds, meadow plants colonising the nutrient-rich mudflats. To survive, some riverbank animals need a continuous corridor of high quality. The river otter (*Lutra canadensis*) in North America has up to 80-km-long home ranges along rivers, and requires clear and cool water, which ordinarily means shaded and uneroded riverbanks and pollutant-free river water (Melquist and Hornocker 1983). Some species demand a corridor wide enough to accommodate their territory or home range. In California, the yellow-billed cuckoo (*Coccyzus americanus occidentalis*) demands riparian corridors wider than 90 m (Reiner and Griggs 1989). However, other species, including the white-footed mouse (*Peromyscus leucopus*), alter their behaviour to live in long and narrow home ranges, which are unlike their elliptical home ranges in the prairies (Merriam and Lanoue 1990).

Many mammals use river corridors as conduits. Mountain lion (*Felix concolor*), bobcat (*Lynx rufus*), grizzly bear (*Ursus horribilis*) and black bear (*U. americanus*) are known to move many kilometres along river corridors (e.g. Noss 1993). If river meanders are present, people and animals (e.g. egrets, kingfishers, and river otters) may move directly between meanders, rather than follow the circuitous channel. Hawks and related birds of prey commonly migrate along windward edges of ridges, where updraughts of air facilitate gliding. Some exotic (non-native) plants use river corridors to spread. In Britain, Japanese knotweed (*Reynontria japonica*) rhizomes are dispersed by floods; and Indian balsam (*Impatiens glandulifera*), introduced from the Himalayas as a garden plant, has spread along riverbanks. In the western United States, two exotic species – tamarisk or salt cedar (*Tamarix hispida rubra*) and Russian olive (*Elaeagnus angustifolia*) – have colonised widely along rivers and cause serious problems in local ecosystems. The tamarisk is a very invasive shrub-tree that was introduced from Eurasia and planted across the western United States by government agencies in the early 1900s in an effort to control soil erosion. The Russian olive is a native tree of Europe and Asia. It was introduced to North America by settlers some 150 years ago, and has since colonised widely along rivers where it is having a severe impact on native birds and fish (e.g. Dixon and Johnson 1999). Similarly, purple loosestrife (*Lythrum salicaria*) was originally introduced into North America from Europe in the early 1800s as an ornamental plant and a contaminant of ship ballast. It was spread by waterborne commerce, and has invaded many wetlands, causing severe problems in some areas (Thompson *et al.* 1987; Mullin 1998).

Networks

Tree networks and life

Corridor networks, tree networks, and circuit networks all influence animals and plants. The influences of corridor and circuit networks are the most studied, but tree networks play a role in determining the distribution of species in landscapes and deserve further investigation. Such studies as have been made hint at relationships between species distribution and stream

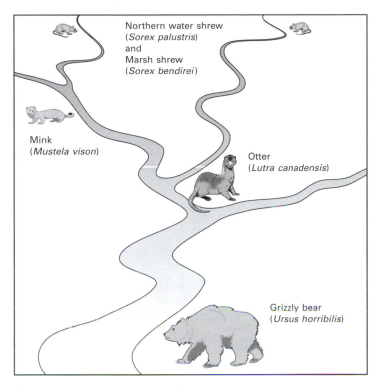

Figure 6.24 The association of different-sized carnivorous mammals with stream order and typical food-particle size according to the river-continuum concept. (Adapted from Harris, 1984, 143.)

networks. Some relationships are broad, as in seasonally migrating mammals using river networks to go from lowlands to uplands. Examples in the Grand Tetons of North America are elk (*Cervus canadensis*), mule deer (*Odocoileus hemionus*), and moose (*Alces americana*), which migrate to subalpine meadows in early summer and back in autumn to winter feeding areas in the plains. Other relationships are more specific and relate species distribution to stream order. Plants in first-order and second-order streams are usually different from plants in high-order streams (e.g. mosses and ferns in headwaters). Because terrain elevation changes with stream order, floodplain vegetation also changes (e.g. Hughes 1988). In Europe, riverine hardwood forests of big river plains contain similar species and genera that often recur in different geographical regions – common hawthorn (*Crataegus monogyna*) or black hawthorn (*C. nigra*), ash (*Fraxinus excelsior*) or narrow-leaved ash (*F. angustifolia*), fly honeysuckle (*Lonicera xylosteum*) or perfoliate honeysuckle (*L. caprifolium*), pedunculate

oak (*Quercus robur*) or holm oak (*Q. ilex*), hoary willow (*Salix eleagnos*) or common osier (*S. viminalis*) (Schnitzler 1994).

Some animals are more abundant in the vicinity of streams. Raccoons (*Procyon lotor*) and opossums (*Didelphis virginiana*) are predators of forest songbird eggs and nestlings. In Missouri, USA, raccoon and opossum abundances were related to, among other things (see p. 223), stream density (Dijak and Thompson 2000). More generally, a sequence of terrestrial mammal species occupying similar functional niches is found along different-sized streams according to the river-continuum concept (Vannote *et al.* 1980). In the western Cascades of North America, a series of carnivorous, amphibious mammals with similar ecological roles eat prey of different sizes, live in streams of different orders, and occur at different elevations in a drainage basin (Figure 6.24). The northern water shrew (*Sorex palustris*) and marsh shrew (*S. bendirei*) live in the headwater and lower-order streams, the mink (*Mustela vison*) lives in slightly higher-order streams, the otter (*Lutra canadensis*) lives in middle-order streams, and the grizzly bear (*Ursus horribilis*) lives in higher-order streams.

Corridor and circuit networks

Many network properties affect the distribution of abundance of animals and plants. Location within a circuit, like location within a river network, is sometimes a significant property and determines the species present. In Brittany, France, the carabid beetle fauna differs in various parts of the hedgerow network. Some species live near the periphery, some nearer the centre, and some in wide corridors within the network (Burel and Baudry 1990). Similarly, while single corridors tend to favour edge species, networks that involve the proximity of two corridors (at right-angles or an acute angle) may be able to support some patch species, even in the absence of a patch. This appears to be the case for kangaroos using roadside strips in Western Australia (Arnold *et al.* 1991). But the main thrust of research into landscape networks concerns connectivity and its effects upon population dynamics. Connectivity has been studied in mathematical models and in the field.

Modelling network connectivity

Of all network properties, connectivity is of particular consequence for animals and plants. It affects the viability of metapopulations by determining how easy or difficult it is for animals to move between resource patches. A mathematical simulation suggested that two- or three-patch metapopulations are doomed to extinction, no matter how much movement there is between patches, when all local populations (subpopulations) are below a minimum viable population size (Wu *et al.* 1993). That finding has management implications: given a set of scattered small populations, augmenting individual populations might be more advisable, rather than trying to bolster migration between patches by maintaining or by building corridors. Another simulation showed that, when at least one subpopulation is larger than the minimum viable population size, there is a critical size for that subpopulation above which the metapopulation as a whole will persist and below which it will collapse. A third simulation indicated that, when a metapopulation comprises two or more patches, metapopulation dynamics and persistence were strongly affected by two factors: (1) the pattern of patch connections; and

(2) the spatial position of the populations above the viable minimum level. All three simulation results suggested that both the number of connections between patches and the magnitude of movements along them are decisive for overall patch connectivity, and the magnitude of migration between patches is positively related to the minimum size of the subpopulation that is above the minimum viable population in both the two-patch and three-patch metapopulation systems, owing to a population sink effect.

Many other mathematical models have been devised (e.g. Wu and Vankat 1991; Wu and Levin 1997). An excellent overview is given by Wu and Loucks (1995; see also Wu 1999).

Connectivity in the field

Theoretical studies are supported by a battery of field investigations that explore the effect of connectivity on species distributions and abundance. Graham W. Arnold and his colleagues have carried out an enormous amount of research on connectivity and its effects on animals in Australia (e.g. Arnold *et al.* 1991, 1993, 1994, 1995; Fortin and Arnold 1997). Other researchers have made weighty contributions to the field (e.g. Bennett *et al.* 1994; Bowne *et al.* 1999).

A very instructive study involved the building of a large-scale experimental landscape in which to examine corridor use and the forest-matrix habitat use by the hispid cotton rat (*Sigmodon hispidus*) (Bowne *et al.* 1999). The experimental landscape was set up at the Savannah River Site in Aiken County, South Carolina, USA. The matrix was a loblolly pine (*Pinus taeda*) forest. Within it, ten 1.64-ha patches, each 128×128 m, and 32-m-wide corridors of differing lengths were cleared of trees and burned. Four of the patches were isolated; six were connected in pairs by the corridors, which were either 128, 256, or 384 m long. Ninety-six cotton rats, which are habitat generalists with a preference for grassland and old-field habitats, were captured within 13.4 km of the experimental study area. They were then released, 12 per release period, at the same time into both an isolated patch and a connected patch. Their movements were monitored by radiotelemetry for 10 days. The results showed that the loblolly pine forest matrix was not a deterrent to cotton-rat movement, with 50 per cent of the released cotton rats moving through the matrix. In addition, there was no significant difference between the number of cotton rats leaving connected patches (60 per cent) and the number leaving isolated patches (50 per cent). Even so, the majority of cotton rats preferred to leave a connected patch via a corridor, rather than venture into the forest matrix. Cotton rats leaving connected and isolated patches were equally good colonists. Interestingly, microhabitat preferences displayed by the cotton rats were not the same in patch and corridor habitats as they were in forest matrix habitats, a finding which indicates that cotton rats have different selection preferences when in transit. In patch and corridor habitats, cotton rats preferred sites with tall (>1 m) shrubs and high percentage vegetation cover. In the forest matrix, cotton rats preferred sites with abundant vine cover and low tree-canopy cover. The work suggests that cotton rats are able to negotiate large-scale spatial structures in the landscape, moving long distances and crossing different habitat types, but in doing so, they display preferences for certain microhabitats. These findings have important implications regarding habitat connectivity for small mammals, for the long-distance movements observed would mean that the level of interaction between habitat patches would be high enough to prevent subpopulations evolving.

Eastern chipmunks (*Tamias striatus*) were studied in a patch-and-corridor network in 200 ha of farmland at Manotick, about 20 km south of Ottawa, Canada (Bennett *et al.* 1994). The eastern chipmunk is a small, diurnal, burrow-dwelling member of the squirrel family. In the Ottawa region, it is active from April to October, hibernating in winter when snow covers the ground. It is a native woodland species, but persists in farm landscapes where some wooded vegetation remains. In the study area the chipmunks lived in a farmland mosaic of patches and corridors. The patches were small woodland remnants, with sugar maple (*Acer saccharum*), white ash (*Fraxinus americana*), eastern white cedar (*Thuja occidentalis*), basswood (*Tilia americana*), and white elm (*Ulmus americana*) being the dominant species. Fields used for pasture and for corn (*Zea mays*) and oats (*Avena sativa*) crops surrounded the patches. Vegetated fencerows that occupied narrow uncultivated strips of land along fence-lines between fields connected the patches. These fencerows provided a corridor network linking the woodland patches (Figure 6.25). The

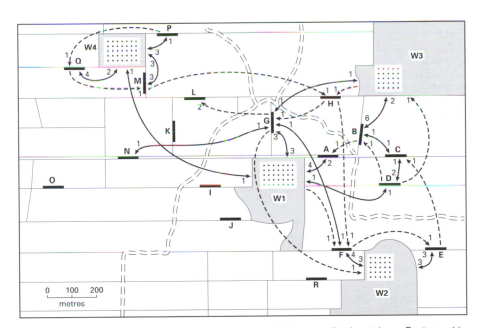

Figure 6.25 Fencerows provide a corridor network linking woodland patches. Eastern chipmunks (*Tamias striatus*) were studied in a patch-and-corridor network in 200 ha of farmland at Manotick, about 20 km south of Ottawa, Canada. Trapping sites are labelled W1–W4 (woods, where trapping sites are shown by circles) and A–R (fencerows). Arrows show the direction of chipmunk movement, and not the route taken. Heavy solid lines show two-way movement, heavy dashed lines show one-way movement, and the digits indicate the number of recorded directional movements between two landscape elements. Fencerows are depicted by thin solid lines, and streams by thin dashed lines. (Adapted from Bennett *et al.*, 1994.)

vegetation in the fencerows was variable, ranging from long grasses to shrubs and vines, or to woodland strips of mature trees. In addition, there were stone walls, rocks, and fence rails within some fencerows. Trapping and radiotelemetry in a separate study showed that the chipmunks confined themselves to wooded or shrubby vegetation, and were seldom recorded in, or moving across, fields of pasture or crops.

In the present study, chipmunks were trapped in four woods and 18 fencerows in four trapping sessions between May and September 1989. They were assigned to one of three residence statuses valid for the season studied. 'Resident' chipmunks were recorded in the same fencerow or wood during two or more trapping sessions. 'Temporary resident' chipmunks were recorded during only one session, even if trapped several times. 'Transient' chipmunks were trapped in a particular fencerow or wood only once. In all, 530 captures of 119 chipmunks were recorded during the study. They resided in all four woods, and were trapped in 14 of the 18 fencerows. Resident individuals lived within and along many fencerows, which enhanced the continuity of the resident population between woods. These residents preferred fencerows with tall trees and a woodland structure. Transient chipmunks seem to use the fencerow network as a pathway across the farmland. They were more abundant in fencerows with high linear continuity (preferring fencerows without too many gaps) and with 'woody' habitat attributes (preferring tall trees, small trees, and tall shrubs). Chipmunks never used grassy vegetation, which, like the surrounding farmland, appears to be an inhospitable habitat.

Mosaics

Studies of landscape mosaics are in their infancy. At least two lines of enquiry have emerged. The first considers population dynamics in landscapes with emphasis on movement between patches, corridors, and matrixes. The main techniques employed in these studies are observation by capture or radio-tracking or mathematical models. Radio-tracking has proved an effective tool to study animals' movements. Small radio transmitters may be fitted to a range of animals – including mice, geese, and lions – which enables them to be tracked with relative ease. Radio-tracking is helping studies of foraging within home ranges, dispersal out of home ranges, and migration between habitats. A key idea in these studies is landscape structure and connectivity. The second line of enquiry focuses around the relation between vegetation and landform and landscape properties, the main tools used in these investigations being GIS, DTMs, and statistical analysis.

Landscape structure and connectivity

Landscape mosaics are specific arrangements of patches, corridors, and matrixes. The arrangement of these three elements affects the ease with which animals and plants may move through a landscape. In its turn, the ease of movement affects population dynamics. The ease of movement is commonly referred to as landscape resistance.

Landscape resistance is a mosaic property. It is determined by the structural landscape properties, including the degree of connectedness, that impede movement. There is no standard way of measuring landscape resistance, though several landscape factors contribute to it, including high fragmentation of patches, gaps in corridors, a lack of corridors between patches, and the presence of main roads. A study in southern Holland found that the movement of woodland birds, including the nuthatch (*Sitta europaea*), was affected by landscape resistance (Harms and Opdam 1990; see also Knaapen *et al*. 1992). In this study, landscape resistance was increased by built area, glasshouses, and busy roads, while it was decreased by wooded vegetation.

An important point to bear in mind is that the resistance offered by a landscape to one species, or group of species, may not be the same as the resistance offered by the same landscape to another species. So, if the spatial structure of a landscape should change, species will fare differently (Henein *et al.* 1998). For instance, species with small home ranges will be affected differently from species with large home ranges. Indeed, many wide-ranging species (including caribou, tigers, black bears, and vultures) are sensitive to the arrangement of regional landscapes, and commonly use two or more landscapes that need to be close together (Forman 1995, 25). In addition, habitat generalists may find it easier to move through heterogeneous landscapes created by habitat fragmentation than may habitat specialists.

Empirical investigations

Most of the empirical data identifying landscape spatial patterns (connectedness) that interfere with individual movements concern walking animals with low powers of dispersal. For flying animals, the distance between patches is measured in almost all cases as a straight-line distance between points. This metric does not account for the behavioural characteristics of species that depend on landscape patterns for their movements (Clergeau and Burel 1997). The presence of the short-toed treecreeper (*Certhia brachydactyla*), for instance, seems to depend on the spatial structure of agricultural landscapes. In western France, around the city of Roz, two adjacent and contrasting rural landscapes – long-established bocage and a recent polder – differ in grain size, in the quality of linear landscape elements delimiting fields, and in their history. The bocage landscape consisted of a mosaic of fields with an average size of less than 2 ha comprising permanent pasture (15 per cent), temporary grassland (20 per cent), maize fields (20 per cent), and scattered small woodlands, with an old hedgerow network. The polder landscape was reclaimed between 1850 and 1930, and since 1980 all the polder fields have been used for market gardening. In the recently reclaimed polder, treecreepers are present only in linear rows of 'well-connected' trees that are long enough to support the home range of this bird. Colonisation of this recent polder landscape from the bocage, which serves as a source of dispersers, required connections between the bocage hedgerows and the planted dykes.

A consideration of the structure and composition, and particularly the connections between corridors and the matrix, in cultivated landscapes has helped to explain seasonal fluctuations in wood mouse (*Apodemus sylvaticus*) populations (Ouin *et al.* 2000). A study conducted in the polders of the Mont Saint-Michel Bay, western France, used GIS to look into the seasonal dispersal of wood mice from hedges to crops at field and landscapes scales. Ninety per cent of the area was under intensive agriculture, with wheat (*Triticum aestivum*), maize (*Zea mays*), peas (*Pisum sativum*), and carrots (*Daucus carotta*) as the main crops. The semi-natural habitats (hedges and grassy linear habitats) were distributed over a dyke network. The results suggested that the summertime drop in hedgerow populations of wood mice resulted from a movement into the crops. Hedgerows serve as a source of wood mice in spring, and the rate at which they colonise fields in summer depends on the quality of the crops, as reflected in the crop cover and seed availability in fields and in landscapes.

Landscape structure affects ecosystem processes, including seed predation. In an agricultural landscape in Ingham County, Michigan, USA, weed seed-predation by invertebrates and vertebrates displayed interesting similarities and differences in simple settings (large crop fields embedded in a matrix of widely scattered woodlots and hedgerows) and complex settings (small crop fields embedded in a matrix of numerous hedgerows and woodlots) (Menalled *et al.* 2000). Structural differences between the two landscapes were evaluated by analysis of black-and-white aerial photographs and digital land-use data. Fields in the complex landscape were 75 per cent smaller and had 63 per cent more wooded perimeter and an 81 per cent wider hedgerow perimeter than fields in the simple landscape. Fields in the simple landscape were surrounded mainly by herbaceous roadside and crops, whereas the complex landscape had fields surrounded primarily by wide hedgerows. Seed predation experiments were conducted in four conventional tillage corn (*Zea mays*) fields within each landscape type. Four common agricultural weed species – crabgrass (*Digitaria sanguinalis*), giant foxtail (*Setaria faberii*), pigweed (*Amaranthus retroflexus*), and velvetleaf (*Abutilon theophrasti*) – were used to examine the effects of agricultural landscape complexity on seed removal after dispersal. Three treatments were used to measure seed predation: (1) no exclosure, which let vertebrates and invertebrates remove seeds; (2) vertebrate exclosure, which allowed invertebrates to remove seeds; and (3) total exclosure, which stopped vertebrates and invertebrates from removing seeds. The treatments were established 27 m from hedgerows, that being the distance from the centre of the smallest field to the hedge. In simple and complex landscapes and for all weed species, a substantial portion of the seed was removed. The complex landscape tended to have higher rates of seed removal, though only on the first trial were the differences statistically significant. Although there were no differences in the rate of seed removal among the four weed species, seed predation showed a high degree of variability within and among fields.

Landscape structure and its effects upon wildlife can be probed using field experiments that manipulate landscape patterns. Two mixed-grass prairie sites in the plains of Front Range, north-central Colorado, USA, situated about 12 km south-east of Boulder, were mowed to produce various arrangements of grassland fragments and corridors (Collinge 1998). Insects were the chosen object of study, and over 500 species were collected over the study area over three growing seasons (1992–1994), with representatives from all major insect orders. Two experiments were conducted, one at each site. The first experiment looked at the influence of fragment size and

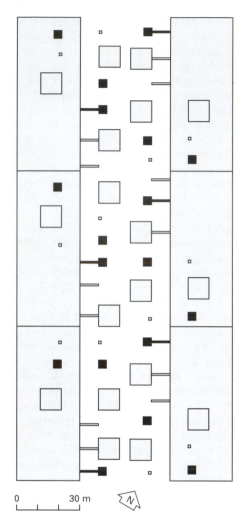

Figure 6.26 Landscape spatial structure in an experimental design of native prairie-grassland fragment size and connectivity, near Boulder, Colorado, USA. Shaded areas depict unmown grassland vegetation; white areas show mown grassland vegetation. Three plot sizes (1 m², 10 m², and 100 m²) were set up randomly in the context of three connectivity treatments – control, corridor, and isolated. The whole study area, as shown in the diagram, covers about 3 ha. (Adapted from Collinge, 1998.)

connectivity on insect-species loss from fragments, insect recolonisation of fragments, and individual insect movements among fragments (Figure 6.26). The second experiment investigated the effects of landscape spatial configuration upon insect species composition. Specifically, it examined the effects of four processes of habitat fragmentation – shrinkage, bisection, fragmentation, and perforation (Figure 6.27). An intriguing finding of the study was that medium-

sized fragments had higher recolonisation rates and lower species-loss rates than small or large fragments. This suggests that corridors would not mitigate against the effects of habitat isolation. It was also found that the spatial structure of land conversion sequences significantly influenced species richness. Total species richness was higher in bisection and fragmentation sequences, and lower in perforation and shrinkage sequences, a pattern that is probably a short-term 'crowding effect'. Large, rare species were more abundant in shrinkage and fragmentation sequences than in bisection and perforation sequences, perhaps because such large species perceive the fragmented habitat patches as one large habitat patch (Collinge 1998). These findings are some of the first experimental evidence obtained under field conditions for the ecological function of corridors, and they show that different spatial patterns of native prairie habitat vary in species composition.

Mathematical models: the case of eastern chipmunks in a fragmented landscape

Models are used to simulate the effects of landscape spatial structure and connectivity upon animal populations. The prediction of species persistence within human-altered landscapes requires a knowledge of the spatial structure of the landscape, the demography of the populations, the movement of individuals within and between patches, and the variation of resource availability over time (Henein *et al.* 1998).

An individual-based, spatially explicit simulation model was built to explore landscape effects on the population size and persistence of two species of woodland small mammals – eastern chipmunks (*Tamias striatus*) and white-footed mice (*Peromyscus leucopus*) – living in an agricultural landscape (Henein *et al.* 1998). These species differ in their behavioural flexibility in the face of landscape change. The chipmunks have an inflexible response to change, being specialists that retain their partiality for wooded habitat and avoid crop fields despite the new larder they provide. White-footed mice are generalists with a fairly flexible, opportunistic response to change, expanding their use of the landscape to include corn and small-grain fields. 'Individual-based' means that each individual is tracked and has its own habitat-specific probabilities of reproducing, dying, and moving at each time step. 'Spatially explicit' means that the geographical relationships between landscape elements are incorporated in the model, which used a 64 × 64 spatial grid cells. The model landscape

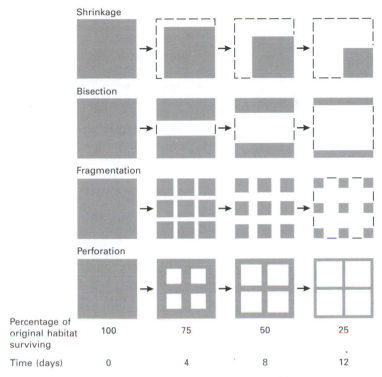

Shrinkage

Bisection

Fragmentation

Perforation

| Percentage of original habitat surviving | 100 | 75 | 50 | 25 |
| Time (days) | 0 | 4 | 8 | 12 |

Figure 6.27 Schematic representation of changing spatial structures in four different sequences of land conversion. The shaded areas are the original habitat type; the white areas are the invading habitat type. In all sequences, the change proceeds from 100 per cent original habitat type to 25 per cent original habitat type. (Adapted from Collinge, 1998.)

represented the 200-ha farm landscape near Manotick, Ontario, Canada, used by Andrew F. Bennett and his colleagues (1994) to study chipmunks. A representative model landscape is depicted in Figure 6.28. Model parameters were derived from field studies, the literature, and if all else failed, best guesses. The simulations followed chipmunk populations, which in all simulations began with 16 chipmunks, through 36 weeks per year (the active period for chipmunks), over 25 years. Thirty-six landscape patterns were simulated. Each landscape pattern was a unique combination of the area of the wooded habitat (10, 30, or 50 per cent of the total area), the subdivision of the wooded habitat (two, four, or eight patches), and the quality of connectivity (high, intermediate, low, or none). White-footed mouse populations were simulated in eight of these landscapes (30 per cent wooded habitat, four patches at each level of connectivity, and 10 per cent wooded habitat, eight patches at each level of connectivity).

The simulation results are important and worth rehearsing at length. First and foremost, connectivity was a decisive factor in chipmunk population persistence. All factors and interactions in the model contributed to population persistence, but unequally. In all landscapes, connectivity (high, mixed, low, or absent) explained over 75 per cent of population persistence. In connected landscapes (i.e. excluding unconnected landscapes), connectivity (high, mixed, or low) and the amount of wooded habitat both explained about 35 per cent of the variance in population persistence. Regardless of the level of connectivity, the degree of habitat subdivision explained a small portion (4 to 15 per cent) of population persistence, and even in better-connected and more-wooded landscapes it contributed little to maximum population size. Indeed, chipmunk population density was higher in well-connected and more-wooded landscapes with fewer patches. Some factors were interrelated. The susceptibility of the chipmunk population to

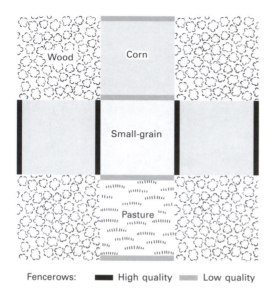

Fencerows: ■ High quality ▨ Low quality

Figure 6.28 Example of simulated landscape. The landscape has four patches of woodland, three corn fields, a small-grain field, a pasture, and a mix of fencerow types. From the eastern chipmunks' viewpoint, the fencerows are high quality (dark bars) or low quality (light bars). (Adapted from Henein *et al.*, 1998.)

lower-quality connections depended partly on the amount of wooded habitat available.

Chipmunk populations with high year-to-year variability in size were at greater risk of extinction, especially when habitat was limited (10 per cent). One chipmunk population went extinct with as many as 65 individuals, although 79 per cent of extinction occurred in populations below 30 individuals, and some populations persisted with as few as 20 individuals. Long survival times and low extinction probabilities for chipmunk populations required over 30 per cent woodland habitat, and high to intermediate landscape connectivity. White-footed mice benefited from being generalists and not confined to forests and fencerows: their populations fared well in all the low to moderately wooded landscapes and sustained no extinctions. The generalist mice out-performed the specialist chipmunks in all the subdivided landscapes in which they were compared, and their populations lasted for the full 25 years of simulation in 99 per cent of the experiments.

A salutary lesson from these simulation experiments is that the survival of behaviourally inflexible species, like the chipmunk, demands that habitat connectivity should be preserved in the face of progressive habitat subdivision and accompanying habitat loss.

Another significant finding is that landscape connectivity regulates population persistence, size, and density more than landscape composition or landscape configuration. Finally, it might be thought reasonable to argue that landscapes rich in remnant habitats would support larger and more persistent populations. However, the simulations showed that the amount of habitat is not the only factor at work – connectivity and patchiness play vital roles. Take chipmunk populations in two model landscapes. Landscape A had 10 per cent wooded habitat, two patches, and high connectivity. Landscape B had 50 per cent wooded habitat, eight patches, and low connectivity. Which landscape had the longer surviving and larger chipmunk population? The answer is counterintuitive: landscape A.

Landscape properties

Vegetation is influenced by many landform and landscape properties. It is reasonable to expect, therefore, that vegetation type and distribution should to some degree be predictable from landform and landscape features. Research efforts in this direction have proved fruitful.

An early piece of research on 'mosaics' used Landsat Thematic Mapper (TM) data in conjunction with topographic and topoclimatic variables to map dominant vegetation communities in the Colorado Rocky Mountain Front Range, USA (Frank 1988). Landsat TM transformations, elevation, aspect, and a slope–aspect index successfully separated alpine and subalpine vegetation types. However, forest vegetation types in the montane zone were not distinguishable. Alpine and subalpine vegetation distributions mapped with Landsat TM and landscape variables agreed favourably with a map of dominant vegetation communities prepared from field observations and large-scale colour and colour–infrared aerial photographs. A later study used an integrated remote sensing–GIS approach to derive critical factors that influence vegetation distribution (Peddle and Duguay 1995). The factors included precipitation, temperature, wind, soil moisture, and snow accumulation. Data were culled from a 16-year archive of digital Landsat imagery, a DEM, and meteorological station data. Five climatic indices were computed: an orogenic precipitation index, a slope–aspect index, a snow probability index, an insolation index, and a growing degree-days index. These indices were tested individually and together with a Landsat TM image and topographic measures from a DEM to assess the signi-

Table 6.9 Pairwise correlation coefficients between 'vegetation cover' and topographic attributes in four Kazakhstan study areas

Topographic attributes	Sekisovka	Berezovka	Ust-Feklistka	Chistopolka
Elevation	**−0.36**	**0.68**	**−0.53**	−0.10
Slope inclination	−0.10	**0.70**	−0.01	0.00
Slope aspect	−0.15	**−0.40**	0.22	**0.54**
Slope curvature	**−0.38**	**0.35**	*−0.26*	0.00
Contour curvature	−0.28	**−0.40**	**−0.50**	−0.11
Mean land-surface curvature	**−0.43**	−0.04	**−0.46**	−0.06
Specific catchment area	0.35	*−0.31*	**0.46**	0.24
Topographic index	**0.51**	**−0.40**	**0.49**	0.13
Stream-power index	−0.11	**0.61**	−0.21	−0.07

Notes: Bold italics, $p = \leq 0.01$; italics, $p = \leq 0.05$.
Source: Adapted from Florinsky and Kuryakova (1996).

ficance of increasing the accuracy and precision of maximum likelihood land-cover classification with respect to the hierarchical Braun-Blanquet vegetation classification system. These results were favourable for the sensor resolutions and classification algorithms used in this complex environment.

Igor V. Florinsky and Galina A. Kuryakova (1996) examined the influence of six local, one non-local, and two combined topographic attributes on vegetation cover, altitude, and density in four areas of the Rudny Altai, Kazakhstan. Local topographic attributes were elevation, slope inclination, slope aspect, slope curvature, contour curvature, and mean land-surface curvature. The non-local topographic attribute was specific catchment area. The two combined topographic attributes were a topographic index, defined as the natural logarithm of the specific catchment area divided by the slope gradient, and a stream-power index, defined as the product of the specific catchment area and the slope gradient. The topographic attributes were calculated by applying a DEM to the digitised contours of 1:50 000-scale topographic maps. Vegetation cover was estimated from a field survey, topographic maps, and aerial scenes. It was characterised according to a rank scale that ranged from field to forest. Four study areas, each 4 × 4 km, were chosen – the Sekisovka area, the Berezovka area, the Ust-Feklistka area, and the Chistopolka area. The results showed a strong dependence of vegetation cover upon topographic attributes, though the most important attributes varied from one study area to another (Table 6.9). Patterns of correlation between vegetation cover and topographic attributes varied considerably between the study areas. Elevation was a highly significant variable at all sites, save Chistopolka,

reflecting altitudinal zonation of vegetation. The correlation is highest in Berezovka where the human impact on vegetation is minimal. Slope inclination did not significantly explain vegetation except in Berezovka, where the correlation coefficient of 0.70 was the highest value in the data set. Slope curvature, contour curvature, and mean land-surface curvature were all significant variables at two sites, though not the same two. Specific catchment area affected all sites and, although it was only highly significant at Ust–Feklistka ($p \leq 0.01$) and significant at Berezovka ($p \leq 0.05$), it was close to the 95 per cent confidence level ($p = 0.06$) at the other two sites. Likewise, the topographic index was highly significant at Sekisovka, Berezovka, and Ust-Feklistka. The stream-power index was highly significantly correlated with vegetation cover at Berezovka, but was not significant at the other three sites. An interesting finding of this study was that two composite topographic attributes – the specific catchment area and the topographic index – explained a fair percentage of the variation in vegetation cover in all sites. Single topographic attributes were significant in some sites, but there was no consistent pattern. However, at least one measure of land-surface curvature was significant at most sites. Indeed, most of the topographic attributes, single and composite, that influence vegetation in this area – land-surface curvatures, specific catchment area, and topographic index – regulated the flow and storage of water, which in turn was a master factor controlling vegetation distribution.

The abundance of some mammals is related to landscape properties. In Missouri, USA, raccoons (*Procyon lotor*), opossums (*Didelphis virginiana*), and striped skunks (*Mephitis mephitis*) prey on forest songbird eggs and nestlings (Dijak and Thompson 2000). Predation risk varies with the abundance of these three species. The predation risk was examined by studying predator abundances at local and landscape scales. Abundances were measured in 25 Missouri counties that took in the northern landscapes dominated by cropland, the south-western landscapes dominated by grassland, and the south-eastern landscapes dominated by forest. Fourteen landscape attributes, largely derived from digitised maps using FRAGSTATS software (McGarigal and Marks 1995), were used to characterise the landscapes produced by forest and agricultural habitats. Multiple linear regression established relationships between species abundances and the landscape attributes. Several landscape attributes were eventually dropped from the analysis owing to multicollinearity. The best predictive regression equation for raccoon abundance, $y_{raccoon}$, was

$$y_{raccoon} = -4.463 + 0.127 \text{ latitude}$$
$$+ 0.097 \text{ stream density}$$
$$+ 1.0E–05 \text{ agriculture mean patch size}$$
$$(r^2 = 0.62, n = 25, p < 0.001) \qquad (6.2)$$

Stream density and latitude explained most of the variation. The best regression equation for opossums, $y_{opossum}$, was

$$y_{opossum} = -5.516 + 0.98 \text{ stream density}$$
$$+ 0.0011 \text{ forest mean nearest-neighbour}$$
$$\text{distance} + 0.155 \text{ latitude} - 0.011 \text{ contagion}$$
$$(r^2 = 0.69, n = 25, p < 0.001) \qquad (6.3)$$

Striped skunk abundance was not significantly predicted by any combination of landscape attributes.

At a local scale, raccoons were more abundant along agricultural and riparian edges than in forest interiors, but their abundance did not differ between forest interior and road or clear-cut edges. Opossum abundance did not differ in forest interior or agricultural, road, or clear-cut edges. In summary, raccoons were more abundant in agricultural landscapes with high stream densities than in forested landscapes with low stream densities. This partly resulted from the raccoons' fondness for foraging along stream courses. Higher raccoon abundance at higher latitudes in Missouri may be explained by landscape characteristics that were not directly measured. Pasture and cropland were not differentiated: pastures and hayfields dominate agricultural landscapes in the southern part of the State while cropland under corn and soybean dominate the northern part of the State. Latitude also affected opossum abundance, perhaps reflecting the opossum's liking for cropland or dislike of grasslands.

An important point to emerge from this study is that findings vary with the scale of investigation. Such local features as proximity to some edge types, as well as such large-scale factors as landscape patterns in land-use, may affect predator abundance and potentially songbird nest-predation rates.

The latest investigations are rather sophisticated. A good example used a geographic information system (GIS) approach in conjunction with forest-plot data to develop an integrated moisture index (IMI), which was then used to predict forest productivity (site index) and species composition for Vinton Furnace Experimental Forest in Vinton County, Ohio, USA (Iverson *et al.* 1996, 1997). The prediction used the correlation between abundance of tree species, as measured by stand basal area, and soil and topographic factors. The causal connection worked mainly through pedological and topographic influences on the soil-water storage. Ohio forests are typical of eastern hardwoods across the Midwest and southern Appalachians in that slope exposure and position (rather than elevation) vary considerably over small areas and strongly influence many ecosystem processes. The area is part of the Allegheny Plateau. The terrain is dissected and the total relief is less than 100 m. The study was designed to do three things: (1) to evaluate the quality of DEM data in deriving topographic and moisture indices at four scales of resolution; (2) to create an easily produced model for estimating the integrated moisture index from DEM and soils data; and (3) to test the value of an integrated moisture index, as calibrated from field data, to estimate an oak site index and tree species composition of the forest that will follow the removal of the 100-year old, oak-dominated canopy growing at present.

Elevational contours, soil series mapping units, and plot locations were digitised and gridded to 7.5-m cells for GIS modelling. Several landscape features (a slope-aspect shading index, cumulative flow of water downslope, curvature of the landscape, and water-holding capacity of the soil) were used to create a single integrated moisture index (Figure 6.29). The integrated moisture index was then statistically related to site-index values and composition data for forest land harvested in the past 30 years. In particular, an oak site-index and the percentage composition of two major species groups in the region – oak (*Quercus* spp.), and yellow poplar (*Liriodendron tulipifera*) plus black cherry (*Prunus serotina*) – were estimated. The derived statistical relationships were then applied in the GIS to create maps of site index and composition, and verified with independent data.

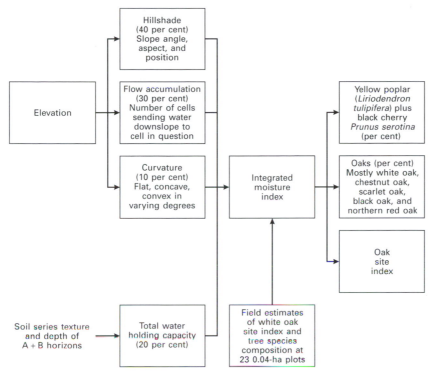

Figure 6.29 Flow diagram of the GIS process to derive an integrated moisture index and compositional maps of forests. (Adapted from Iverson *et al.*, 1997.)

The maps showed that oaks will dominate on dry, ridge-top positions (where the site-index is low), while the yellow poplar and black cherry will predominate on mesic sites (where the site-index is high).

The GIS-derived integrated moisture index approach is simple and easily transferred to other areas. It has the advantage of using readily available GIS information that does not change significantly with time and that requires no time-consuming and expensive fieldwork. The integrated moisture index has practical applications. As Louis R. Iverson and his colleagues (1997) explained, it may be used to improve the management of forests where moisture is a limiting factor, and to predict change in such forests under assorted forms of ecosystem management.

More generally, GIS models are being used, with statistical packages, to estimate regional biodiversity (e.g. Iverson and Prasad 1998a), and to predict changes of plant abundance and composition resulting from climatic change, and especially the likely changes that will occur as the globe warms up through the twenty-first century (e.g. Iverson and Prasad 1998b).

References

Allainé, D., Rodrigue, I., Le Berre, M., and Ramousse, R. (1994) Habitat preferences of alpine marmots, *Marmota marmota*. *Canadian Journal of Zoology* **72**, 2193–8.

Allainé, D., Graziani, L., and Coulon, J. (1998) Postweaning mass gain in juvenile alpine marmots *Marmota marmota*. *Oecologia* **113**, 370–6.

Anderson, D. G. and Bliss, L. C. (1998) Association of plant distribution patterns in microenvironments on patterned ground in a polar desert, Devon Island, N.W.T., Canada. *Arctic and Alpine Research* **30**, 97–107.

Anderson, S. H., Mann, K., and Shugart, H. H. (1977) The effect of transmission-line corridors on bird populations. *American Midland Naturalist* **97**, 216–21.

Andrén, H. (1995) Effects of landscape composition on predation rates at habitat edges. In L. Hansson, L. Fahrig, and G. Merriam (eds) *Mosaic Landscapes and Ecological Processes*, pp. 225–55. New York: Chapman & Hall.

Andrén, H. and Angelstam, P. (1988) Elevated predation rates as an edge effect in habitat islands: experimental evidence. *Ecology* **69**, 544–7.

Andrén, H. and Angelstam, P. (1993) Moose browsing on Scots pine in relation to stand size and distance to forest edge. *Journal of Applied Ecology* **30**, 133–42.

Armand, A. D. (1992) Sharp and gradual mountain timberlines as a result of species interaction. In A. J. Hansen and F. di Castri (eds) *Landscape Boundaries: Consequences for Biotic Diversity and Ecological Flows* (Ecological Studies, Vol. 92), pp. 360–78. New York: Springer.

Arnold, G. W., Weeldenburg, J. R., and Steven, D. E. (1991) Distribution and abundance of two species of kangaroo in remnants of native vegetation in the central wheatbelt of Western Australia and the role of native vegetation along road verges and fencelines as linkages. In D. A. Saunders and R. J. Hobbs (eds) *Nature Conservation 2: The Role of Corridors*, pp. 273–80. Chipping Norton, Australia: Surrey Beatty & Sons.

Arnold, G. W., Steven, D. E., Weeldenburg, J. R., and Smith, E. A. (1993) Influences of remnant size, spacing pattern and connectivity of population boundaries and demography in euros *Macropus robustus* living in a fragmented landscape. *Biological Conservation* **64**, 219–30.

Arnold, G. W., Steven, D. E., and Weeldenburg, J. R. (1994) Comparative ecology of western grey kangaroos (*Macropus fuliginosus*) and euros (*M. robustus erubescens*) in Durokoppin Nature Reserve, isolated in the wheatbelt of Western Australia. *Wildlife Research* **21**, 307–22.

Arnold, G. W., Weeldenburg, J. R., and Ng, V. M. (1995) Factors in the distribution and abundance of Western grey kangaroos (*Macropus fuliginosus*) and euros (*M. robustus*) in a fragmented landscape. *Landscape Ecology* **10**, 65–74.

Ashton, D. H. and Kelliher, K. J. (1996) Effects of forest soil dessication on the growth of *Eucalyptus regnans* F. Muell. seedlings. *Journal of Vegetation Science* **7**, 487–96.

Bennett, A. F. (1990) Habitat corridors and the conservation of small mammals in a fragmented forest environment. *Landscape Ecology* **4**, 109–22.

Bennett, A. F. (1991) Roads, roadsides and wildlife conservation: a review. In D. A. Saunders and R. J. Hobbs (eds) *Nature Conservation 2: The Role of Corridors*, pp. 99–118. Chipping Norton, Australia: Surrey Beatty & Sons.

Bennett, A. F., Henein, K., and Merriam, G. (1994) Corridor use and the elements of corridor quality: chipmunks and fencerows in a farmland mosaic. *Biological Conservation* **68**, 155–65.

Billings, W. D. (1954) Temperature inversions in the piñon–juniper zone of a Nevada mountain range. *Butler University Botanical Studies* **11**, 112–18.

Billings, W. D. (1990) The mountain forests of North America and their environments. In C. B. Osmond, L. F. Pitelka, and G. M. Hidy (eds) *Plant Biology of the Basin and Range* (Ecological Studies, Vol. 80), pp. 47–86. Berlin: Springer.

Bliss, L. C. (1971) Arctic and alpine plant life cycles. *Annual Review of Ecology and Systematics* **2**, 405–38.

Bowne, D. R., Peles, J. D., and Barrett, G. W. (1999) Effects of landscape spatial structure on movement patterns of the hispid cotton rat (*Sigmodon hispidus*). *Landscape Ecology* **14**, 53–65.

Braun-Blanquet, J. (1932) *Plant Sociology: The Study of Plant Communities*. Authorised translation of *Pflanzensoziologie*. Translated, revised and edited by George D. Fuller and Henry S. Conard. New York and London: McGraw-Hill.

Bright, P. W. (1998) Behaviour of specialist species in habitat corridors: arboreal dormice avoid corridor gaps. *Animal Behaviour* **56**, 1485–90.

Bunnell, S. D. and Johnson, D. R. (1974) Physical factors affecting pika density and dispersal. *Journal of Mammalogy* **55**, 866–9.

Burel, F. and Baudry, J. (1990) Hedgerow network patterns and processes in France. In I. S. Zonneveld and R. T. T. Forman (eds) *Changing Landscapes: An Ecological Perspective*, pp. 99–120. New York: Springer.

Burhans, D. E. and Thompson, F. R., III (1999) Habitat patch size and nesting success of yellow-breasted chats. *Wilson Bulletin* **111**, 210–15.

Cantlon, J. E. (1953) Vegetation and microclimates on north and south slopes of Cushetunk Mountain, New Jersey. *Ecological Monographs* **23**, 241–70.

Clergeau, P. and Burel, F. (1997) The role of spatio-temporal patch connectivity at the landscape level: an example in a bird distribution. *Landscape and Urban Planning* **38**, 37–43.

Clifford, H. T. (1959) Seed dispersal by motor vehicles. *Journal of Ecology* **47**, 311–15.

Cloudsley-Thompson, J. L. (1951) On the responses to environmental stimuli, and the sensory physiology of millepedes (Diplopoda). *Proceedings of the Zoological Society of London* **121**, 253–77.

Collinge, S. K. (1998) Spatial arrangement of habitat patches and corridors: clues from ecological field experiments. *Landscape and Urban Planning* **42**, 157–68.

Cone, C. D., Jr (1962) Thermal soaring of birds. *American Scientist* **50**, 180–209.

Corbit, M., Marks, P. L., and Gardescu, S. (1999) Hedgerows as habitat corridors for forest herbs in central New York, USA. *Journal of Ecology* **87**, 220–32.

Crome, F. H. J. and Richards, G. C. (1988) Bats and gaps: microchiropteran community structure in a Queensland rainforest. *Ecology* **69**, 1960–9.

Daniels, R. J. R. (1992) Geographical distribution patterns of amphibians in the Western Ghats, India. *Journal of Biogeography* **19**, 521–9.

Daubenmire, R. (1954) Alpine timberlines in the Americas and their interpretation. *Butler University Botanical Studies* **11**, 119–36.

Degórski, M. L. (1994) Analysis of a catenary complex of soils and forest plant communities in the South Mazovia Uplands (Central Poland). *Ekologia Polska* **42**, 263–88.

Diamond, J. M. (1973) Distributional ecology of New Guinea birds. *Science* **179**, 759–69.

Dijak, W. D. and Thompson, F. R., III (2000) Landscape and edge effects on the distribution of mammalian predators in Missouri. *Journal of Wildlife Management* **64**, 209–16.

Dixon, M. D. and Johnson, W. C. (1999) Riparian vegetation along the Middle Snake River, Idaho: zonation, geographical trends, and historical changes. *Great Basin Naturalist* **59**, 18–34.

Donovan, T. M., Thompson, F. R., III, Faaborg, J., and Probst, J. (1997) Reproductive success of neotropical migrant birds in habitat sources and sinks. *Conservation Biology* **8**, 1380–95.

Dos Anjos, L. and Boçon, R. (1999) Bird communities in natural forest patches in southern Brazil. *Wilson Bulletin* **111**, 397–414.

Druitt, D. G., Enright, N. J., and Ogden, J. (1990) Altitudinal zonation in the mountain forests of Mt Hauhungatahi, North Island, New Zealand. *Journal of Biogeography* **17**, 205–20.

Ehrenfeld, J. G. (1995a) Microtopography and vegetation in Atlantic white cedar swamps: the effects of natural disturbances. *Canadian Journal of Botany* **73**, 474–84.

Ehrenfeld, J. G. (1995b) Microsite differences in surface substrate characteristics in *Chamaecyparis* swamps of the New Jersey Pinelands. *Wetlands* **15**, 183–9.

Ferreyra, M., Cingolani, A., Ezcurra, C., and Bran, D. (1998a) High-Andean vegetation and environmental gradients in northwestern Patagonia, Argentina. *Journal of Vegetation Science* **9**, 307–16.

Ferreyra, M., Clayton, S., and Ezcurra, C. (1998b) La flora altoandina de los sectores este y oeste del Parque Nacional Nahuel Huapi, Argentina. *Darwiniana* **36**, 65–79.

Finney, H. R., Holowaychuk, N., and Heddleson, M. R. (1962) The influence of microclimate on the morphology of certain soils of the Allegheny Plateau of Ohio. *Soil Science Society of America Proceedings* **26**, 287–92.

Florinsky, I. V. and Kuryakova, G. A. (1996) Influence of topography on some vegetation cover properties. *Catena* **27**, 123–41.

Forman, R. T. T. (1991) Landscape corridors: from theoretical foundations to public policy. In D. A. Saunders and R. J. Hobbs (eds) *Nature Conservation 2: The Role of Corridors*, pp. 71–84. Chipping Norton, Australia: Surrey Beatty & Sons.

Forman, R. T. T. (1995) *Land Mosaics: The Ecology of Landscapes and Regions*. Cambridge: Cambridge University Press.

Forman, R. T. T., Galli, A. E., and Leck, C. F. (1976) Forest size and avian diversity in New Jersey woodlots with some land use implications. *Oecologia* **26**, 1–8.

Forseth, I. N. and Ehrelinger, J. R. (1982) Ecophysiology of two solar tracking desert winter annuals. *Oecologia (Berlin)* **54**: 41–9.

Fortin, D. and Arnold, G. W. (1997) The influence of road verges on the use of nearby small shrubland remnants by birds in the central wheatbelt of Australia. *Wildlife Research* **24**, 679–89.

Frank, T. D. (1988) Mapping dominant vegetation communities in the Colorado Rocky Mountain Front Range with Landsat Thematic Mapper and digital terrain data. *Photogrammetric Engineering and Remote Sensing* **54**, 1727–34.

French, D. D., Miller, G. R., and Cummins, R. P. (1997) Recent development of high-altitude *Pinus sylvestris* scrub in the northern Cairngorm Mountains, Scotland. *Biological Conservation* **79**, 133–44.

Fuller, R. J. (1988) A comparison of breeding bird assemblages in two Buckinghamshire clay vale woods with different histories of management. In K. J. Kirby and F. J. Wright (eds) *Woodland Conservation and Research in the Clay Vale of Oxfordshire and Buckinghamshire*, pp. 53–65 (Research and Survey in Nature Conservation No. 15). Peterborough: Nature Conservancy Council.

Fuller, R. J. (1992) Effects of coppice management on woodland breeding birds. In G. P. Buckley (ed.) *Ecology and Management of Coppice Woodlands*, pp. 169–92. London: Chapman & Hall.

Galli, A. E., Leck, C. F., and Forman, R. T. T. (1976) Avian distribution patterns in forest islands of different sizes in central New Jersey. *Auk* **93**, 356–64.

García, F. J., Díaz, M., de Alba, J. M., Alonso, C. L., Carbonell, R., de Carrión, M. L., Monedero, C., and Santos, T. (1998) Edge effects and patterns of winter abundance of wood mice *Apodemus sylvaticus* in Spanish fragmented forests. *Acta Theriologica* **43**, 225–62.

Gardner, A. S. and Fisher, M. (1996) The distribution and status of the montane juniper woodlands of Oman. *Journal of Biogeography* **23**, 791–803.

Gates, B. D. and Gysel, L. W. (1978) Avian nest dispersion and fledgling success in field–forest ecotones. *Ecology* **59**, 871–83.

Gates, J. E. (1991) Powerline corridors, edge effects, and wildlife in forested landscapes of the Central Appalachians. In J. E. Rodiek and E. G. Bolen (eds) *Wildlife and Habitats in Managed Landscapes*, pp. 13–32. Washington, DC: Island Press.

Gerrard, A. J. (1990) *Mountain Environments: An Examination of the Physical Geography of Mountains*. London: Belhaven Press.

Getz, L. L., Cole, F. R., and Gates, D. L. (1978) Interstate roadsides as dispersal routes for *Microtus pennsylvanicus*. *Journal of Mammalogy* **59**, 208–12.

Giblin, A. E., Nadelhoffer, K. J., Shaver, G. R., Laundre, J. A., and McKerrow, A. J. (1991) Biogeochemical diversity along a riverside toposequence in Arctic Alaska. *Ecological Monographs* **61**, 415–35.

Godfryd, A. and Hansell, R. I. C. (1986) Predictions of bird-community metrics in urban woodlots. In J. Verner, M. L. Morrison, and C. J. Ralph (eds) *Wildlife 2000: Modeling Habitat Relationships of Terrestrial Vertebrates*, pp. 321–6. Madison, WI: University of Wisconsin Press.

Goldblum, D. and Beatty, S. W. (1999) Influence of an old field/forest edge on a northeastern United States deciduous forest understorey community. *Journal of the Torrey Botanical Society* **126**, 335–43.

Greenberg, C. H. and McNab, W. H. (1998) Forest disturbance in hurricane-related downbursts in the Appalachian mountains of North Carolina. *Forest Ecology and Management* **104**, 179–91.

Grubb, P. J. (1971) Interpretation of the 'Massenerhebung' effect on tropical mountains. *Nature* **229**, 44–6.

Hack, J. T. and Goodlett, J. C. (1960) *Geomorphology and forest ecology of a mountain region in the central Appalachians*. US Geological Survey Professional Paper 347. Washington, DC: US Government Printing Office.

Hamazaki, T. (1996) Effects of patch shape on the number of organisms. *Landscape Ecology* **11**, 299–306.

Hansson, L. (2000) Edge structures and edge effects on plants and birds in ancient oak–hazel woodlands. *Landscape and Urban Planning* **46**, 203–7.

Harms, W. B. and Opdam, P. (1990) Woods as habitat patches for birds: application in landscape planning in The Netherlands. In I. S. Zonneveld and R. T. T. Forman (eds) *Changing Landscapes: An Ecological Perspective*, pp. 73–97. New York: Springer.

Harper, J. L. and Sagar, G. R. (1953) Some aspects of the ecology of buttercups in permanent grassland. *Proceedings of the British Weed Control Conference 1958*, pp. 256–65. London: British Weed Control Council.

Harris, L. D. (1984) *The Fragmented Forest: Island Biogeography Theory and the Preservation of Biotic Diversity*. With a Foreword by Kenton R. Miller. Chicago and London: Chicago University Press.

Harris, L. D., Maser, C., and McKee, A. (1982) Patterns of old growth harvest and implications for Cascades wildlife. *Transactions of the North American Wildlife and Natural Resources Conf.* **47**, 374–92.

Heikkinen, O., Obrębska-Starkel, B., and Tuhkanen, S. (1995) Introduction: the timberline – a changing battlefront. *Zeszyty Naukowe Universytetu Jagiellonskiego, Prace Geograficzne* **98**, 7–16.

Helliwell, D. R. (1976) The effects of size and isolation on the conservation value of wooded sites in Britain. *Journal of Biogeography* **3**, 407–16.

Henein, K., Wegner, J., and Merriam, G. (1998) Population effects of landscape model manipulation on two behaviourally different woodland small mammals. *Oikos* **81**, 168–86.

Hinsley, S. A., Bellamy, P. E., Enoksson, B., Fry, G., Gabrielsen, L., McCollin, D., and Schotman, A. (1998) Geographical and land-use influences on bird species richness in small woods in agricultural landscapes. *Global Ecology and Biogeography Letters* **7**, 125–35.

Holland, P. G. and Steyn, D. G. (1975) Vegetational responses to latitudinal variations in slope angle and aspect. *Journal of Biogeography* **2**, 179–83.

Horrocks, M. and Ogden, J. (1998) The effects of the Tampo tephra eruption of c. 1718 BP on the vegetation of Mount Hauhungatahi, central North Island, New Zealand. *Journal of Biogeography* **25**, 649–60.

Horrocks, M. and Ogden, J. (2000) Evidence for Lateglacial and Holocene tree-line fluctuations from pollen diagrams from the subalpine zone on Mount Hauhungatahi, Tongariro National Park, New Zealand. *The Holocene* **10**, 61–73.

Hsieh, C.-F., Chen, Z.-S., Hsu, Y.-M., Yang, K.-C., and Hsieh, T.-H. (1998) Altitudinal zonation of evergreen broad-leaved forest on Mount Lopei, Taiwan. *Journal of Vegetation Science* **9**, 201–12.

Hughes, F. M. R. (1988) The ecology of African floodplain forests in semi-arid and arid zones: a review. *Journal of Biogeography* **15**, 127–40.

Hustich, I. (1979) Ecological concepts and biogeographical zonation in the North: the need for a generally accepted terminology. *Holarctic Ecology* **2**, 208–17.

Inger, R. F., Shaffer, H. B., Koshy, M., and Badke, R. (1987) Ecological structure of a herpetological assemblage in south India. *Amphibia–Reptilia* **8**, 189–202.

Isard, S. A. (1986) Factors influencing soil moisture and plant community distributions on Niwot Ridge, Front Range, Colorado, USA. *Arctic and Alpine Research* **18**, 83–96.

Iverson, L. R. and Prasad, A. (1998a) Estimating regional plant biodiversity with GIS modelling. *Diversity and Distribution* **4**, 49–61.

Iverson, L. R. and Prasad, A. (1998b) Predicting abundance of 80 tree species following climate change in the eastern United States. *Ecological Monographs* **68**, 465–85.

Iverson, L. R., Scott, C. T., Dale, M. E., and Prasad, A. (1996) Development of an integrated moisture index for predicting species composition. In M. Kohl and G. Z. Gertner (eds) *Caring for the Forest: Research in a Changing World. Statistics, Mathematics and Computers* (Proceedings of the International Union of Forestry Research Organizations S4.11-00, Tamper, 1995), pp. 101–16. Birmensdorf, Switzerland: Swiss Federal Institute for Forest, Snow and Landscape Research.

Iverson, L. R., Dale, M. E., Scott, C. T., and Prasad, A. (1997) A GIS-derived integrated moisture index to predict forest composition and productivity of Ohio forests (U.S.A.). *Landscape Ecology* **12**, 331–48.

Ives, J. D. (1992) The Andes: geoecology of the Andes. In P. B. Stone (ed.) *The State of the World's Mountains: A Global Report*, pp. 185–256. London and New Jersey: Zed Books.

Kershaw, K. A. (1964) *Quantitative and Dynamic Ecology*. London: Edward Arnold.

Kitayama, K. (1992) An altitudinal transect study of the vegetation on Mount Kinabalu, Borneo. *Vegetatio* **102**, 149–71.

Knaapen, J. P., Scheffer, M., and Harms, W. B. (1992) Estimating habitat isolation in landscape planning. *Landscape and Urban Planning* **23**, 1–16.

Knight, R. L. and Kawashima, J. Y. (1993) Responses of raven and red-tailed hawk populations to linear right-of-ways. *Journal of Wildlife Management* **57**, 266–71.

Knutson, R. (1987) *Flattened Fauna: A Field Guide to Common Animals of Roads, Streets and Highways*. Berkeley, CA: Ten Speed Press.

Korb, J. and Linsenmair, K. E. (1998) Experimental heating of *Macrotermes bellicosus* (Isoptera, Macrotermitinae) mounds: what role does microclimate play in influencing mound architecture? *Insectes Sociaux* **45**, 335–42.

Lapointe, S. L., Serrano, M. S., and Jones, P. G. (1998) Microgeographic and vertical distribution of *Acromyrmex landolti* (Hymenoptera: Formicidae) nests in a neotropical savanna. *Environmental Entomology* **27**, 636–41.

Leuschner, C. (1996) Timberline and alpine vegetation on the tropical and warm-temperate oceanic islands of the world: elevation, structure and floristics. *Vegetatio* **123**, 193–206.

Leuschner, C. (1998) Vegetation an der Waldgrenze auf tropischen und subtropischen Inseln. *Geographische Rundschau* **50**, 690–7.

Levenson, J. B. (1981) Woodlots as biogeographic islands in southeastern Wisconsin. In R. L. Burgess and D. M. Sharpe (eds) *Forest Island Dynamics in Man-dominated Landscapes*, pp. 13–39. New York: Springer.

Lidicker, W. Z., Jr (1999) Responses of mammals to habitat edges: an overview. *Landscape Ecology* **14**, 333–43.

Lindenmayer, D. B., Mackey, B. G., Cunningham, R. B., Donnelly, C. F., Mullen, I. C., McCarthy, M. A., and Gill, A. M. (2000) Factors affecting the presence of cool temperate rain forest tree myrtle (*Nothofagus cunninghamii*) in southeastern Australia: integrating climatic, terrain and disturbance predictors of distribution patterns. *Journal of Biogeography* **27**, 1001–9.

Lynch, J. F. and Saunders, D. A. (1991) Responses of bird species to habitat fragmentation in the wheatbelt of Western Australia: interiors, edges and corridors. In D. A. Saunders and R. J. Hobbs (eds) *Nature Conservation 2: The Role of Corridors*, pp. 143–58. Chipping Norton, Australia: Surrey Beatty & Sons.

Mack, R. N. (1981) Invasion of *Bromus tectorum* L. into western North America: an ecological chronicle. *Agro-Ecosystems* **7**, 145–65.

Mader, H.-J. (1984) Animal habitat isolation by roads and agricultural fields. *Biological Conservation* **29**, 81–96.

Mader, H. J., Schell, C., and Kornacker, P. (1990) Linear barriers to arthropod movements in the landscape. *Biological Conservation* **54**, 209–22.

Manson, R. H., Ostfeld, R. S., and Canham, C. D. (1999) Responses of a small mammal community to heterogeneity along forest–old-field edges. *Landscape Ecology* **14**, 355–67.

Martin, A. A. and Tyler, M. J. (1978) The introduction into western Australia of the frog *Limnodynastes tasmaniensis*. *Australian Zoologist* **19**, 320–44.

Matthews, A., Dickman, C. R., and Major, R. E. (1999) The influence of fragment size and edge on nest predation in urban bushland. *Ecography* **22**, 349–56.

Matthiae, P. E. and Stearns, F. (1981) Mammals in forest islands in southwestern Wisconsin. In R. L. Burgess and D. M. Sharpe (eds) *Forest Island Dynamics in Mandominated Landscapes*, pp. 55–66. New York: Springer.

McCollin, D. (1993) Avian distribution patterns in a fragmented wooded landscape (North Humberside, U.K.): the role of between-patch and within-patch structure. *Global Ecology and Biogeography Letters* **3**, 48–62.

McCollin, D. (1998) Forest edges and habitat selection in birds: a functional approach. *Ecography* **21**, 247–60.

McGahan, J. (1968) Ecology of the golden eagle. *Auk* **85**, 1–12.

McGarigal, K. and Marks, B. J. (1995) *FRAGSTATS: Spatial Pattern Analysis Program for Quantifying Landscape Structure*. US Forest Service General Technical Report PNW-GTR-351. Portland, OR: Pacific Northwest Research Station, USDA Forest Service.

Meiners, S. J. and Pickett, S. T. A. (1999) Changes in community and population responses across a forest–field gradient. *Ecography* **22**, 261–7.

Melquist, W. E. and Hornocker, M. G. (1983) Ecology of river otters in west central Idaho. *Wildlife Monographs* **83**, 5–60.

Menalled, F. D., Marino, P. C., Renner, K. A., and Landis, D. A. (2000) Post-dispersal weed seed predation in Michigan crop fields as a function of agricultural landscape structure. *Agriculture, Ecosystems and Environment* **77**, 193–202.

Merriam, C. H. (1894) Laws of temperature control of the geographic distribution of terrestrial animals and plants. *National Geographic Magazine* **6**, 229–38.

Merriam, G. and Lanoue, A. (1990) Corridor use by small mammals: field measurement for three experimental types of *Peromyscus leucopus*. *Landscape Ecology* **4**, 123–31.

Morkill, A. E. and Anderson, S. H. (1991) Effectiveness of marking powerlines to reduce sandhill crane collisions. *Wildlife Society Bulletin* **19**, 442–9.

Mosher, J. A. and White, C. M. (1976) Directional exposure of golden eagle nests. *Canadian Field Naturalist* **90**, 356–9.

Mowbray, T. B. and Oosting, H. J. (1968) Vegetation gradients in relation to environment and phenology in a southern Blue Ridge gorge. *Ecological Monographs* **38**, 309–44.

Mueller-Dombois, D. and Bridges, K. W. (1981) Introduction. In D. Mueller-Dombois, K. W. Bridges, and H. L. Carson (eds) *Island Ecosystems: Biological Organization in Selected Hawaiian Communities* (US/IBP Synthesis Series, Vol. 15), pp. 35–76. Stroudsberg, PA, and Woods Hole, MA: Hutchinson Ross.

Mueller-Dombois, D., *et al.* (1981) Altitudinal distribution of organisms along an island mountain transect. In D. Mueller-Dombois, K. W. Bridges, and H. L. Carson (eds) *Island Ecosystems: Biological Organization in Selected Hawaiian Communities* (US/IBP Synthesis Series, Vol. 15), pp. 77–180. Stroudsberg, PA, and Woods Hole, MA: Hutchinson Ross.

Mullin, B. H. (1998) The biology and management of purple loosestrife (*Lythrum salicaria*). *Weed Technology* **12**, 397–401.

Murcia, C. (1995) Edge effects in fragmented forest: implications for conservation. *Trends in Ecology and Evolution* **10**, 58–62.

Neilson, R. P. and Wullstein, L. H. (1983) Biogeography of two southwest American oaks in relation to atmospheric dynamics. *Journal of Biogeography* **10**, 275–97.

Neilson, R. P. and Wullstein, L. H. (1986) Microhabitat affinities of Gambel oak seedlings. *Great Basin Naturalist* **46**, 294–8.

Newbey, B. J. and Newbey, K. R. (1987) Bird dynamics of Foster Road reserve, near Ongerup, Western Australia. In D. A. Saunders, G. W. Arnold, A. A. Burbidge, and A. J. M. Hopkins (eds) *Nature Conservation: The Role of Remnants of Native Vegetation*, pp. 341–3. Chipping Norton, Australia: Surrey Beatty & Sons.

Noss, R. (1993) Wildlife corridors. In D. S. Smith and P. C. Hellmund (eds) *Ecology of Greenways: Design and Function of Linear Conservation Areas*, pp. 43–68. Minneapolis, MN: University of Minneapolis: Press.

Ogle, C. C. (1987) The incidence and conservation of animal and plant species in remnants of native vegetation within New Zealand. In D. A. Saunders, G. W. Arnold, A. A. Burbidge, and A. J. M Hopkins (eds) *Nature Conservation: The Role of Remnants of Native Vegetation*, pp. 79–87. Chipping Norton, Australia: Surrey Beatty & Sons.

Ogle, C. C. and Wilson, P. R. (1985) Where have all the mistletoes gone? *Forest and Bird* **16**, 10–13.

Oksanen, L., Moen, J., and Helle, T. (1995) Timberline patterns in northernmost Fennoscandia: relative importance of climate and grazing. *Acta Botanica Fennica* **153**, 93–105.

Oliveira-Filho, A. T., Curi, N., Vilela, E. A., and Carvalho, D. A. (1997) Tree species distribution along soil catenas in a riverside semideciduous forest in southeastern Brazil. *Flora* **192**, 47–64.

Ortega, Y. K. and Capen, D. E. (1999) Effects of forest roads on habitat quality for ovenbirds in a forested landscape. *Auk* **116**, 937–46.

Osterkamp, W. R., Hupp, C. R., and Schening, M. R. (1995) Little River revisited – thirty-five years after Hack and Goodlett. *Geomorphology* **13**, 1–20.

Ouin, A., Paillat, G., Butet, A., and Burel, F. (2000) Spatial dynamics of wood mouse (*Apodemus sylvaticus*) in an agricultural landscape under intensive use in the Mont Saint Michel Bay (France). *Agriculture, Ecosystems and Environment* **78**, 159–65.

Oxley, D. J., Fenton, M. B., and Carmody, G. R. (1974) The effects of roads on populations of small mammals. *Journal of Applied Ecology* **11**, 51–9.

Pears, N. (1985) *Basic Biogeography*, 2nd edn. Harlow: Longman.

Pearson, S. M., Smith, A. B., and Turner, M. G. (1998) Forest patch size, land use, and mesic forest herbs in the French Broad River Basin, North Carolina. *Castanea* **63**, 382–95.

Peddle, D. R. and Duguay, C. R. (1995) Incorporating topographic and climatic GIS data into satellite image analysis of an alpine tundra ecosystem, Front Range, Colorado Rocky Mountains. *Geocarto International* **10**, 43–60.

Pennycuick, C. J. (1973) The soaring flight of vultures. *Scientific American* **229** (December), 102–9.

Pérez, F. L. (1995) A high-Andean toposequence: the geoecology of caulescent paramo rosettes. *Mountain Research and Development* **15**, 133–52.

Potzger, J. E. (1939) Microclimate and a notable case of its influence on a ridge in central Indiana. *Ecology* **20**, 29–37.

Redford, K. and Fonseca, G. (1986) The role of gallery forests in the zoogeography of the Cerrado's non-volant mammalian fauna. *Biotropica* **18**, 126–35.

Reh, W. (1989) Investigations into the influence of roads on the genetic structure of populations of the common frog *Rana temporaria*. In T. E. S. Langton (ed.) *Amphibians and Roads*, pp. 101–3. Shefford, Bedfordshire: ACO Polymer Products.

Reh, W. and Seitz, A. (1990) The influence of land use on the genetic structure of populations of the common frog *Rana temporaria*. *Biological Conservation* **54**, 239–49.

Reiner, R. and Griggs, T. (1989) TNC undertakes riparian restoration projects in California. *Restoration and Management Notes* **7**, 3–8.

Remillard, M. M., Gruendling, G. K., and Bogucki, D. J. (1987) Disturbance by beaver (*Castor canadensis* Kuhl) and increased landscape heterogeneity. In M. G. Turner (ed.) *Landscape Heterogeneity and Disturbance*, pp. 103–22 (Ecological Studies, Vol. 64). New York: Springer.

Richards, P. W. (1996) *The Tropical Rain Forest: An Ecological Study*. With contributions by R. P. D., 2nd edn. Cambridge: Cambridge University Press.

Roberts, C. and Norment, C. J. (1999) Effects of plot size and habitat characteristics on breeding success of scarlet tanagers. *Auk* **116**, 73–82.

St Hilaire, L. R. and Leopold, D. J. (1995) Conifer seedling distribution in relation to microsite conditions in a central New York forested minerotrophic peatland. *Canadian Journal of Forest Research* **25**, 261–9.

Saracco, J. F. and Collazo, J. A. (1999) Carolina bottomland hardwood forest. *Wilson Bulletin* **111**, 541–9.

Saunders, D. A. (1990) Problems of survival in an extensively cultivated landscape: the case of Carnaby's cockatoo, *Calyptorhynchus funereus latirostris*. *Biological Conservation* **54**, 277–90.

Schnitzler, A. (1994) European alluvial hardwood forests of large floodplains. *Journal of Biogeography* **21**, 605–23.

Shaver, G. R., Laundre, J. A., Giblin, A. E., and Nadelhoffer, K. J. (1996) Changes in live plant biomass, primary production, and species composition along a riverside toposequence in Arctic Alaska, U.S.A. *Arctic and Alpine Research* **28**, 363–79.

Singer, F. J. and Doherty, J. L. (1984) Movements and habitat use in an unhunted population of mountain goats (*Oreamnos americanus*). *Canadian Field Naturalist* **99**, 205–17.

Spencer, J. W. and Kirby, K. J. (1992) An inventory of ancient woodland for England and Wales. *Biological Conservation* **62**, 77–93.

Stevens, G. C. and Fox, J. F. (1991) The causes of treeline. *Annual Review of Ecology and Systematics* **22**, 177–91.

Stevens, S. M. and Husband, T. P. (1998) The influence of edge on small mammals: evidence from Brazilian Atlantic forest fragments. *Biological Conservation* **85**, 1–8.

Storm, G. L., Andrews, R. D., Phillips, R. L., Bishop, R. A., Siniff, D. B., and Tester, J. R. (1976) Morphology, reproduction, dispersal and mortality of midwestern red fox populations. *Wildlife Monographs* **49**, 5–82.

Stoutjesdijk, P. and Barkman, J. J. (1992) *Microclimate, Vegetation and Fauna*. Knivsta, Sweden: Opulus Press.

Suckling, G. C. (1984) Population ecology of the sugar glider *Petaurus breviceps* in a system of fragmented habitats. *Australian Wildlife Research* **11**, 49–75.

Swanson, D. K. (1996) Soil geomorphology on bedrock and colluvial terrain with permafrost in central Alaska, USA. *Geoderma* **71**, 157–72.

Terborgh, J. and Weske, J. S. (1975) The role of competition in the distribution of Andean birds. *Ecology* **56**, 562–76.

Thompson, D. Q., Stuckey, R. L., and Thompson, E. B. (1987) *Spread, Impact, and Control of Purple Loosestrife* (Lythrum salicaria) *in North American Wetlands*. Washington, DC: United States Department of Agriculture, Fish, and Wildlife Service.

Tinner, W., Ammann, B., and Germann, P. (1996) Treeline fluctuations recorded for 12 500 years by soil profiles, pollen, and plant macrofossils in the central Swiss Alps. *Arctic and Alpine Research* **28**, 131–47.

Troll, C. (1973a) The upper timberlines in different climatic zones. *Arctic and Alpine Research* **5**, 3–18.

Troll, C. (1973b) High mountain belts between the polar caps and the equator: their definition and lower limit. *Arctic and Alpine Research* **5**, 19–28.

Turesson, G. (1914) Slope exposure as a factor in the distribution of *Pseudotsuga taxifolia* in arid parts of Washington. *Bulletin of the Torrey Botanical Club* **41**, 337–45.

Unsworth, J. W., Kuck, L., Scott, M. D., and Garton, E. O. (1993) Elk mortality in the Clearwater drainage of north-central Idaho. *Journal of Wildlife Management* **57**, 495–502.

Vannote, R. L., Minshall, G. W., Cummins, K. W., Sedell, J. R., and Cushing, C. E. (1980) The river continuum concept. *Canadian Journal of Fisheries and Aquatic Sciences* **37**, 130–7.

Vaughan, T. A. (1954) Mammals of the San Gabriel Mountains of California. *University of Kansas Publications, Museum of Natural History* **7**, 513–82.

Veblen, T. T., Ashton, D. H., Schlegel, F. M., and Veblen, A. T. (1977) Plant succession in a timberline depressed by vulcanism in south-central Chile. *Journal of Biogeography* **4**, 275–94.

Vitt, D. H. and Slack, N. G. (1984) Niche diversification of *Sphagnum* relative to environmental factors in northern Minnesota peatlands. *Canadian Journal of Botany* **62**, 1409–30.

Wace, N. M. (1977) Assessment of dispersal of plant species – the car-borne flora in Canberra. *Proceedings of the Ecological Society of Australia* **10**, 167–86.

Wagner, J. and Reichegger, B. (1997) Phenology and seed development of the alpine sedges *Carex curvula* and *Carex firma* in response to contrasting topoclimates. *Arctic and Alpine Research* **219**, 291–9.

Walker, D. A. and Everett, K. R. (1991) Loess ecosystems of northern Alaska: regional gradient and toposequence at Prudhoe Bay. *Ecological Monographs* **61**, 437–64.

Walker, D. A., Binnian, E., Evans, B. M., Lederer, N. D., Nordstrand, E., and Webber, P. J. (1989) Terrain, vegetation and landscape evolution of the R4D research site, Brooks Range Foothills, Alaska. *Holarctic Ecology* **12**, 238–61.

Ward, J. P. and Anderson, S. H. (1988) Influences of cliffs on wildlife communities in southcentral Wyoming. *Journal of Wildlife Management* **52**, 673–8.

Ward, J. P. and Anderson, S. H. (1989) Evaluation of wildlife response to a retained mine highwall in south central Wyoming. *Great Basin Naturalist* **49**, 449–55.

Wardle, P. (1974) Alpine timberlines. In J. D. Ives and R. G. Barry (eds) *Arctic and Alpine Environments*, pp. 370–402. London: Methuen.

Wardle, P. (1993) Causes of alpine timberline: a review of the hypotheses. In J. Alden, J. L. Mastrantonio, and S. Ødum (eds) *Forest Development in Cold Climates* (NATO Advanced Science Institutes Series, Series A: Life Sciences, Vol. 244), pp. 89–103. New York and London: Plenum Press.

Weste, G. (1977) Future forests – to be or not to be? *Victoria's Resources* **19**, 26–7.

Wilcox, B. A. (1989) Migration and control of purple loosestrife (*Lythium salicaria* L.) along highway corridors. *Environmental Management* **13**, 365–70.

Woodell, S. R. J. (1974) Anthill vegetation in a Norfolk saltmarsh. *Oecologia* (*Berlin*) **16**, 221–5.

Wu, J. (1999) Hierarchy and scaling: extrapolating information along a scaling ladder. *Canadian Journal of Remote Sensing* **25**, 367–80.

Wu, J. and Levin, S. A. (1997) A patch-based spatial modeling approach: conceptual framework and simulation scheme. *Ecological Modelling* **101**, 325–46.

Wu, J. and Loucks, O. L. (1995) From balance of nature to hierarchical patch dynamics: a paradigm shift in ecology. *The Quarterly Review of Biology* **70**, 439–66.

Wu, J. and Vankat, J. L. (1991) An area-based model of species richness dynamics of forest islands. *Ecological Modelling* **58**, 249–71.

Wu, J., Vankat, J. L., and Barlas, Y. (1993) Effects of patch connectivity and arrangement on animal metapopulation dynamics: a simulation study. *Ecological Modelling* **65**, 221–54.

CHAPTER 7

Humans

Jean Bruhnes (1952, 36) once wondered what signs of human civilisation observers in a balloon or an aeroplane could see. If the observers were close to the ground, they could see humans themselves as 'a mobile covering of the surface, differing greatly in density at different points'. Although the covering is mobile, 'great patches of living humanity appear for a long time in the same place'. Wherever humans live, Bruhnes averred, six essential phenomena or 'facts' of human geography would be manifest. First, they could see human abodes – houses, shelters, habitations, and human constructions, or what Bruhnes dubbed the 'superficial excrescence'. Second, they could espy the routes linking the abodes, or the 'lines of passage sacrificed to movement'. Third and fourth, they could see cultivated fields and domesticated animals – the 'productive use of the land'. Fifth, they could see the physical evidence of 'economic plunder' – quarries, tip-heaps, smoke stacks, and so on. And sixth, they could see the 'facts of a destructive economy' – destroyed forests, battlefields, and, though perhaps not obvious from an aerial survey, depleted fisheries and recently extinct birds and beasts. These are Bruhnes's six essentials of human geography, each a visible expression of a material aspect of civilisation. Bruhnes realised that other 'facts' also exist, unseen by observers in the skies – human tools, for example, and economic, political, and social relationships.

Aerial survey has advanced almost beyond recognition since Bruhnes was writing, and discerning his 'essential facts' of human geography is now far easier. High-flying satellites can spy on any part of the Earth and pick out details. From the perspective of the present book, a cardinal question bears on the extent to which topography influences these 'facts'. Geographers have long recognised that topographical elements do wield an influence upon the location of settlements, roads, agriculture, industry, and so on.

The degree of this influence has always been debatable, and once was heatedly discussed in arguments for and against environmental determinism. Up to the middle of the twentieth century, school textbooks in particular pedalled many examples of crude environmental determinism with no mention of social, cultural, and political factors. Clement Cyril Carter's (1931) *Land-forms and Life: Short Studies on Topographical Maps* is a charmingly outdated example of this practice. From about 1950 onward, physical geography has, with a few exceptions, become more refined and sharply focused around a handful of paradigms, even if in doing so it has lost its core. Over the same time, human geography has become diffuse, a collection of several competing explanatory paradigms, their only common thread being a flirtation with post-modernism. In consequence, physical geography and human geography have an uneasy relationship, though interactions between humans and their environment would make a natural core for the geographical discipline (see Huggett and Robinson 1996).

This last chapter will investigate topographic influences on settlements and routes, agriculture, environmental destruction and conservation, and, finally, human individuals and populations. In doing so, it will show that, although crude environmental determinism may be dead, a modern environmental 'interactionalism' is alive and thriving.

Settlements

The nature of settlement

Human settlements are the places where people live – isolated dwellings, farmsteads, buildings, hamlets, villages, towns, cities, and metropolises. They are the physical manifestation of the social organisation of space (Roberts 1996). Before the advent of agriculture, hunting and gathering economies forced people to move with the seasons. Settlements were ephemeral and temporary, except where fishing was a primary source of food. Agriculture paved the way for permanent settlements. A few settlements, such as camps and caravan sites, still are temporary, but the great majority are permanent and have been for centuries or even millennia. The major centres of population in Greece have remained in the same location in many parts of the country: a few small sites, probably remnants of single-family farmsteads, appeared and disappeared but the great majority of settlements have stayed rooted to the same spot (Davis 1991). Settlements, especially larger ones, tend to persist because their form changes less quickly than their function, and human activities today often take place within buildings and streets that were generated by human needs decades or centuries ago (Roberts 1987, 6). Nonetheless, settlement abandonment is common. It may arise in times of crisis, witness the deserted medieval villages in England, or through a more gradual change in the environment. In Poland, an abandoned settlement was discovered in the spring of 1982, 7 km south-east of Elbląg, in fields adjacent to the village of Janów Pomorski (Jagodziński and Kasprzycka 1991). The settlement was formerly a flourishing trading port established on a once much-larger Lake Drużno that provided contact with the Baltic Sea. When access to the sea was cut off, about 3000 years ago, the town declined and was eventually deserted. Recent abandonment is also known. In the lower Yukon Delta, Alaska, two Eskimo villages were abandoned during the twentieth century owing to disasters (Griffin 1996). Pastuliarraq, reportedly the oldest and largest Eskimo village in the lower Yukon Delta, was abandoned after the 1918 influenza epidemic, most of its residents moving to Caniliaq. In 1962, Caniliaq was abandoned due to flooding, and its residents moved to Kotlik.

All settlements, temporary and permanent, have a form and a site. Settlement form is the layout of the buildings and structures within a farm, a hamlet, a village, a town, or a city. A settlement site is the place where a settlement is founded. It is an absolute location that can be expressed unromantically by a grid reference. Each settlement site also has a situation, that is, the local area surrounding the site. Settlement forms are dotted over landscapes and their spatial distribution is called a settlement pattern. Settlement patterns are discerned by comparing the relative locations of several settlements in relation to one another. In rural settlements, two chief types of settlement pattern are recognised – nucleated and dispersed (e.g. Houston 1963; Jones 1967; Roberts 1996). The contrast is fundamental. On the one hand are isolated farmsteads where the individual houses lie scattered. One the other hand are villages where the dwellings and farmyards cluster together in an arrangement incorporating streets, lanes, and open spaces (Roberts 1987, 14). In England, one has only to think of the scattered hamlets of Kent, the nucleated villages of the Midlands, the forest hamlets of Shropshire, the fen villages of south Lincolnshire, and the farms and hamlets of the Lake District – these are all distinct settlement patterns. Nucleation and dispersion are extreme ends of a spectrum of possible pattern and intermediate cases are common (Figure 7.1).

Settlement distribution encompasses wider aspects of settlement, such as the location of settled and unsettled areas and the limits of settlement. In many cases, settlement distribution looks at sites on a broad scale – what, for example, is the upper altitudinal limit of settlements in England? That is a question of distribution that can only be answered by looking at a collection of individual sites. Interestingly, settlement distribution has little bearing on settlement patterns, but it is dependent on the geography of sites. Moreover, settlement sites may bear little relationship to settlement patterns.

Key questions are why settlements are founded in particular places, why some settlement patterns are nucleated while other are dispersed, and why some areas are settled and others are not. Three main factors help to answer these deceptively simple questions: the natural environment, and especially topography; social and cultural factors; and agricultural economic factors. Figure 7.2 portrays the factors that have the potential to influence the siting of traditional rural settlements. Factors intrinsic to the site include water supply, free drainage, flat land, shelter, aspect, local accessibility, and culturally perceived qualities. Factors extrinsic to the site and pertaining to the local situation include communications, water bodies, agriculture, woods, turbaries (places where peat may be dug), and quarries.

Physical influences

Physical conditions undoubtedly affect the siting of settlements. Some past geographers may perhaps have taken physical determinism too far in explaining settlement features. Indeed, the interaction between humans

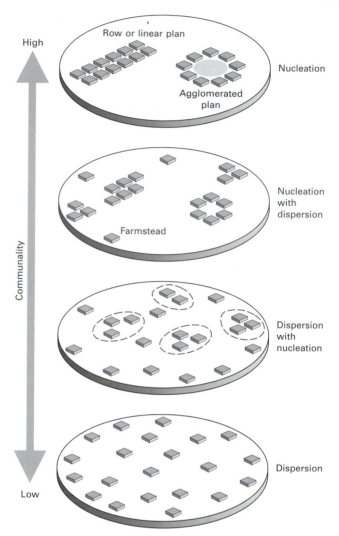

Figure 7.1 Nucleation, dispersion, and other combinations of settlement patterns. Nucleated and dispersed tendencies may change with time. The individual blocks are farmsteads. (After Roberts, 1987, 17.)

and the physical environment is two-way: the environment constrains human actions, but humans can alter their local environment and often modify the landscape wherever they settle. Slash-and-burn cultivators create temporary forest gaps. In the New World, prairie conversion to agriculture has produced a grid-iron pattern of roads, farmsteads, and towns. In parts of Poland, a moderate level of human activity has made landscape patches more elongated, convoluted, and variable in shape, while intense human activity has produced the least elongated, convoluted, and

variable patches (Hammett 1992). In cities, humans have altered almost all the natural environment, creating characteristic urban topography. Nevertheless, environmental conditions do exert a powerful influence on settlement features.

Settlement sites

The location of settlements is sensitive to factors in the physical landscape. As Paul Vidal de la Blache wrote:

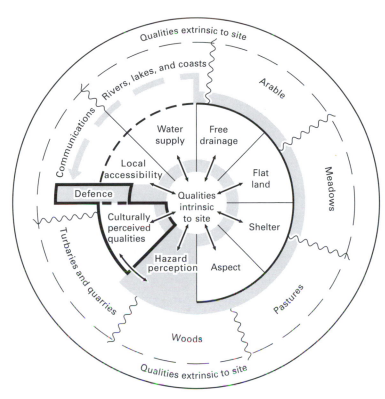

Figure 7.2 The separate but interrelated factors that influence traditional rural settlements: factors intrinsic to the site, and factors extrinsic to the site and associated with the wider situation. (After Roberts, 1996, 31.)

. . . whatever makes for variety of landscape, whatever makes it easier to foregather, has its influence on the distribution of establishments. In the mountains, slopes of moraines and alluvial fans are favourite sites for hamlets or villages. Along lake shores, lines of dunes are skirted by buildings upon their slopes. In the river-plains of France, the terraces which were the banks of ancient streams are adorned with houses or villages. Isolated hillocks in the swampy regions of the *Marschen* or *Polders* of Prussia were the sites of the earliest settlements. In Finland, stony ridges (*osar*) between clayey depressions have a magnetic attraction. In the Sahara, mountain massifs (Aïr, Haggar), cloud-condensers as they are, are the only sites suitable for permanent occupancy. (Vidal de la Blache 1926, 288–9)

Settlement sites are commonly located near resources, and especially, of necessity, near a supply of potable water and of food. Prehistoric settlements in the Hiawatha National Forest Region, Michigan, were sited in sturgeon spawning areas, areas with open water, beaver, moose, deer, and plant resources, and bedrock outcrops for chert procurement (Franzen 1986). A similar dependence upon resources explains settlement sites in the historical period in the south-eastern USA, where marine resources added to the terrestrial resources and most inland settlements have been located near running water for the last 5000 years (Hammett 1992). Prehistoric coastal sites along the North American Great Lakes were located near spring spawning habitats for fish, including areas of shallow water in complex shoreline reaches (Anderton 1993). Eskimo sites dating from the third century AD on the Chukotka peninsula, Siberia, which lies to the west of the Bering Strait, were largely coastal (Gusev *et al.* 1999). Few sites along this rocky coastline were suitable for settlement, and almost every rocky cape or sandy spit at altitudes of 13–18 m bore one (lower altitudes were subject to storm surges). Kaniskak and Nun'amo are examples, which, like other settlements at the time, had easy access to a lagoon or a small river. From the ninth century to the fourteenth century

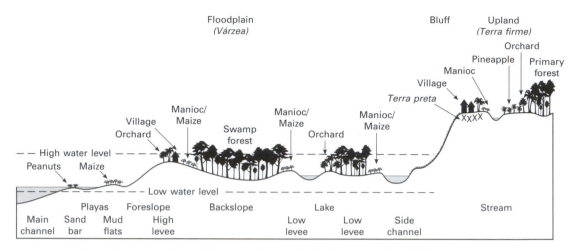

Figure 7.3 Schematic cross-section of the upper Amazon floodplain near Iquitos. Notice the main channel, high-water level, bluff, villages, and *terra preta* sites. (Adapted from Denevan, 1996.)

AD, settlements were sited on sandy spits near the shoreline and close to lagoons (e.g. Un'olen'on) or small rivers (e.g. Leimin), on the remnants of glacial terraces. From the fourteenth to the seventeenth century, settlement sites were chosen more for their ability to be defended. They were located on high rocky shores on the base of the Dezhnev Mountain Range, some 20–40 m above sea level, on cliffs that protected them on the mainland side. Settlements at Nunak and Imaklïkh are examples, and probably resulted from Chukchee people arriving in the area and endeavouring to oust the Eskimos.

Nucleated rural settlement sites may be associated with favourable water supply. In seasonal or habitually dry regions, such as parts of the Mediterranean area and India, access to a well or spring is a sufficient spur for nucleation. Even in well-watered areas, villages may grow around intermittent stream courses, such as the 'winterbournes' of Dorset, England, and many villages are located near springs and wells. In Northamptonshire, England, 200 out of 290 villages, towns, and hamlets are situated at the contact of two geological outcrops (Houston 1963, 97). Contact springs in the Northampton Sands account for the siting of one-third of the villages in the county. In the wolds of east Yorkshire, England, the permeability of the underlying chalk has invariably made surface water scarce (Hayfield and Wagner 1995). This paucity of water has constrained settlement and husbandry from prehistory to the present day. Originally, people depended wholly upon natural water sources, but later developments (rainwater traps, wells, artificial ponds,

and pumps) have progressively eased the water-supply problem. Along the Wold scarp, township units were apparently planned in segments to include, within each, an area of Wold land, a section down the scarp, and a part of the vale below. Significantly, this arrangement provided each township with a share of the spring line.

In Amazonia, prehistoric settlement and agriculture were especially concentrated along the major rivers. Traditionally, this concentration has been explained by the superior soil and wildlife resources of the *várzea* (floodplain) compared to the *terra firme* (upland or interfluve) (e.g. Steward 1948; Meggers 1971; Denevan 1966). However, the floodplain is a high-risk habitat because of regular and periodic extreme flooding of even the highest terrain. A new explanation of the concentration identifies valley-side bluffs (cliffs) as favoured settlement sites (Denevan 1996). A bluff model is proposed, arguing that most prehistoric 'riverine' settlement was not in the floodplain but rather on the fringing bluff tops lying adjacent to active river channels (Figure 7.3). Subsistence was a multiple strategy involving the seasonal utilisation of floodplain *playas* (beaches) and levee soils and aquatic wildlife, in conjunction with more permanent bluff-edge gardens, agroforestry, and hunting. All the same, bluff occupation was spatially sporadic rather than continuous, with large settlements mostly located where main river channels impinged against bluffs. This pattern persisted with colonial missions, and it continues today. Figure 7.4 depicts the Amazon floodplain between Manaus and Monte

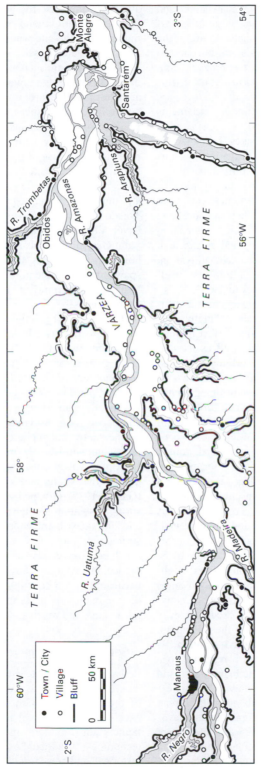

Figure 7.4 The Amazon floodplain between Manaus and Monte Alegre, Brazil. Notice the channels, islands, bluffs, and recent settlement locations. *Terra firme* is the stippled area and *várzea* is white. (Adapted from Denevan, 1996.)

Alegre, Brazil. It shows bluffs, channels, islands, and the sites of 153 villages, towns, and large cities. Of the settlements, 106 are on bluff edges; 35 are in the floodplain, most of these on islands and all but two being small villages; and 12 are on *terra firme* within 10 km of the floodplain. Additional towns occur further from the bluffs, mostly along roads, but these are not shown. Most of the 106 bluff-edge settlements are adjacent to major channels, and of the remainder, most are next to large floodplain lakes. Evidence for permanent and semi-permanent cultivation systems on the poor bluff soils comes from archaeology, ethnic history, palaeoecology, and the location of anthropogenic soils or *terra preta*. An interesting point is that continuing cultivation was able to support large and fairly permanent villages. Even more interesting is the suggestion that, although upland Amazonian soils are too infertile to support permanent cultivation, permanent or semi-permanent agriculture may itself have created fertile soils (see Denevan 1998).

Each town is, in some ways, unique and it is difficult to generalise about the influence of landform on town sites. To be sure, landform may affect the original siting and plan of a town, as well as its subsequent development. Towns were originally created for a variety of reasons. Sites were often chosen by physical criteria – strongholds, fords, bridging points, river confluences, road junctions, and so forth. Many Mediterranean towns were sited near a fortified hill that was neither too high nor inaccessible. Athens started on the Acropolis site and spread all round. Many other towns have grown around a fortified hill that survives at the core. Edinburgh is a case in point. In the thirteenth century, the town clustered on the steep castle slope and was well protected. Other towns are located between a fortified hill and a nearby sheltered coastal haven. Nice is an example, being positioned between the Château and the harbour of St Lambert. Most Bronze Age sites in the Central Highlands of Yemen are situated either on rocky hilltops or on plateaux (Wilkinson and Edens 1999). However, Hawagir, the largest site, is an exception. It is situated on the eastern loam plains of the QāʻJahrān, and lacks an outer wall and a defensive position. Its development appears to have been a response to a large area of cultivated land on the loam plains and its position on a major north–south route. The implications are that a relocation from hills to plain to reap the benefits of increased agricultural potential was possible, and that the loss of a defensible site cannot have been a top priority at that time (later second millennium BC). This example shows why the situation – in this case the potential of the surrounding land for agriculture and communication – may be as important as the site when considering why a settlement was founded (see Figure 7.2).

Many towns are sited at bridge points or ford crossings. Famous bridge towns include Geneva, London, and Dresden. Hertford, England, is at a crossing place on the River Lea. Other towns are sited at the confluences of rivers – Granada, Milan, and Lyons. Where a river is broad, twins towns may develop on either bank, as in the case of Buda and Pest on the Danube. Some towns were sited to take advantage of the natural water defences offered by meander spurs, islands, and marshes. Meander spurs are the original sites of Toledo on the River Tagus and of Prague on the River Vltava. Island sites on rivers include the Île Saint Géry of Brussels and the Stadholmen of Stockholm. Many low-lying sites defended by marshes have spawned towns – the north Italian towns of Cremona, Mantua, and Ferrara, and the English town of Oxford are examples.

Road sites and junctions tend to attract towns. The Roman road network in Europe has fixed many important town sites. In Italy, the Emilian, Flaminian, and Egnatian roads are lined with market towns and regional centres, and Milan is a route centre, at the crossing place of road and rail routes between central Europe, through Alpine passes, to the Mediterranean, and between the Adriatic and western Europe.

The sunny and sheltered slopes of valleys (south-facing and north-facing) influence settlement patterns. In Alpine valleys, the duration of sunshine greatly affects the siting and distribution of settlements (Garnett 1937). Grouped settlements are located where spring insolation is high on high hillside positions, and the oldest hamlets are placed where, as well as spring insolation, there is a long period of winter sunshine. Conversely, in Corsica and the Cévennes, France, many villages are located in the shade, mainly because they are built in chestnut forests, which traditionally provide food, and sweet chestnut trees cannot tolerate too much sunshine.

Settlement patterns and distributions

Perhaps the most interesting questions about settlement patterns revolve around their evolution. Some of the best work on this topic takes advantage of excellent archaeological and historical records of settlement sites, modern imaging techniques, and GIS technology. A splendid example related prehistorical, historical, and present settlement patterns to landscape

Figure 7.5 Study area in which prehistorical, historical, and present settlement patterns are related to landscape features, Eastern Upper Peninsula of Michigan, Great Lakes region, USA.

features in the Eastern Upper Peninsula of Michigan, USA (Figure 7.5) (Silbernagel *et al.* 1997). Patterns of settlement in the three time periods were 'reconstructed' from the Michigan Bureau of History data of archaeological sites for the Middle or Late Woodland Period (about 3000 to 300 years ago), the General Land Office Survey notes of the 1850s for the pre-European settlement, and present 'urban' sites from Landsat Thematic Mapper imagery. ARC/INFO was used to examine the spatial patterns and dynamics of settlement areas in each time period at three geographical scales – subregional ecosystems, landscape ecosystems, and terrain characteristics. The results revealed a tendency for settlements in all three time periods to occur on bedrock and lowland landscapes near the Great Lakes shoreline, mostly on slopes of less than 2 per cent (Figure 7.6). These sites would have ready access to combinations of coastal and wetland resources. A slightly larger percentage of historical settlements occurred on the transition and clay plains group with no documented occurrences of the outwash group (Figure 7.6). Prehistorical and present sites favoured south-facing sites, historical villages tending to face east and not south (Figure 7.6). This difference may arise from the relocation of historical people from preferred site locations, or their acceptance of new sites in the face of changing trade needs. Fur trapping was at its peak in this period and a wave of 'Americanisation' spread northwards that involved the setting up of Indian

reservations. Elevation data is complex because some 4500 years ago, glacial Lake Nipissing's shoreline stood at about 192 m, and is seen in the present landscape as a prominent bench around much of Lakes Michigan and Superior. Prehistorical settlements were sited just below the Nipissing bench at around 186 m, close to the then shoreline (Figure 7.6). Historical settlements were located a little higher at around 200 m. Prehistorical and historical settlements show a secondary cluster of sites at higher elevations – around 234 m and 255 m respectively. These were probably inland winter hunting camps or maple-sugaring camps (Franzen 1986). Present settlements are located around 192 m and decrease in frequency as elevation increases, with no secondary cluster. Settlements in all three periods favour lower slope gradients (Figure 7.6). Perhaps surprisingly, the present distribution of settlements accorded with the prehistorical distributions in terms of slope aspect and geographical subregion areas in the approximate range 25 to 25 000 km^2: prehistorical and present sites were mainly south-facing and often found along Green Bay and Lake Michigan shorelines.

Rural settlement patterns are influenced by landform. It has long been observed that dispersed patterns are commonly associated with mountainous terrain and other areas of broken relief, while settlements are often nucleated in areas where subdued relief allows a widespread and uniform agricultural practice (e.g. Vidal de la Blache 1926, 271–318).

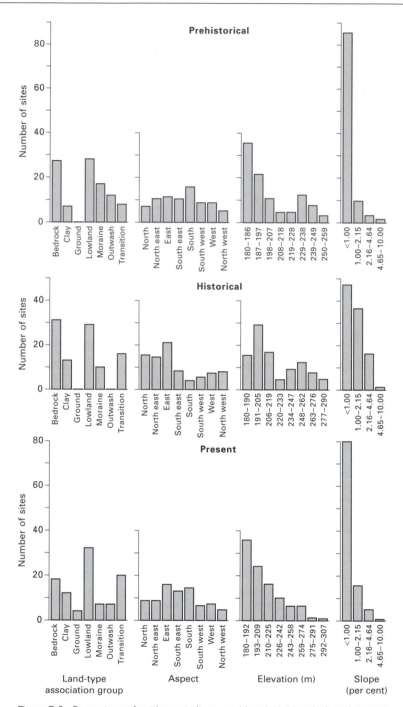

Figure 7.6 Percentage of settlement sites – prehistorical, historical, and present – associated with different land-type associations, aspects, elevations, and slopes. (Adapted from Silbernagel *et al.*, 1997.)

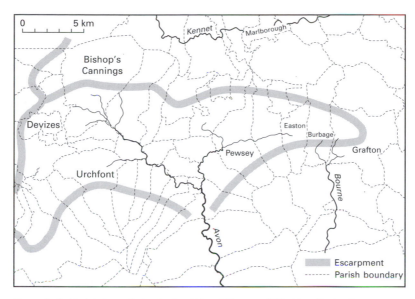

Figure 7.7 Parish boundaries in the Vale of Pewsey, Wiltshire, England.

In France, the scattered hamlets and farmsteads of the mountainous Vosges give way to the nucleated settlements of the Plain of Burgundy. In Hungary, hamlets mark the limits of the Carpathian foothills while villages are found in the Danube lowlands. But rules of thumb are not rigid laws – in Mediterranean regions, nucleated villages occur in both hilly and flat terrain, except where irrigation water favours dispersion, as in the *huertas* of Spain (Houston 1963, 99).

Soils affect the distribution of settlements. In Eurasia, three broad patterns are recognised associated with loess soils, clay soils, and infertile sands and leached soils (Houston 1963, 99–100). Loess soils occupy a broad swath from East Anglia, in England, across northern France, Germany, and eastern Europe to Ukraine. They favour cereal cultivation and are associated with nucleated villages and an open-field system. Soils of clay lowlands are heavy and were not settled until the Middle Ages. In many areas they have remained wooded and are associated with dispersed settlements. Infertile sands and leached soils have had a negative influence upon settlement.

Another facet of settlements concerns the division of areas between them. A farmstead will have an associated set of fields, a village a parish, a county town, a county, and so forth. Field systems and administrative boundaries produce distinct spatial patterns that vary from place to place. These patterns are

fashioned in part by the physical constraints dictated by topography. Manors and parishes on the chalklands of north Wiltshire, England, are related to the physical conditions that originally supplied their needs (Carter 1931, 189–92). Scarp settlements were sited near the springs and, from these centres, manors and parishes ran up and over the scarp to the chalk grassland and down over the wet and wooded clay plain. The result was long and narrow shapes set transversely to the scarp across the bands of differing soils, allowing each manor and parish to gain a share of land for sheep, cattle, arable farming, and wood. In the Vale of Pewsey, parish boundaries adjust to topography (Figure 7.7). At the narrow head of the Vale, parishes, including Grafton, Burbage, and Easton, cross the full width, from one side to another. Moving westwards, the Vale widens and northern and southern scarps each have their own lines of parishes, Urchfont and Bishop's Cannings being examples. Also in the broader west, the northern and southern series are parted by a narrow band of small central parishes.

Cultural, social, and economic influences

Although topography and other environmental factors influence settlement features, cultural, social, and economic factors may be equally or more important, and in some cases of overarching importance.

Although beyond the scope of this book, these non-physical factors demand a little explanation because it seems common sense to suppose that social, cultural, economic, climatic, pedological, and topographic factors will work together over time to fashion settlement sites, patterns, and distributions in a particular region. Two examples will support this supposition. First, Rani T. Alexander's (1997, 1998, 1999) research on community organisation in the *Parroquia de Yaxcabá*, Yucatán, Mexico, from 1750 to 1847, demonstrates how the spatial organisation of settlements, and of house lots, varied according to variations in population growth, tax structure, and land stress. The variations accompanied changes in the colonial regime, resulting from the Bourbon reforms and Mexican Independence from Spain. The second example is Inge Schjellerup's (1999) study of native peoples living in the upper lowlands of Amazonia, north-eastern Peru. Before 1532, before the Spaniards arrived, people in this region consisted of many indigenous groups speaking different languages. Most of them lived in small and independent communities or in dispersed settlements suitable for slash-and-burn cultivation and hunting and fishing. Under Spanish colonial domination, groups were forced to live together in nuclear settlements for conversion to Christianity and the collection of tributes. For instance, the Lamistas, a native group who spoke Quechua, the language of the Incas, became established as an ethnic group in the upper Huallaga catchment near the Río Mayo during the Spanish Colonial Period. In the twentieth century, and especially in the last 30 years, drastic environmental changes have occurred in the region as the forest has been transformed to an open savannah with scattered forest trees. The change was caused by agrarian reform. A 'development wave' spread into Amazonia as 'colonos', people from highland departments, were given free land and cheap loans to encourage them to settle in Amazonia and to raise cattle and grow rice, sugar cane, and coffee. Other parts of the forest have been cleared for timber. Soil degradation and impoverishment reduced agricultural productivity and land was abandoned. Local peasants then moved in and colonised the cleared land. The colonisation process was accelerated in the 1990s by an increase in demand for coca, and drug dealers encouraged local people to grow coca leaves and process cocaine paste for North American and European markets. So, despite the cultural and environmental upheaval, the Lamistas have preserved their culture and identity through strategies that have affected their settlement patterns and land tenure.

Routes

The presence of a route-way is fixed by human needs; its location is guided by topography.

> Ships slide through the water, the cleft waves roll together, and all trace of the passage is blotted out. But land preserves traces of the routes early travelled by mankind. The road is branded on the soil. It sows the seeds of life – houses, hamlets, villages, and towns. Even what would seem at first glance to be haphazard trails, the random tracks of hunters and shepherds, leave their mark. (Vidal de la Blache 1926, 370)

Topographic obstacles are hard to overcome and force routes to follow a particular direction. Rivers, marshes, and mountains may require construction work or assistance in crossing. Mountains are traversed at passes or tunnels.

Maintaining routes

All routes are as vulnerable to the elements as any other parts of landscapes. They may be blocked by floods, snow, landslides, and fallen trees, or broken by earth movements associated with earthquakes.

Snow is particularly troublesome in mountainous areas. Drifting snow may form cornices and snowdrifts. Cornices can pose an avalanche hazard, while snowdrifts on roads disrupt communications and transport. A survey of drifting snow in France, carried out in 1991 and based on replies from local branches of the Road Administration, revealed that snowdrifts were dealt with by snow clearing (77 per cent of replies), snow fences (14 per cent), planting trees (7 per cent), and modification of road profiles (2 per cent) (Naaim-Bouvet and Mullenbach 1998a, 1998b). Since then, the use of windbreaks to control avalanches has become far more widespread, especially in new road projects that take account of the environment in road design. The windbreaks are normally built of trees and are referred to as 'living' snow fences. The first living snow fences were placed alongside some German railways as early as 1852 (Nordling 1864), but their general use was limited until the 1990s. To be useful, natural plant barriers need to 'store' snow. Plainly, they take time to reach an effective height after planting and may be difficult to establish on windy sites at high altitudes. Research on these matters has confirmed the value of living snow fences and discovered the best means of building them.

In the Col de Manse, which lies at 1200 m in the French Alps, four types of hedge were tested: a 1.7-m-high hedge of 20 pruned spruce trees spaced at 1-m intervals (pruned means the lower branches removed to leave a 50-cm gap between the lowest foliage and the ground); a 1.7-m-high hedge of 20 unpruned spruce trees spaced at 1-m intervals; a 1.7-m-high hedge of 60 rowan trees spaced at 0.33-m intervals; and a 1.7-m-high artificial snow fence, with a 50 per cent porosity and a bottom gap of 50 cm, constructed of horizontal wooden slats (Naaim-Bouvet and Mullenbach 1998a, 1998b). The wooden snow fence is of the kind often used to keep snowdrifts from roads and has optimal design features, maximising snow storage. It is therefore a reference fence against which to gauge the effectiveness of the living snow fences. The results showed that deciduous trees, such as the rowan, trap sizeable snowdrifts, and have about 80 per cent of the snow storage capacity of artificial snow fences. Pruning low branches was ineffective owing to the weight of snow on the lower branches. The living fences must be kept to a height of about 1.5 m to avoid damage to branches and trunks by the pressure of snow. It might be better, therefore, to install several rows of living snow fence. The problem of establishment time is a problem – a tall windbreak cannot be planted owing to cost. It can be partly circumvented by planting in the lee of an artificial snow fence, the protection afforded reducing the mortality of young plants for about 22 m.

Environmental impacts of routes

The trampling of humans (walking or riding) and other animals along trails may lead to soil erosion (e.g. Kuss 1983; Jacoby 1990). Anyone who has walked along footpaths, especially those in hilly terrain, is bound to have first-hand experience of the problem. The problem has become acute over the last 20 or 30 years as the number of people using mountain trails, either on foot or on horseback, has risen sharply (e.g. Godin and Leonard 1979). A study in Costa Rican forest confirmed that trails generated runoff more quickly, and are eroded sooner, than is the case in off-trail settings (Wallin and Harden 1996). This finding, which is typical of trail erosion studies in all environments, underscores the need for careful management of ecotourism in trail-dependent activities. Strategies for combating trail erosion can work. Smedley Park lies in the Crum Creek watershed, Delaware County, near Media, Pennsylvania, USA. The trails in the park pass through several areas with fragile environments (Lewandowski and McLaughlin 1995). A strategy was devised using network analysis, that altered the efficiency of the trail system by more fully connecting sites with robust environments and reducing the potential for visitors to use environmentally fragile sites. Some of the severest erosion is associated with logging trails. In the Paragominas region of eastern Amazonia, tree damage in unplanned and planned logging operations was associated with each of five logging phases: tree felling, machine manoeuvring to attach felled boles to chokers, skidding boles to log landings, constructing log landings, and constructing logging roads (Johns et al. 1996).

Trail use affects soils and vegetation. In the northern Rocky Mountains, Montana, USA, trails across meadow vegetation bear signs of damage – bare soil and eroded areas – through human use (Weaver and Dale 1978). The meadows were principally Idaho fescue–Kentucky bluegrass (*Festuca idahoensis–Poa pratensis*) communities. Experiments were run on meadows underlain by deep, sandy-loam soils at 2070 m near Battle Ridge US Forest Ranger Service Station, in the Bridge Range. They involved getting hikers, horse riders, and a motorcyclist to pass up and down slopes of 15°. The hikers weighed 82–91 kg and wore hiking boots with cleated soles; the horses weighed 500–579 kg and had uncleated shoes; the motorcycle was a Honda 90 running in second gear at speeds below 20 km/hr. The experiments showed that horses and motorcycles do more damage (as measured by percent bare area, trail width, and trail depth) on these trails than do hikers (Figure 7.8). Hikers, horses, and motorcycles all do more damage on sloping ground than on level ground. Hikers cause their greatest damage going downhill. Horses do more damage going uphill than downhill, but the difference is not great. Motorcycles do much damage going downhill and uphill, but cut deep trails when going uphill. In the alpine environments of the Central Plateau, Tasmania, the passage of 20–30 horses damaged vegetation (Whinam et al. 1994). Damage was severest in shrubland, where the southern or alpine grevillea (*Grevillea australis*), a prostrate shrub, lost a large part of its biomass. Damage was also observed in herb-field and bolster heath. Waymarked trails may be necessary in this environment to limit damage in the eroded shrubland, while herb-field and bolster heath are best avoided by horse-riders.

The comparative impact of hikers, horses, motorcycles, and off-road bicycles on water runoff and sediment yield was investigated on two trails – the Emerald Lake Trail in the Gallatin National Forest,

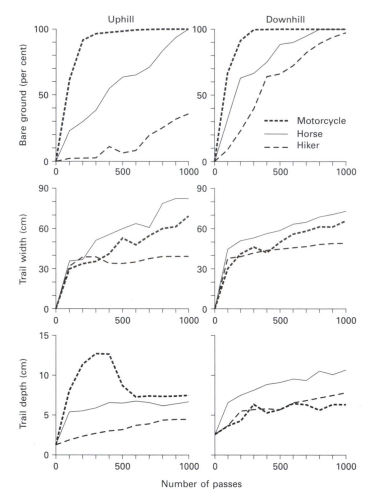

Figure 7.8 Experimental damage done by hikers, bikers, and horses (moving uphill and downhill) on trails in the Bridge Range, Montana, on a sloping (15°) meadow site. (Adapted from Weaver and Dale, 1978.)

Montana, USA, and the New World Gulch Trail just outside the same forest (Wilson and Seney 1994). A modified Meeuwig drip-type rainfall simulator was used to reproduce natural rainstorm events. Treatments of 100 passes were applied to 54 sample plots on each of the trails. Twelve sample plots were used per mode of travel with two antecedent moisture conditions (dry and pre-wetted) and two slope gradient classes (0–6 and 8–21 per cent) with two replications. Control sites required just six plots. The results confirmed the complex interactions that occur between topographic, soil, and geomorphic variables, and the difficulty of interpreting their impact on existing trails. Multiple-comparison test results showed that horses and hikers (hooves and feet) made more sediment available than wheels (motorcycles and off-road bicycles), with horses producing the most sediment, and that sediment production was greater on pre-wetted trails. The higher susceptibility to erosion of wet sites was an especially interesting finding, and merits further study.

Agriculture

The term agriculture covers a gamut of human activities concerned with food procurement and production. Of the multifarious activities, crop production is perhaps the most sensitive to topographic factors, but livestock farming is also subject to topographic influences. These effects will be considered under the headings aspect, shelterbelts, land-use catenas, and agricultural mosaics.

Aspect

Considering that many plant species are sensitive to slope exposure, it is not surprising that agricultural crops respond to differences in aspect. Slope exposure modifies surface water and energy balances, which in turn alter the productivity of crops. In the Central Chernozem Region, Russia, soils of north-facing slopes are potentially more productive than soils of south-facing slopes (Nakonechnaya and Yavtushenko 1989). At Coopers Creek, North Canterbury, New Zealand, rainfall, soil moisture, and soil and air temperatures were measured over five years on sheep-grazed pasture slopes of about 25° with north and south aspects, and on the exposed intervening ridge crest. Over a year, potential evapotranspiration on the north-facing slope was almost double that on the south-facing slope (Radcliffe and Lefever 1981). Soils on the north-facing slope were invariably drier than those on the south-facing slope throughout the growing season, and sometimes approached wilting point between December and March. During the study period, mean annual herbage accumulation and the green content of the herbage were greater on pasture of southerly aspect than on pasture of northerly aspect (Radcliffe 1982). On northerly-aspect pasture, herbage production was greatest in areas where moisture was retained – hummocky spurs and seepage areas. On southerly-aspect pasture, most herbage grew on the ridge crest and in hollows of hummocky spurs. Variation in herbage production was always greater on northerly-aspect pasture, owing to windier, warmer, and drier conditions stretching the gap between wetter and drier sites.

Shelterbelts

Windbreaks bring a variety of benefits, including an increase in productivity. Productivity in a field varies due to, among other things, distance from a hedge, with the most productive parts often lying within the shelterbelt of a windbreak. The fetch (or run) of the wind (distance without obstructions) increases with increasing field size. The wind tends to blow more strongly across large fields and may exacerbate soil erosion and reduce productivity. In windy climates, wind fetch may also affect livestock (Caborn 1965). The reasons for increased productivity within shelterbelts are complicated. Growth depends very much on leaf temperature. Reduced windspeed in the lee of windbreaks produces less turbulence, less evapotranspiration, and warmer leaves. In temperate and cold climates this effect will increase productivity, but in hot climates it could prove lethal. In addition, plants in the shelterbelt suffer less shaking and mechanical abrasion by blowing particles, which also promotes growth.

Much research attests to the benefits of windbreaks. A field of spring wheat (*Triticum aestivum*) in the Canadian prairies and northern plains of the USA gave greater yields between $1h$ and about $17h$ downwind of windbreaks (where h is the height of the windbreak) than in the open portion (Kort 1988). Maximum crop production occurred about $3h$ downwind. In fields at Ridgetown, Ontario, Canada, soybean (*Glycine max*) yields from 1980 to 1982 were lower from $0–1h$ downwind and higher from $2–9h$ downwind, with a maximum level at about $4h$ downwind (Baldwin 1988). As well as boosting crop productivity, windbreaks shelter buildings, reduce stress on livestock, reduce soil erosion, protect crops, trap snow, and provide wildlife habitats. In the Great Plains of North America, the trees forming the windbreaks are subject to low rainfall, cold winters, attack by insects and diseases, and exposure to herbicides (Dow *et al.* 1998). Owing to the severity of conditions in windbreaks, only a few species are planted in the Great Plains. Ponderosa pine (*Pinus ponderosa*) and Scots pine (*P. sylvestris*) are the main species, with limited numbers of jack pine (*P. banksiana*) and red pine (*P. resinosa*). Fifteen-year tests on lodgepole pine (*P. contorta*) showed that, due to its extensive genetic variability, this species makes a good shelterbelt tree. Once established, lodgepole pines in the Northern Great Plains were hardy and drought resistant, with no mortality after 6 years (Dow *et al.* 1998).

The benefits of windbreaks are felt worldwide, though in the tropics the chief benefit is shading rather than protection from the wind. Consider coffee plantations on the Tay Nguyen Plateau, central Vietnam (Błazejczyk *et al.* 1995). The principal advantage of shelterbelts in this monsoon area is that shading enables coffee bushes to survive the hot and dry season, when incoming levels of solar radiation are very high and winds may be very strong. The shading also makes conditions more bearable for people working on the plantations. Golden shower (*Cassia seamia*), grown as 15–17-m dense canopies, was the most effective shade plant, reducing the solar radiation received at the coffee canopies by 41 per cent. White lead-tree (*Leucaena leucocephala*), grown as 5–7-m high trees with sparse canopies, achieved just a 12 per cent reduction in incoming solar radiation. Sunn hemp (*Crotalaria arecta*) bushes, planted in rows between coffee bushes, caused a slight increase in solar radiation receipt on coffee bushes by reflecting light. Golden shower and breadfruit (*Artocarpus communis*) were planted around all the fields to reduce windspeed. In fields without shading plants, the surrounding trees reduced windspeed by about 42 per cent. Wind-protecting trees and shading plants combined reduced windspeeds by between 74 and 87 per cent.

Land-use catenas

Some land-use systems form agricultural toposequences that often display subtle adaptations to slope, climate, and soils. In Leyte, the Philippines, several ecosystems were all used traditionally by the farm households to ensure their subsistence (Sauerborn 1994). These ecosystems run from the sea, through the coast, to the hillsides (Figure 7.9). Traditionally, they were

exploited by Kaingin farming, which is shifting cultivation combined with forest-product collecting activities. The aquatic ecosystems yielded protein, the coastal and hillside agro-ecosystems provided starch, and the forest ecosystem supplied the families with firewood and timber. This traditional system was adapted to the given natural resources, with long fallow periods of 15 years allowed to restore soil productivity. Owing to a growing population, the supplementary role of hillside farms has evolved into a primary source of subsistence, followed by resource depletion. More productive and sustainable land-use systems need to be introduced and adapted which are able to restore soil productivity in the same efficient way as does the secondary forest during long fallow periods.

In the Karnataka uplands, southern India, a steep environmental gradient exists on the backslope of the Western Ghats that in large measure controls the agricultural and land-use patterns within the region (Gunnell 1999). The rainfall along the gradient ranges from a mean annual figure of over 3000 mm on the Ghats to around 1000 mm in the Tamilnad Plain lying nearly 400 km to the east. The Indian rural landscape patterns along the environmental gradient are finely tuned to variations in rainfall, soil, and

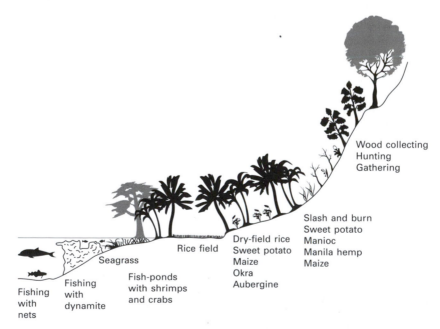

Figure 7.9 A land-use catena in Leyte, the Philippines. The ecosystems along the catena were traditionally used by families to secure a livelihood. (Adapted from Sauerborn, 1994.)

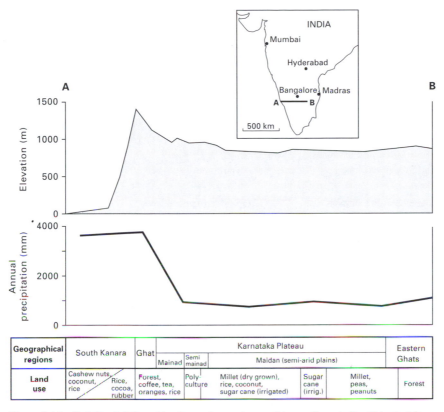

Figure 7.10   Relief, rainfall and land-use along a transect in southern India. (Adapted from Gunnell, 1999.)

landforms (Figure 7.10). Between the Ghats and the Indian Ocean, rice, cashew nuts, and coconut are grown on lateritic mesas, and rice, rubber, and cocoa on piedmont hills. The mountains and hills immediately to their east at the western edge of the Karnataka Plateau (the *mainad*) support forest, but coffee, oranges, tea, and rice are cultivated. The broad expanse of the semi-arid plains of the Karnataka Plateau, called the *maidan*, supports millet production in the eastern region, though rice, coconut, and sugar cane are grown under irrigation. Millet, peas, and peanuts are grown in the western region of the plateau. Interestingly, the compact chain of agroclimatic zones promotes simultaneously varying degrees of economic complementarity and of self-sufficiency.

Soil erosion is best combated by an understanding of catenary and landscape processes. Soil and vegetation cover degradation is prevalent in many areas of Sudano-Sahelian West Africa. A successful programme for curing the problem, which takes into account the diversity of ecological conditions and socio-economic conditions of the farmers, analyses traditional technologies of water and soil fertility management, infiltration rates, organic matter and nutrient balances, and the crop yields and net income (Roose *et al.* 1992; Dugué *et al.* 1993). The programme was developed for farmers in Yatenga, a north-west province of Burkina Faso (formerly Upper Volta). It goes beyond simple soil conservation, which demands considerable labour for farmers but produces a negligible increase in soil fertility and therefore no economic return for the farmer, and combines soil conservation practices with improved management of water resources and soil nutrients, which does boost yields and economic returns. The measures adopted depend on soil type, which in turn varies with catenary position (Figure 7.11). Farmers are encouraged to redevelop their individual plots first, which occupy the best land on tropical ferruginous soils, and then the communal areas of the upper catena (gravelly lithosols) and the valley bottom (hydromorphic and sandy soils that are flooded during heavy

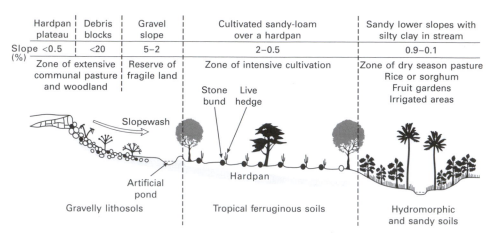

Figure 7.11 Land-use catena on a granitic substrate on the Mossi Plateau, central Yatenga, Burkina Faso. (Adapted from Roose *et al.*, 1992.)

storms). Soil conservation structures include stone bunds, live hedges of African vetiver (*Andropogon guyanus*), a perennial grass, and shrubs (*Acacia, Ziziphus,* etc.), and lines of balsam spurge (*Euphorbia balsamifera*), a dwarf shrub. The measures are integrated within drainage basins (Figure 7.12). On the gravelly and sandy hills, defences against soil erosion are built and grasses and shrubs planted, which helps to regenerate pasture. Soil erosion defences include structures that collect runoff and include half-moon structures, artificial ponds, and, where topography permits, lakes. This stored water is used for livestock and extra irrigation for small gardens of fruit trees and crops. On the main cultivated areas in the central portion of the hillslopes, groves of trees, live hedges, tree lines along roads and farm boundaries are all planted, and stone bunds constructed. Completely degraded soils are restored using the traditional technique of Zaï, which involves digging holes and filling them with compost to add organic material to the soil and build up fertility, in conjunction with four other measures. The other measures are redeveloping gullies and valley bottoms; building stone barriers in villages to prevent gullying; planting orchards and vegetable gardens in the dry season; and fixing gullies at road-crossing points.

Catenary position affects the yields of some crops. Yields vary from summit to footslope in accordance with changes in soil type and soil properties. In some parts of the world, farmers are aware of these differences and plant different crops along catenas. For example, in the West African savannah, upper and middle slopes were under millet, cowpea, and ground-

nut, lower slopes were under sorghum and maize, and valley floors were given over to rice (Van Staveren and Stoop 1985). This pattern emerged from farmers' experience. In south-west Nigeria, yields of maize (*Zea mays*) and soybean (*Glycine max*) were compared at four catenary positions (upper slope, middle slope, lower slope, and valley floor) and under two management regimes (traditional and 'improved') (Ogunkunle and Onasanya 1993). Yields varied significantly ($p \leq 0.05$–0.01) according to topographic position, which produced differences of 0.5 to 2.4 t/ha for maize and 0.3 to 1.3 t/ha for soybean. In the highlands of Rwanda, soil parameters change in a characteristic way along toposequences running from hilltops and upper slopes to lower slopes and valley bottoms (Steiner 1998). Variations in soil parameters are reflected in crop yields at different positions on the toposequence. On average, the lower slopes yield 20–50 per cent less compared with upper slopes, depending on the season and crop rotation. Results of surveys and on-farm studies revealed farmers' detailed knowledge of the soils. Farmers know that soil suitability is closely related to relief. They classify soils by a number of criteria and choose crops according to soil suitability.

Agricultural mosaics

Most agricultural landscapes comprise mosaics of agricultural and 'natural' patches, corridors, and matrixes. Agricultural production in part depends on the spatial structure of the agricultural mosaic. This important relationship is exemplified by some applications

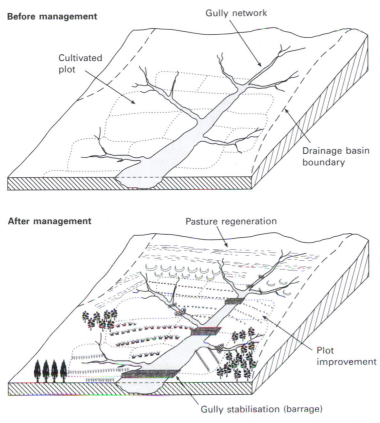

Figure 7.12 Integrated soil conservation and cropping measures in a drainage basin, central Yatenga, Burkina Faso. (Adapted from Dugué *et al.*, 1993.)

of agroforestry, an agricultural system designed to mitigate against, or even reverse, the harmful effects of intensive farming practices and the concomitant land degradation. Agroforestry is a collective name for land-use systems involving trees, shrubs, and other woody perennials grown in association with herbaceous plants (in crops and pastures) or livestock (or both). The growing is done in a spatial arrangement or a rotation, or both, and there are ecological and economic interactions between the tree and herbaceous plant–livestock components of the system (Young 1989).

An interesting point is to what extent the spatial arrangement of the agricultural mosaic in an agroforestry system influences crop production. This issue was tackled in a study of agroforestry in a 21-km² area within the coffee zone on the southern footslopes of Mount Kenya, Kenya (Olson 1998). The land is used mainly for maize, coffee, bananas, and silk oak (*Grevillea robusta*) trees, which are the domin-

ant hedgerow species. The research tested two hypotheses: maize productivity is related to the spatial structure of (1) fields (e.g. patch size, shape, and adjacency relationships) and (2) landscapes (density of trees and hedgerows). A Compact Airborne Spectrographic Imager (CASI) was used to acquire a detailed digital image of the site at a high spatial resolution, with a pixel size of 1.75 m. Chlorophyll density, crop type, and crop vigour (as measured by the normalised difference vegetation index, NDVI) were determined from six spectrally narrow width-bands. It was found that maize production is not significantly related to the attributes measured at the landscape scale. However, significant results did emerge from the field studies. Crop vigour was inversely and curvilinearly related to distance from shelter elements (hedgerows). The curvilinear relationship meant that crop vigour beyond about 25 m from the hedge remained unchanged. Converting the crop vigour figures into actual yields gives an increase from 6800 to 9000 kg/ha for a site at the Kenyan Agricultural Research Institute. This means that maize would benefit from trees within 25 m of a hedge and, at optimal distances (about 3–25 m) within the sheltered zone, would increase production by 32 per cent. So, it would be advantageous to design the landscape to maximise the shelter effect of trees. The optimum distance between hedgerows is about 40 m. Changing the landscape to accommodate optimum hedgerow spacing would lead to a net loss of crop area but that would be made good by an increase in the yield of maize and of wood products. On the basis of the findings, the maize yields and wood production in the existing agricultural landscape were compared with two possible alternative landscapes. The first alternative landscape was one in which 50 per cent of the silk oaks in the area were destroyed by a pest outbreak. Such a catastrophic disturbance would halve tree production and reduce maize production by 5 per cent. The second alternative landscape kept existing coffee-growing areas and enclosed maize fields in an optimal manner and identified under-utilised planting niches (e.g. field boundaries and trails) throughout the landscape. The effect of this planned change was a modest increase (over 4 per cent) in maize production and a doubling of wood production. Plainly, this kind of investigation holds out much promise for establishing sustainable land-use patterns that avoid soil degradation on smallholder farms throughout the tropics.

Environmental destruction and conservation

Conservation practice needs to take account of topography's role as an environmental factor. Taking topographic study in its broad sense to include all landscape features, it becomes tantamount to landscape ecology, which considers spatial heterogeneity and spatial processes as they affect ecological processes. Many principles and theories from landscape ecology are used to inform conservation practice. Six major themes in landscape ecology have relevance to conservation plans: reserve distribution, reserve shape, conservation corridors, reserve edges, reserve functioning, and regional settings (cf. Kupfer 1995). These are not the only topics that could be considered under the conservation rubric. For example, the conservation of soil catenas has been suggested as being important in building planning (Reinirkens and Vartmann 1996). However, this section will stick to the six major themes.

Reserve distribution

A major conservation debate concerns two seemingly simple questions: is it better to have a large patch or a small patch (LOS)? And, is it better to have a single large or several small patches (SLOSS)? This long-running debate was sparked off in the mid-1970s (e.g. Simberloff and Abele 1976a, 1976b; Diamond 1976; Terborgh 1976; Whitcomb et al. 1976; see Shafer 1990, 79–82). To an extent, arguments over the connectivity of nature reserves, the worth of wildlife corridors, and the problems of scale have superseded the debate, but its arguments still have currency.

Large patches have several qualities of ecological value (Forman 1995, 47). The water quality of aquifers and lakes is safeguarded. Fish and terrestrial mammal populations are aided by the connectivity of a low-order stream network. Patch-interior species are sustained by a large area of suitable habitat. Vertebrates with large home ranges are supported by core habitat and ample escape cover. Large patches are a source of species dispersing into the surrounding matrix, they maintain near-natural disturbance regimes, which is advantageous to those species that have evolved with, and require, disturbance, and they provide a buffer against extinction during environmental change. Small patches also have qualities of ecological value (Forman 1995, 47). They provide much-needed habitats and stepping-stones for species dispersal, and for recolonisation after local extinction of interior species. They support high densities of species and large populations of edge species. Small patches augment matrix heterogeneity, so providing escape cover for prey species and decreasing wind fetch and erosion. They provide habitats for small-patch-restricted species (p. 201), and they preserve scattered small habitats and rare species. Nature reserves in specific areas may possess other qualities of ecological worth. Large patches of natural vegetation in suburban areas may house a limited number of interior species, but these patches can play a major role in ameliorating microclimates, making the downwind neighbourhood cooler and moister (e.g. Gilbert 1991). They also soak up rainwater, so reducing floods. On balance, large benefits tend to accrue from large reserves, and small benefits from small reserves. Oddly, few benefits arise in middling-sized reserves. Disadvantages, such as an increase in the proportion of edge species, are described mainly for small reserves.

Accepting that large patches are ordinarily preferable to small patches, the question then becomes how many large patches are needed to achieve the desired ecological outcomes, if we might adapt a jargonistic phrase from education? This question is not well researched, but there are snippets of studies that offer some clues. In New Jersey woodlots, USA, more than three 24-ha woods would be needed to maximise bird species diversity at the landscape scale (Forman *et al.* 1976). Another way of approaching the question is to look at the species pool (the total species living in a habitat type within a landscape). If a large patch should contain the entire pool, then no more patches would be required since they would add no extra species. If a large patch contains 90 per cent of the species pool, then a second patch may add the other 10 per cent, though it is more likely that three or more patches would be needed to include the 'tail-end', rare species. The minimum number point is useful as a conservation target. It is the number of patches beyond which the species diversity–patch number curve gains less than 5 per cent of the species pool with the addition of another patch. It thus includes nearly all the characteristic species of the patch-type in a landscape. Plainly, the lower the number of species present in the first patch, the more patches are required. In northern Sweden since 1945, boreal forests have been hit by large-scale clear-cutting. To boost biodiversity in this region, the preservation of small patches of old forest, less than 1 ha in size, has been advocated. However, a study of birds in Grandanlet, a 280-km^2 landscape mosaic of virgin boreal forest and mire, suggested that such small forest patches would provide some habitat for more generalist species, but to maintain a diverse bird fauna requires larger forest patches, preferably larger than 10 ha, to be set aside (Edenius and Sjöberg 1997).

Recent fieldwork on reserve size reveals the problem with generalisations. For instance, the finding that patch-size–species diversity relationships vary with species even in the same area has management implications. A recent study tackled the SLOSS question head-on, and also considered the effect of reserve shape (Virolainen *et al.* 1998). The subject of the study was vascular plant species living on boreal spruce and pine mires. Species richness, species rarity, and taxonomic diversity in the reserves were all investigated. The results showed that species richness was not related to the spruce mire size, but it did increase in relation to the pine mire size. In contrast, the species-rarity score increased in relation to the area of spruce mires, but it was not related to the area of pine mires. Taxonomic diversity was unrelated to reserve size in the case of spruce-mire reserves, but

it increased with the size of pine-mire reserves. By comparing large and small reserves, it was found that several small mire reserves contained more vascular plant species than a large mire reserve of equal area. Several small mire reserves also had higher rarity scores and higher taxonomic diversities than a single large mire reserve. Indeed, species richness, rarity score, and taxonomic diversity all increased in relation to the number of small mires in both spruce and pine groups. Mire shape had no influence upon species richness, rarity score and taxonomic diversity, which brings us to the next theme.

Reserve shape

It is reasonable to think that reserve shape, particularly in smaller reserves, will affect population dynamics on a reserve. Highly convoluted shapes have a greater length exposed to surrounding matrix, and this increased exposure may raise susceptibility to external stresses or disturbances arising from domesticated animals, poaching, and so on (Schonewald-Cox 1988). They also have proportionately more edge than less-convoluted shapes, which brings a range of attendant consequences. It means that edge effects are more pronounced than in a less-convoluted patch of the same size, which may explain some area-related extinctions (Schonewald-Cox and Bayless 1986). Convoluted reserves also contain less interior habitat than a less-convoluted reserve of the same size, and this may interfere with reserve dynamics (Laurance and Yensen 1991). Despite these reasons for supposing that reserve shape would play a role in reserve dynamics, research into the effects of reserve shape has produced a mixed message. A study of species richness in English farm woodlands showed, after controlling for the effects of patch area, that patch shape explained no significant variation in species richness (Usher *et al.* 1992). Conversely, in urban woodlots in four adjacent cities in southern Ontario, Canada, edge length (which is related to shape) was a superior predictor of bird species richness than patch area (see p. 204; Gotfryd and Hansell 1986). The relationship between reserve shape and species richness may be inconsistent (cf. p. 207), but the relationship between reserve shape and the abundance of specific species adds another and little-researched dimension to the problem (Kupfer 1995). It is likely that small reserves with convoluted edges will house mainly edge species, whereas more circular reserves will house a larger number of rare, interior species. However, far more work needs to be done on reserve shape and its effect upon species composition and dynamics before secure conclusions are reached.

Conservation corridors

Corridors have captured the public imagination and a range of 'greenways', 'greenbelts', 'linear reserves', 'lifelines', 'buffer strips', and 'wildlife corridors' are built into management plans and planning strategies (Bennett *et al.* 1994; Bennett 1999). The idea that corridors may negate the isolation of populations resulting from habitat fragmentation carries an intuitive appeal, and the building of corridors to ease the movement of endangered animals is often within the scope and resources of conservation-minded groups.

Corridors possess several advantages in conservation, but there are potential disadvantages, too. Corridors are thought to boost dispersal between reserves, which compensates the damaging effects of small populations and restricted gene-pools by expediting interbreeding between populations. Inter-reserve dispersal also influences species demography positively by increasing population growth, facilitating recolonisation of reserves following local extinctions, and promoting the survival of metapopulations by raising the chances of 'rescue' (Merriam 1991). Other advantages of corridors are an increase in the foraging area of wide-ranging species; the provision of temporary refuges from predators for species moving between patches; the provision of a mix of accessible habitats and successional stages for species demanding a diversity of habitats for different activities or stages of their life cycles; the provision of alternative shelter from large distances, such as a refuge from fire; and the provision of 'greenbelts' to limit urban sprawl, abate pollution, furnish recreational opportunities, and enhance scenery and land values (Noss 1987). In addition, corridors may assist large-scale species migrations necessitated by climatic change (Hobbs and Hopkins 1991). A negative aspect of corridors is that they may hasten the spread of diseases, exotic predators, and other disturbance while failing to enhance the survival or dispersal of the targeted species (Simberloff and Cox 1987). Other possible disadvantages are that corridors may increase species flow between reserves, so impoverishing genetic variability through the loss of local genotypes; permit the spread of fire and other 'contagious' disturbances; increase the exposure of wildlife to hunters, poachers, and other predators; in the case of riparian strips used as conduits, block the dispersal or survival of upland species; and conflict with strategies to preserve the habitat of endangered species, when the inherent quality of a corridor habitat is low (Simberloff and Cox 1987; Noss 1987; see also Simberloff *et al.* 1992).

Despite the arguments against them, the consensus is that, if conservation corridors are designed well, they are efficacious in managing species within fragmented landscapes. And there is evidence that several reserves linked by corridors helps to maintain regional-scale landscapes and conserve many species. In 14 western North American parks, 13 had lost 43 per cent of their historical lagomorph, carnivore, and ungulate species (Newmark 1987). The Kootenay–Banff–Jasper–Yoho park system, which has significant connections between reserves, maintained all its original mammal fauna. Nonetheless, it is well to bear in mind the makeshift nature of conservation corridors:

> Corridors, regardless of how effective they may be, can never replace large reserves for the protection of ecosystems and species. This is because corridors, by definition, contain little habitat, and because they are intrinsically 'edgy'. This edginess of corridors means that these landscape links are hazardous for edge-sensitive and predation-sensitive species, but very suitable habitat indeed for many weedy species and pathogens. Corridors are bandages for a wounded natural landscape, and at best can only partly compensate for the denaturing activities of humans (Soulé and Gilpin 1991, 8)

Road corridors often serve as conservation areas. The paradox of road systems is that, on the one hand, they split and isolate wildlife populations, cause disturbance and pollution in the adjacent environment, and lead to a heavy toll of dead animals; on the other hand, roadside habitats attract a wide range of animals and plants, may preserve the only remnants of natural habitat in a region, and may act as conduits for animal passage through forbidding terrain (Bennett 1991). To resolve this paradox, road systems must be viewed in their regional setting. New road proposals should consider impacts on the regional landscape. For instance, where a road is to pass through forest, wildlife management priorities should be to minimise the efficiency of the road as a barrier, to limit road kills, and to keep noise and pollution to the lowest possible levels. And when a road is to pass through agricultural landscapes, efforts should be made to retain and expand wildlife habitats along the roadside, to enhance the continuity of roadside vegetation so as to build natural corridors across the landscape, and to monitor road kills (Bennett 1991). In some agricultural areas of the world, such as the central wheatbelt of Western Australia, native vegetation survives as small remnants, the use of which depends on connections through road verges (Fortin and Arnold 1997). This underscores the importance of roadside habitats in conservation.

Reserve edges

As well as reducing the size of patches, habitat fragmentation also increases the proportion of edge to interior. Although forest edges were once considered a boon by wildlife mangers, mainly because they are often species-rich and attract many gamebirds, current thinking is less sanguine. For reasons given in an earlier section (p. 207), edge habitats are not necessarily desirable. For conservation purposes, for example, their propensity to attract weedy species, rather than species that need protecting, is hardly a blessing. With these drawbacks in mind, conservation schemes try to build in edge processes. In addition, edges tend now to be treated as dynamic, functioning habitats that interact with the matrix on either side. The pressing needs are to re-evaluate edges and achieve a fuller comprehension of their roles in landscape processes, and to examine the significance of reserve boundaries to reserve functioning and viability (Kupfer 1995).

Reserve functioning

The animal and plant species that are preserved within a reserve do not live in isolation. They are components of the reserve ecosystem and interact with the reserve's physical environment. Traditionally, reserves were managed to protect a particular species or to sustain species richness. Reserves for single species are inefficient and philosophically questionable. Piecemeal species management and piecemeal land management may pervert ecosystem integrity and precipitate severe ecosystem changes (cf. Merriam 1991). Management plans now recognise that some kind of ecological completeness is desirable and they commonly include the conservation of ecosystem processes within a reserve.

Upsets of reserve ecosystems arise from many processes, but habitat fragmentation comes high on the list, if not at the top. Effects of upsets are varied and fall into biotic and abiotic categories. Biotic effects resulting from habitat fragmentation include 'extinction cascades'. If a top carnivore goes extinct because there is not enough of its habitat left to support it, repercussions may be felt right down the food web. Fragmentation also leads to the incursion of edge habitats into interior habitats. Concomitant changes

in ecosystem process ensue, including higher nest predation rates. Human modification of habitats in reserves may alter the age structure of habitat patches. Logging, for instance, tends to induce a faster turnover of forest patches and maintain a young forest. This change of forest age, with a drastic reduction in old trees, has injurious effects on such species as the northern spotted owl (*Strix occidentalis caurina*) and the red-cockaded woodpecker (*Picoides borealis*). Abiotic effects resulting from disruption of a reserve ecosystem involve changes in sediment budgets and the land-surface water cycle. Both geomorphic and hydrological changes may have knock-on effects in animal and plant communities on a reserve. Microclimates are also altered by ecosystem modifications. Near-surface wind, temperature, and relative humidity all change following deforestation.

Disturbance is a major ecological factor that needs to be included in reserve design. Soon after disturbance became a popular topic of research in the 1970s, reserve functioning often focused on the connection between reserve size and the scale of the natural disturbance regime (Kupfer 1995). Particular attention was paid to 'minimum dynamic area', defined as the smallest area under a natural disturbance regime that keeps sources for recolonisation and so minimises extinctions (e.g. Pickett and Thompson 1978). Subsequently, the value of the minimum dynamic area concept has been questioned (e.g. Baker 1989, 1992). Despite these problems, the notion of reserve functioning now recognises the important relationships between changing landscape structures and altered landscape functioning (Kupfer 1995).

Regional reserve setting

The regional setting of a reserve affects its local dynamics. For this reason, ecologists, planners, and wildlife managers should heed large-scale ecological processes and look beyond the bounds of an individual reserve (e.g. Noss 1983). Of course, the process is two-way: local dynamics affect regional processes. A danger in metapopulations is that local extinctions in patchy landscapes will accumulate into landscape extinctions, and even regional and subcontinental extinctions (cf. Merriam 1991). The acknowledgement of regional processes and their interaction with local ecosystem dynamics has at least two implications for conservation (Kupfer 1995). First, it encourages a switch towards landscape-level population management and strengthens the case for studying large-scale population dynamics within spatially heterogeneous landscapes. Reserves do not stand in isolation; they interact with surrounding ecosystems and the interaction affects species within the reserve. Second, the recognition of regional processes suggests a need to include spatial factors and scale effects in conservation plans. David B. Lindenmayer's (2000) study of factors affecting the distribution of Leadbeater's possum (*Gymnobelideus leadbeateri*), an endangered arboreal marsupial, and their implications for animal conservation, is a superb example of this practice.

Lindenmayer's research has been running for over a decade (e.g. Lindenmayer 1991, 1997, 2000; Lindenmayer *et al.* 1993). It has involved field and modelling studies at four spatial scales – tree level, stand level, landscape level, and regional-climate level. Leadbeater's possum is largely confined to ash-type eucalypt forests in the Central Highlands of Victoria, south-eastern Australia. Most of these forests are designated for timber harvesting and pulpwood production. The possums live in colonies of up to 12 individuals that shelter and nest in hollows within large trees. They eat arthropods, sap from wattle (*Acacia* spp.) and eucalypt manna, and honeydew from phloem-feeding insects. The total population is predicted to fall by over 90 per cent by 2020–2030, owing to the natural collapse of existing large trees containing hollows, low recruitment of new hollow-bearing trees, and timber extraction. Studies on the possums at the four scales cover a variety of themes (see Lindenmayer 2000, Table 1). Tree-level studies include models of den-tree attributes and tree occupancy, performance tests of nest tree models, den-tree use, long-term occupancy of nest sites, cavity development in trees, and fall rate of hollow trees. Stand-level studies involve habitat requirement models, performance tests of habitat models, and applications of habitat models. Landscape-level studies consider corridor use and population viability models. Regional-scale studies focus on possum-based climate analysis and tree-distribution climate analysis. The research findings strongly suggest that the conservation of Leadbeater's possum must embody the needs of the animals at different spatial scales. Plans must include large reserves, landscape-scale strategies such as wildlife corridors, and stand level operation, such as maintaining structural attributes and vegetation complexity. By adopting strategies at several ecological and management scales, it should be possible to ensure that appropriate conditions for the possum will exist somewhere in part of its range, including areas that are used for wood production.

Although the possum work has focused on one species, the implications of its findings have universal pertinence. Lindenmayer identifies five generic implications:

1. Data gathered at each scale can inform processes and conservation strategies at other scales.
2. The conservation implications for different scales may be disparate.
3. Conservation strategies may be more readily accepted if the scales required for their employment are congruent with the scales used in the management and planning of natural resources.
4. An array of conservation strategies is needed, even for a single species, that cannot stem from a single study at one spatial scale.
5. Implementing a range of conservation strategies across the scales spreads the risk involved and means that, if a strategy at one scale should prove to be faulty, other strategies are in place that should work and so the overall success of the conservation exercise is not fatally jeopardised.

Human individuals and populations

This chapter has not covered all possible influences of topography on humans. Indeed, it has deliberately omitted any discussion of human beings themselves. This is partly because this is not a rich area of research. It is through the material expressions of human cultures – settlements, communications and movement networks, agricultural systems, and industries – that topographic constraints exercise themselves. That is not to say that humans as individuals and humans as populations are unaffected by topography. As individuals, humans are adapted to live at different altitudes and different environments. Some examples that could be quoted here might be redolent of the extreme environmental determinism once rife in geography. Nonetheless, physiological changes in adapting to life at high altitudes cannot be denied. Nor can adaptations of skin colour and stature in adapting to different climatic zones. And as populations, some human actions are directly or indirectly related to topography. Terrain may directly affect military campaigns. It may indirectly influence a host of human activities through an intermediary factor such as climatic change: topographic attributes may modify regional climatic trends to produce spatial differences that impress themselves upon agriculture, tourism, and many other facets of socio-economic systems. To an extent, topography lies at the heart of ecological, economic, social, and cultural demands on environments and is an indispensable part of practices that try to harmonise conflicting demands. Agroforestry and ecosystem management are such harmonising practices. They take cognisance of the physical environment, but they may shape that environment according to the agreed wishes of different human factions to create a novel topography. This final section will briefly explore these three themes: the effects of terrain on warfare; topography as a mediator of climatic change and the resulting influence on human activities; and the reconciliation of conflicting human use of landscapes in agroforestry and ecosystem management practices. It will do so by taking case studies.

Terrain and military history

People have always been able to recognise defensible sites for their forts, castles, or towns. If called upon to engage in an offensive action, they have usually had the common sense to use terrain to the best advantage, though some have used it better than others and there have been some fearful blunders. Military geography was once a popular subject, especially in the first half of the twentieth century (e.g. Hogarth 1915; Leaf 1916; Strahan 1919; Beeby-Thomson 1924; Cole 1953; see also Rose and Rosenbaum 1993). In 1906, Colonel D. Johnston declared to Scottish geographers that 'Topographic maps are . . . of utmost military value even in peace time for training and manoeuvre purposes and for military problems connected with the defence of the country' (Johnston 1906, 18). Two years later, Major E. H. Hills announced that 'We may first note the somewhat curious fact that the production of a map of a country, useful as a work is for many purposes, has almost always been embarked upon because the imperative necessity of maps of the theatre of operations in war has been brought home to the people and government of a nation' (Hills 1908, 506). The chief champion of geography's military potential was Sir Thomas H. Holdich, who argued for a school of practical geography to service the military's topographical corps (Holdich 1902), and, in *Political Frontiers and Boundary Making* (1916), asserted that war is the greatest civilising agent! Lieutenant-Colonel Charles A. Court Repington, too, argued that geography should form part of a military training, an understandable viewpoint given the parlous state of military cartography and geographical knowledge at the time (see Stoddart 1992; Livingstone 1992, 241–53).

Military applications of geography made a partial comeback when terrain evaluation emerged as a substantial research area within geomorphology in the 1960s: plainly, terrain can be evaluated for military as well as scientific purposes (e.g. Mitchell 1991, 322–41). During a conflict, the terrain is considered in planning strategy and in tactical operations. Strategic planners need to appreciate the distribution of major geographical features – mountains, seas, plains, plateaux, and so forth – in order to know where to deploy troops and resources. During action, tactics should be informed by terrain conditions to advance the strategic aims of the campaign to best advantage on the ground. Five types of problem normally arise in the tactical evaluation of terrain: position, mobility, ground conditions, resource provision, and hazard mitigation (Mitchell 1991, 322–3). Position means providing vantage points and refuges from the enemy's vantage points. Mobility involves identifying such terrain barriers as rivers and impassable slopes, and the capacity of surfaces – the 'going' surfaces – to support troop, machinery, and animal movements. Ground conditions are important in deciding where to pitch tents, dig trenches, dugouts, and so forth. Resource provision means locating sites of potable water, and building materials for roads and defensive works. Hazard mitigation embraces the identification of potential floods and mass movements that could imperil the lives of the troops and threaten the supply and communications infrastructure, and of prevailing winds that might affect the use of chemical weapons. Intelligence of these terrain attributes has reached a high level of sophistication, but older battles were generally based on a limited knowledge of terrain. Of course, there are exceptions. In the American Civil War, General Ulysses S. Grant's appreciation of the unique geography of river systems in the western theatre was a crucial factor in his victorious campaigns of 1862 and 1863. In 1862 he captured Forts Henry and Donelson with the Illinois Volunteers and ran the successful Vicksburg campaign (1862–63).

Signs of a resurgence of military geography are a book on the subject (Collins 1996), a general interest in the geography of conflict and peace (e.g. Kliot and Waterman 1991; O'Loughlin 1996), and a conference on 'Terrain in Military History', which was organised by the University of Greenwich, England, in association with the Imperial War Museum, London, and the Royal Engineer's Museum, Chatham. The conference examined the impact of terrain on the conduct of military campaigns. It arose from the assertion that terrain is a crucial factor in the running

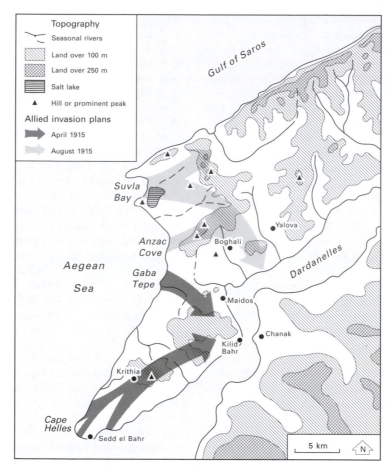

Figure 7.13 Topographic map of the Gallipoli Peninsula and the Dardanelles, showing militarily significant physical features, with Allied Invasion Plans for the Gallipoli Campaign in 1915. (Adapted from Doyle and Bennett, 1999.)

of any military campaign, and it assembled leading practitioners of modern armed forces, terrain experts, geographers, geologists, archaeologists, and military historians. Lecture topics included 'Perception of space through military experience', 'Topographic training in the British Army in the nineteenth century', 'Terrain and guerilla warfare in Navarre', 'The effective use of terrain: the Battle of Fredericksberg, December, 1862', 'A geographic analysis of the Battle of Shiloh', and 'An archaeological approach to the landscape of the Anglo-Zulu War'.

The influence of terrain on the Gallipoli Campaign, 1915 (Doyle and Bennett 1999), epitomises the kind of geographical analysis to which military campaigns may be subjected, often with instructive outcomes. A land-system analysis of the Gallipoli Peninsula was performed. Five land systems were identified, based on aspects of geology, geomorphology, hydrogeology, and vegetation. The landings of 25 April 1915 were made at Cape Helles and Anzac Cove. The objective was to capture high ground in order to command the Dardanelles Straits and to remove the threat of fixed and mobile coastal batteries on the shore of the Dardanelles (Figure 7.13). The land-system analysis brought out two key points that should have been considered in planning the invasion, but were not necessarily available in 1915. First, with the exception of the high ground surrounding the Suvla Plain, the terrain possessed inadequate water supplies, and to succeed, any attack would need to capture high ground speedily. Second, the landing sites were far from ideal: the terrain was steep, with deeply incised slopes, and the beaches were narrow. Better landing sites were available. A later landing at Suvla Bay in August of the same year benefited from better terrain, with wide landing beaches and locally available water supplies. Few Turkish troops held the high ground, but the attack was not pushed forward and failed to exploit the tactical advantages of a lightly held terrain. The Gallipoli Campaign failed primarily because the Allied High Commands in London and Paris planned inadequately and were indecisive. Not enough troops were deployed and munitions and communications were inadequate. In addition, the Turkish armies were efficient at siting defensive positions according to the terrain, while the British and ANZAC troops lacked information about terrain and geology, and had no local groundwater supplies. An important question to emerge from the analysis is why a possible landing at Suvla Bay, with its more suitable beaches and locally available supplies of water, was not fully explored at the outset. Had this option been pursued, Turkey might have been knocked out of the war well before 1918, which might have changed the face of the First World War (Doyle and Bennett 1999).

Topography, climatic change, and tourism

In Scotland, current climatic changes are affecting the pattern of winter visitor activity and threatening the viability of tourism-related enterprises (Harrison *et al.* 1999). Indications are that winters are becoming milder and summers drier. What are the implications of these changes to tourist activities, especially skiing and hill-walking? It was explained in Chapter 3 (p. 58) how changes in snowfall patterns in Scotland were related to the frequency of easterly and westerly airstreams, and that regression equations were established relating the number of days when snow lay (sometimes simply called snow-lie) to altitude, latitude, and longitude. Applying these statistical models to a DEM enabled predictions of changes in snow-lie to be extended to the whole of Scotland using a grid size of about 1 × 1 km. The predictions suggested that, if the winter climate should shift to a more westerly maritime character, then lowlands below 100 m would have more-frequent snow-free winter months. The abatement in snow-lie becomes more pronounced with increasing altitude, especially in the range 500–1000 m, making slopes in the elevational band largely useless for skiing. Surprisingly, perhaps, upland above 1000 m may see little change in the number of days with snow lying, and Scotland would not be blighted by snowless mountains. Even so, the skiing industry may have to relocate to higher altitudes or latitudes, or make more subtle shifts decided by changes in the direction of approach of snow-bearing weather systems. Another potential problem for tourism in Scotland is a change towards wetter summers. Such a change would invariably lead to lower levels of participation in outdoor activities such as hill-walking, and to more Scots taking holidays abroad. A series of poor summers may also greatly reduce the gates at amusement parks (Moutinho 1988). Other predictions of 'summer quality' suggested that summers may become drier, with a welcome reduction of dull and damp summer days (Harrison *et al.* 1999). That is good news for hill-walkers wishing to view the Scottish scenery, but it is worrying for questions of water supply and fire hazard, especially in eastern Scotland where summer drought is already a threat.

Agroforestry and ecosystem management

A good example of agroforestry in action comes from the Winnebago Reservation, in the north-eastern corner of Nebraska, USA (Szymanski *et al.* 1998a, 1998b; Szymanski and Colletti 1999). Agroforestry normally involves creating windbreaks and riparian buffers to reduce soil loss. On communally owned tribal lands, agroforestry systems can be extended to address social and cultural needs, such as passing down knowledge from one generation to the next, involving youth, and building small family gardens. The Winnebago Tribe was originally granted its reservation in 1865, since when as much as three-quarters of it has been 'lost' to non-Indian people. In the 1990s, gambling casino revenue has allowed the Winnebago Tribe to acquire lands that were previously part of the reservation. The tribe is currently deciding which tracts of land to buy and which specific needs should be addressed. As part of the decision process, a feasibility study was conducted in 1991 to ascertain land-use alternatives for Big Bear Hollow, a 1250-acre tract of tribal land in the Missouri River corridor. The option included intensive irrigated and non-irrigated agriculture, a mix of intensive and less intensive agriculture and agroforestry, and a total

Table 7.1 General land-use preferences expressed by members of the Winnebago Tribe, Nebraska

Land-use category	Number of responses	Per cent of responses
The village		
Housing	187	78
Education	122	51
Medical facility	122	51
Nursing home	120	50
Group home for youth	188	49
Other	—	<45
The Missouri River area		
Wildlife	64	28
Spiritual	50	22
Recreation	46	20
Housing	23	10
Forestry	21	10
Roads	11	5
Agriculture	11	5

Source: Adapted from Szymanski *et al.* (1998a).

conversion of the area to a mixed bottomland forest. By consulting with the tribe, the best land-use options that emerged were agroforestry systems yielding a diverse mixture of forest and agricultural and horticultural crops. A 55-acre area was chosen for an agroforestry demonstration. The first step in the demonstration was taken in 1994 with the planting of a windbreak of mixed shrubs on 2.5 acres of the north-

ern edge of the demonstration area, which is a riparian zone. In subsequent years, three crops will be rotated on the land – clover (*Melitotus officinalis*), soybeans (*Glycine max*), and Indian or flint corn (*Zea mays*). Black walnut (*Juglans nigra*) will be grown for nuts, wildlife habitat, and veneer. Wild plums (*Prunus* sp.), wild raspberries (*Rubus* sp.), and milkweed (*Asclepias syriaca*) were deemed to be secondary food crops. The tribe banned all use of chemicals in the agroforestry demonstration after 1995. Various 'community surveys' using participatory rural appraisal techniques revealed land-use preferences that have little to do with natural resources or agriculture and much to do with wildlife, spiritual factors, and recreation (Table 7.1). This is not surprising as the Winnebago Tribe has strong extended families and a strong tradition of connection with the earth. Indian corn topped the list of specific land-use preferences (68 per cent, or 156 responses), chiefly because of the strong cultural value attached to its cultivation. Interest in horticultural activities was quite strong (84 per cent, or 205 responses), with more interest in family gardens and cottage industries. Trees were rated very important for cultural and wildlife reasons, and less important for economic reasons. The proposed components of the agroforestry system and the importance attached to them are listed in Table 7.2. Note the importance attached to cultural and spiritual factors. The demonstration is far from completed, but it shows that, if properly planned, agroforestry systems may meet production needs while maintaining a people's spiritual and cultural values through their connection to the land.

Table 7.2 Preference ratings for socio-cultural values attached to the proposed components of the agroforestry system by 30 members of the Winnebago Tribe. Ratings are on a scale of 10, least value = 1, greatest value = 10

Crop	Cultural importance	Spiritual importance	Food source	Opportunities to teach youth
Primary products				
Indian corn	10	7	8	6
Black walnut	5	2	6	4
Soybeans	3	1	2	2
Clover	3	1	2	2
Secondary products				
Berries	4	2	5	3
Mixed shrubs	5	2	3	4
Milkweed	7	4	7	5

Source: Adapted from Szymanski *et al.* (1998a).

References

Alexander, R. T. (1997) Haciendas and economic change in Yucatán: entrepreneurial strategies in the Parroquia de Yaxcabá, 1775–1850. *Journal of Archaeological Method and Theory* **4**, 331–51.

Alexander, R. T. (1998) Community organization in the Parroquia de Yaxcabá, Yucatan, Mexico, 1750–1847. *Ancient Mesoamerica* **9**, 39–54.

Alexander, R. T. (1999) Mesoamerican house lots and archaeological site structure: problems of interference in Yaxcaba, Yucatan, Mexico, 1750–1847. In P. Allison (ed.) *The Archaeology of Household Activities*. London: Routledge.

Anderton, J. B. (1993) *Paleoshoreline Geoarchaeology in the Northern Great Lakes. Hiawatha National Forest*. United States Forest Service, Heritage Program Monograph No. 1. Hiawatha National Forest. Escanaba, MI: USDA Forest Service.

Baker, W. L. (1989) Landscape ecology and nature reserve design in the Boundary Waters Canoe Area, Minnesota. *Ecology* **70**, 23–35.

Baker, W. L. (1992) The landscape ecology of large disturbances in the design and management of nature reserves. *Landscape Ecology* **7**, 181–94.

Baldwin, C. S. (1988) The influence of field windbreaks on vegetable and speciality crops. In J. R. Brandle and D. L. Hintz (eds) *Windbreak Technology*, pp. 191–203. Amsterdam: Elsevier (reprinted from *Agriculture, Ecosystems and Environment* Vols 22 and 23).

Beeby-Thomson, A. (1924) *Emergency Water Supplies for Military, Agricultural and Colonial Purposes*. London: Crosby Lockwood & Son.

Bennett, A. F. (1991) Roads, roadsides and wildlife conservation: a review. In D. A. Saunders and R. J. Hobbs (eds) *Nature Conservation 2: The Role of Corridors*, pp. 99–118. Chipping Norton, Australia: Surrey Beatty & Sons.

Bennett, A. F. (1999) *Linkages in the Landscape: The Role of Corridors and Connectivity in Wildlife Conservation*. Gland, Switzerland: IUCN (International Union for the Conservation of Nature).

Bennett, A. F., Henein, K., and Merriam, G. (1994) Corridor use and the elements of corridor quality: chipmunks and fencerows in a farmland mosaic. *Biological Conservation* **68**, 155–65.

Błażejczyk, K., Krawczyk, B., and Skoczek, J. (1995) Bioclimatic conditions of coffee plantation in central Vietnam. In K. U. Sirinanda (ed.) *Climate and Life in the Asia Pacific: Proceedings*, pp. 310–18. Brunei: The Department of Geography, Universiti Brunei Darussalam.

Bruhnes, J. (1952) *Human Geography*. Abridged version by M. Jean-Bruhnes Delamare and Pierre Deffontaines, translated by Ernest F. Row. London: Harrap.

Caborn, J. M. (1965) *Shelterbelts and Windbreaks*. London: Faber & Faber.

Carter, C. C. (1931) *Land-forms and Life: Short Studies on Topographical Maps*. London: Christophers.

Cole, D. H. (1953) *Imperial Military Geography: the Geographical Background of the Defence Problems of the British Commonwealth*, 11th edn. London: Sifton, Pread.

Collins, J. M. (1996) *Military Geography*. London: Chrysalis Books.

Davis, J. L. (1991) Contributions to a Mediterranean rural archaeology: historical case studies from the Ottoman Cyclades. *Journal of Mediterranean Archaeology* **4**, 131–216.

Denevan, W. M. (1966) A cultural–ecological view of the former aboriginal settlement in the Amazon Basin. *Professional Geographer* **18**, 346–51.

Denevan, W. M. (1996) A bluff model of riverine settlement in prehistoric Amazonia. *Annals of the Association of American Geographers* **86**, 654–81.

Denevan, W. M. (1998) Comments of prehistoric agriculture in Amazonia. *Culture & Agriculture* **20**, 54–9.

Diamond, J. M. (1976) Island biogeography and conservation: strategy and limitations. *Science* **193**, 1027–9.

Dow, B. D., Cunningham, R. A., and Krupinsky, J. M. (1998) Fifteen-year provenance tests of lodgepole pine (*Pinus contorta*) in North Dakota. *Western Journal of Applied Forestry* **13**, 5–11.

Doyle, P. and Bennett, M. R. (1999) Military geography: the influence of terrain in the outcome of the Gallipoli Campaign, 1915. *The Geographical Journal* **165**, 12–36.

Dugué, P., Roose, É., and Rodriguez, L. (1993) L'aménagement de terroirs villageois et l'amélioration de la production agricole au Yatenga (Burkina Faso). *Cahier Orstrom, Série Pédologie* **28**, 385–402.

Edenius, L. and Sjöberg, K. (1997) Distribution of birds in natural landscape mosaics of old-growth forests in northern Sweden: relations to habitat area and landscape context. *Ecography* **20**, 425–31.

Forman, R. T. T. (1995) *Land Mosaics: The Ecology of Landscapes and Regions*. Cambridge: Cambridge University Press.

Forman, R. T. T., Galli, A. E., and Leck, C. F. (1976) Forest size and avian diversity in New Jersey woodlots with some land use implications. *Oecologia* **26**, 1–8.

Fortin, D. and Arnold, G. W. (1997) The influence of road verges on the use of nearby small shrubland remnants by birds in the central wheatbelt of Western Australia. *Wildlife Research* **24**, 679–89.

Franzen, J. (1986) *Prehistoric Settlement on the Hiawatha National Forest, Michigan: A Preliminary Locational Model*. Culture Resource Management Report No. 4. Hiawatha National Forest.

Garnett, A. (1937) *Insolation and Relief: Their Bearing on the Human Geography of Alpine Regions*. The Institute of British Geographers Publication No. 5. London: George Philip & Son.

Gilbert, O. L. (1991) *The Ecology of Urban Habitats*. London: Chapman & Hall.

Godin, V. B. and Leonard, R. E. (1979) Management problems in designated wilderness areas. *Journal of Soil and Water Conservation* **34**, 141–3.

Gotfryd, A. and Hansell, R. I. C. (1986) Prediction of bird-community metrics in urban woodlots. In J. Venner, M. L. Morrison, and C. J. Ralph (eds) *Wildlife 2000: Modeling*

Habitat Relationships of Terrestrial Vertebrates, pp. 321–6. Madison, WI: University of Wisconsin Press.

Griffin, D. (1996) A culture in transition: a history of acculturation and settlement near the mouth of the Yukon River, Alaska. *Arctic Anthropology* **33**, 98–115.

Gunnell, Y. (1999) Systèmes agraires et facettes écologiques au Karnataka (Inde du Sud): portraits d'une organisation humaine autour d'un gradient bioclimatique exceptionnel. *Annales de Géographie* **605**, 46–66.

Gusev, S. V., Zagoroulko, A. V., and Porotov, A. V. (1999) Sea mammal hunters of Chukotka, Bering Strait: recent archaeological results and problems. *World Archaeology* **30**, 354–69.

Hammett, J. E. (1992) The shapes of adaptation: historical ecology of anthropogenic landscapes in the southeastern United States. *Landscape Ecology* **7**, 121–35.

Harrison, S. J., Winterbottom, S. J., and Sheppard, C. (1999) The potential effects of climate change on the Scottish tourist industry. *Tourism Management* **20**, 203–11.

Hayfield, C. and Wagner, P. (1995) From dolines to dewponds: a study of water supplies on the Yorkshire Wolds. *Landscape History* **17**, 49–64.

Hills, E. H. (1908) The survey of the British Empire. *Scottish Geographical Magazine* **24**, 505–19.

Hobbs, R. J. and Hopkins, A. J. M. (1991) The role of conservation corridors in a changing climate. In D. A. Saunders and R. J. Hobbs (eds) *Nature Conservation 2: The Role of Corridors*, pp. 281–90. Chipping Norton, Australia: Surrey Beatty & Sons.

Hogarth, D. G. (1915) Geography of the war theatre in the Near East. *The Geographical Journal* **66**, 457–67.

Holdich, T. H. (1902) The progress of geographical knowledge. *Scottish Geographical Magazine* **18**, 504–25.

Holdich, T. H. (1916) *Political Frontiers and Boundary Making*. London: Macmillan.

Houston, J. M. (1963) *A Social Geography of Europe*. London: Gerald Duckworth.

Huggett, R. J. and Robinson, M. E. (1996) General introduction. In I. Douglas, R. J. Huggett, and M. E. Robinson (eds) *Companion Encyclopedia of Geography: The Environment and Humankind*, pp. 1–8. London: Routledge.

Jacoby, J. (1990) Mountain bikes: a new dilemma for wildlife recreation managers. *Western Wetlands* **16**, 25–8.

Jagodziński, M. and Kasprzycka, M. (1991) The early medieval craft and commercial centre at Janow Pomorski near Elblag on the south Baltic coast. *Antiquity* **65**, 696–715.

Johns, J. S., Barreto, P., and Uhl, C. (1996) Logging damage during planned and unplanned logging operations in the eastern Amazon. *Forest Ecology and Management* **89**, 59–77.

Johnston, D. (1906) A brief description of the Ordnance Survey and some notes on the advantages of a topographical survey of South Africa. *Scottish Geographical Magazine* **22**, 18–27.

Jones, E. (1967) *Human Geography*. London: Chatto & Windus.

Kliot, M. and Waterman, S. (1991) *The Political Geography of Conflict and Peace*. London: Belhaven Press.

Kort, J. (1988) Benefits of windbreaks to field and forage crops. In J. R. Brandle and D. L. Hintz (eds) *Windbreak Techno-

logy*, pp. 165–90. Amsterdam: Elsevier (reprinted from *Agriculture, Ecosystems and Environment* Vols 22 and 23).

Kupfer, J. A. (1995) Landscape ecology and biogeography. *Progress in Physical Geography* **19**, 18–34.

Kuss, F. R. (1983) Hiking boot impacts on woodland trials. *Journal of Soil and Water Conservation* **38**, 119–21.

Laurance, W. F. and Yensen, E. (1991) Predicting the impacts of edge effects in fragmented habitats. *Biological Conservation* **55**, 77–92.

Leaf, W. (1916) The military geography of the Troad. *The Geographical Journal* **67**, 401–21.

Lewandowski, J. P. and McLaughlin, S. P. (1995) Managing visitors' environmental impacts in a system of sites: a network approach. *Pennsylvania Geographer* **33**, 43–58.

Lindenmayer, D. B. (1991) A note on the occupancy of nest trees by Leadbeater's possum in the montane ash forests of the Central Highlands of Victoria. *Victorian Naturalist* **108**, 128–9.

Lindenmayer, D. B. (1997) Differences in the biology and ecology of arboreal marsupials in forests of southeastern Australia. *Journal of Mammalogy* **78**, 1117–27.

Lindenmayer, D. B. (2000) Factors at multiple scales affecting distribution patterns and their implications for animal conservation – Leadbeater's possum as a case study. *Biodiversity and Conservation* **9**, 15–35.

Lindenmayer, D. B., Cunningham, R. B., and Donnelly, C. F. (1993) The conservation of arboreal marsupials in the montane ash forests of the Central Highlands of Victoria, south-east Australia, IV. The presence and abundance of arboreal marsupials in retained linear habitats (wildlife corridors) within logged forest. *Biological Conservation* **66**, 207–21.

Livingstone, D. N. (1992) *The Geographical Tradition: Episodes in the History of a Contested Discipline*. Oxford: Blackwell.

Meggers, B. (1971) *Amazonia: Man and Culture in a Counterfeit Paradise*. Chicago: Aldine Press.

Merriam, G. (1991) Corridors and connectivity: animal populations in heterogeneous environments. In D. A. Saunders and R. J. Hobbs (eds) *Nature Conservation 2: The Role of Corridors*, pp. 133–42. Chipping Norton, Australia: Surrey Beatty & Sons.

Mitchell, C. W. (1991) *Terrain Evaluation: An Introductory Handbook to the History, Principles, and Methods of Practical Terrain Assessment*, 2nd edn. Harlow: Longman.

Moutinho, L. (1988) Amusement park visitor behaviour – Scottish attitudes. *Tourism Management* **9**, 292–300.

Naaim-Bouvet, F. and Mullenbach, P. (1998a) Haies parecongère: étude expérimentale *in situ*. *Revue Forestière Française* **3**, 263–76.

Naaim-Bouvet, F. and Mullenbach, P. (1998b) Field experiments on 'living' snow fences. *Annals of Glaciology* **26**, 217–20.

Nakonechnaya, M. A. and Yavtushenko, V. Ye. (1989) Differences in agroecological conditions on north- and south-facing slopes of the Central Chernozem Region. *Soviet Soil Science* **21**, 12–21.

Newmark, W. D. (1987) A land-bridge island perspective on mammalian extinctions in western North American parks. *Nature* **325**, 430–2.

Nordling, M. (1864) Sur les moyens de prévenir les amoncellements de neige sur les chemins de fer. *Annales des Ponts et Chaussées* **105**, 1–17.

Noss, R. F. (1983) A regional landscape approach to maintain diversity. *BioScience* **33**, 700–6.

Noss, R. F. (1987) Corridors in real landscapes: a reply to Simberloff and Cox. *Conservation Biology* **1**, 159–64.

Ogunkunle, A. O. and Onasanya, O. S. (1993) Relative influence of management and topographic position on maize and soybean yields at some sites in south-western Nigeria. *International Journal of Tropical Agriculture* **11**, 155–62.

O'Loughlin, J. (1996) The geography of conflicts and the prospects for peace. In I. Douglas, R. J. Huggett, and M. E. Robinson (eds) *Companion Encyclopedia of Geography: The Environment and Humankind*, pp. 353–69. London: Routledge.

Olson, J. D. (1998) A digital model of pattern and productivity in an agroforestry landscape. *Landscape and Urban Planning* **42**, 169–89.

Pickett, S. T. A. and Thompson, J. N. (1978) Patch dynamics and the design of nature reserves. *Biological Conservation* **13**, 23–37.

Radcliffe, J. E. (1982) Effects of aspect and topography on pasture production in hill country. *New Zealand Journal of Agricultural Research* **25**, 485–96.

Radcliffe, J. E. and Lefever, K. R. (1981) Aspect influences on pasture microclimate at Coopers Creek, North Canterbury (New Zealand). *New Zealand Journal of Agricultural Research* **24**, 55–66.

Reinirkens, P. and Vartmann, C. (1996) Die Catena als Kriterium fur den Bodenschutz in der Bauleitplanung. [The catena as a criterion for soil conservation in building planning.] *Naturschutz und Landschaftsplanung* **28**, 5–9.

Roberts, B. K. (1987) *Rural Settlement*. Basingstoke and London: Macmillan Education.

Roberts, B. K. (1996) *Landscapes of Settlement, Prehistory to the Present*. London and New York: Routledge.

Roose, É., Dugué, P., and Rodriguez, L. (1992) La G. C. E. S. [Gestion Conservatoire de l'Eau, de la biomasse et de la fertilité des Sols]. Une nouvelle stratégie de lutte anti-érosive appliquée à l'aménagement de terroirs en zone soudono-sahélienne du Burkina Faso. *Revue des Bois et Forêts des Tropiques* No. 233, 49–63.

Rose, E. P. F. and Rosenbaum, M. S. (1993) British military geologists: the formative years to the end of the First World War. *Proceedings of the Geologists' Association, London* **104**, 41–50.

Sauerborn, J. (1994) Aspekte zu Bodennutzungssystemen Entlang einer Catena auf Leyte, Philippinen. [Aspects of land use systems along a catena on Leyte, Philippines.] *Giessener Beitrage zur Entwicklungsforschung, Reihe I* (Symposien) **21**, 129–38.

Schjellerup, I. (1999) Wayko–Lamas: a Quechua community in the Selva Alta of North Peru under change. *Geografisk Tidsskrift, Danish Journal of Geography, Special Issue 1*, 199–207.

Schonewald-Cox, C. M. (1988) Boundaries in the protection of nature reserves: translating multidisciplinary knowledge into practical conservation. *BioScience* **38**, 480–6.

Schonewald-Cox, C. M. and Bayless, J. W. (1986) The boundary model: a geographical analysis of the design and conservation of nature reserves. *Biological Conservation* **38**, 305–22.

Shafer, C. L. (1990) *Nature Reserves: Island Theory and Conservation Practice*. Washington and London: Smithsonian Institution Press.

Silbernagel, J., Martin, S. R., Gale, M. R., and Chen, J. (1997) Prehistoric, historic, and present settlement patterns related to ecological hierarchy in the Eastern Upper Peninsula of Michigan, U.S.A. *Landscape Ecology* **12**, 223–40.

Simberloff, D. S. and Abele, L. G. (1976a) Island biogeography theory and conservation practice. *Science* **191**, 285–6.

Simberloff, D. S. and Abele, L. G. (1976b) Island biogeography and conservation: strategy and limitations. *Science* **193**, 1032.

Simberloff, D., and Cox, J. (1987) Consequences and costs of conservation corridors. *Conservation Biology* **6**, 493–504.

Simberloff, D., Farr, J. A., Cox, J., and Mehlman, D. W. (1992) Movement corridors: conservation bargains or poor investments? *Conservation Biology* **6**, 493–504.

Soulé, M. E. and Gilpin, M. E. (1991) The theory of wildlife corridor capability. In D. A. Saunders and R. J. Hobbs (eds) *Nature Conservation 2: The Role of Corridors*, pp. 3–8. Chipping Norton, Australia: Surrey Beatty & Sons.

Steiner, K. G. (1998) Using farmers' knowledge of soils in making research results more relevant to field practice: experiences from Rwanda. *Agriculture Ecosystems and Environment* **69**, 191–200.

Steward, J. H. (1948) Culture areas of the tropical forests. In J. H. Steward (ed.) *Handbook of South American Indians* (Bureau of American Ethnology, Bulletin 143), 5, 883–99. Washington, DC: Smithsonian Institution.

Stoddart, D. R. (1992) 'Geography and War': the new geography and the new army in England, 1899–1914. *Political Geography* **11**, 87–99.

Strahan, A. (1919) Introduction. Work in connection with the war. *Memoirs of the Geological Survey. Summary of the Progress of the Survey of Great Britain and the Museum of Practical Geology for 1918*, pp. 1–4. London: HMSO.

Szymanski, M. and Colletti, J. (1999) Combining the socio-economic-cultural implications of community owned agroforestry: the Winnebago Tribe of Nebraska. *Agroforestry Systems* **44**, 227–39.

Szymanski, M., Colletti, J., and Whitewing, L. (1998a) Meeting the Winnebago Tribe's needs through agroforestry. *Journal of Forestry* **96**, 34–8.

Szymanski, M., Whitewing, L., and Colletti, J. (1998b) The use of participatory rural appraisal methodologies to link indigenous knowledge and land use decisions among the Winnebago Tribe of Nebraska. *Indigenous Knowledge Monitor* **6**, 3–6.

Terborgh, J. (1976) Island biogeography and conservation: strategy and limitations. *Science* **193**, 1029–30.

Usher, M. B., Brown, A. C., and Bedford, S. E. (1992) Plant species richness in farm woodlands. *Forestry* **65**, 1–13.

Van Staveren, J. P. and Stoop, W. A. (1985) Adaptation of sorghum, millet and maize to different land types in a common toposequence in West Africa: implications for the testing and introduction of new sorghum varieties in the local agriculture. *Field Crops Research* **11**, 13–35.

Vidal de la Blache, P. (1926) *Principles of Human Geography*. Edited by Emmanuel de Martonne. Translated from the French by Millicent Todd Bingham. London: Constable.

Virolainen, K. M., Suomi, T., Suhonen, J., and Kuitunen, M. (1998) Conservation of vascular plants in single large and several small mires: species richness, rarity and taxonomic diversity. *Journal of Applied Ecology* **35**, 700–7.

Wallin, T. R. and Harden, C. P. (1996) Estimating trail-related soil erosion in the humid tropics: Jatun Sacha, Ecuador, and La Selva, Costa Rica. *Ambio* **25**, 517–22.

Weaver, T. and Dale, D. (1978) Trampling effects of hikers, motorcycles and horses in meadows and forests. *Journal of Applied Ecology* **15**, 451–7.

Whinam, J., Cannell, E. J., and Kirkpatrick, J. B. (1994) Studies on the potential impact of recreational horse-riding on some alpine environments of the Central Plateau, Tasmania. *Journal of Environmental Management* **40**, 103–17.

Whitcomb, R. F., Lynch, J. F., Opler, P. A., and Robins, C. S. (1976) Island biogeography and conservation: strategy and limitations. *Science* **193**, 1030–2.

Wilkinson, T. J. and Edens, C. (1999) Survey and excavation in the Central Highlands of Yemen: results of the Dhamār survey project, 1996 and 1998. *Arabian Archaeology and Epigraphy* **10**, 1–33.

Wilson, J. P. and Seney, J. P. (1994) Erosional impact of hikers, horses, motorcycles, and off-road bicycles on mountain trails in Montana. *Mountain Research and Development* **14**, 77–88.

Young, A. (1989) *Agroforestry for Soil Conservation*. Nairobi: International Centre for Research in Agroforestry and CAB International.

Index